Natural Resources and the Environment Series

Volume 1

The Environmental Impacts of Production and Use of Energy

NATURAL RESOURCES AND THE ENVIRONMENT SERIES

Volume 1: THE ENVIRONMENTAL IMPACTS OF PRODUCTION AND USE OF ENERGY

Essam El-Hinnawi United Nations Environment Programme.
ISBN 907567 00 2. 330 pages, figures and tables

Volume 2: RENEWABLE NATURAL RESOURCES AND THE ENVIRONMENT

Kenneth Ruddle and Walther Manshard, United Nations University.
ISBN 0 907567 01 0. 404 pages, figures and tables.

Volume 3: ASSESSING TROPICAL FOREST LANDS: THEIR SUITABILITY FOR SUSTAINABLE USES

Richard A. Carpenter, Editor East West Center.
ISBN 0 907567 02 9. 256 pages, figures and tables.

Volume 4: RURAL ENERGY PRODUCTION AND SUPPLY

Ross and Morgan, United Nations University. Publication August '81.
ISBN 0 907567 03 7. 192 pages approx.

Volume 5: ECONOMIC APPROACHES TO NATURAL RESOURCES AND ENVIRONMENTAL QUALITY ANALYSIS.

Hufschmidt and Hyman, Editors, East-West Center. Publication August '81.
ISBN 0 907567 04 5. 256 pages approx.

Volume 6: INTERNAL AND EXTERNAL CONSTRAINTS ON TECHNOLOGICAL CHOICE IN DEVELOPING COUNTRIES

Freedman and Lucas, Editors, Publication September, 1981.
ISBN 0 907567 05 3.

The Environmental Impacts of Production and Use of Energy

An Assessment prepared by the United Nations Environment Programme

ESSAM E. EL-HINNAWI

study director

Published for the UNITED NATIONS ENVIRONMENT PROGRAMME

by the TYCOOLY PRESS LTD.

ISBN 907567 00 2

Typeset by Brunswick Press Ltd., 17 Gilford Road, Sandymount, Dublin 4.
Printed by Irish Elsevier Printers Limited, Shannon.

CONTENTS

TABLES

FIGURES

Foreword

THE 1970s brought into focus two global issues that are of primary importance in determining future world development, namely "Environment" and "Energy". In the early years of the decade, the environmental movement was approaching its peak and an "energy crisis" was in the making. The emergence of the two issues at the same time was not mere coincidence. As the last decade has shown, both issues are closely related, and have marked socio-economic as well as political dimensions.

It has been realized that some fossil fuels, especially oil and natural gas, are finite in extent and should be regarded as depleting assets, and the term "energy crisis" has become a common catchword all over the world. However, it is difficult to define this "crisis". Does the problem lie in the scarcity, or the vulnerability, of supply? Or is it the rising cost of energy resources and its implicit relation to the world-wide inflation and recession? Or is the crisis defined by an "excessive" demand for energy, or by widespread habits of energy waste? Of course, it may be a combination of all or most of these things.

At local, national and in some cases regional levels, the environmental aspects of energy production and use have become the subject of wide-ranging debate. Environmental awareness and anti-pollution campaigns have affected the formulation of energy policies in many countries, and it has recently been realized that nations are not isolated in this respect; the actions of one country may affect the environment in a neighbouring one. Nowadays, energy policy decisions are dictated less by technological than by social, environmental and political factors.

The production, transportation and use of different sources of energy raise a number of important environmental issues. Among these, questions of the availability and allocation of resources are likely to play as important a role as pollution problems proper. Land and water use, emissions (including thermal discharges) and their impact on ecosystems and human health are but examples of the problems encountered. The 1970's, brought into focus some related socio-economic and geo-political problems. For example, nuclear power development raises problems related to ultimate disposition of nuclear wastes, possibilities of diversion of nuclear material for military purpose or terrorists action and to the public acceptance of this source of energy. On the other hand, concern has been recently voiced at the implications of extensive coal utilization, especially its possible impact on climate. The atmosphere is believed to show a warming primarily due to the greenhouse effect of increasing carbon dioxide emissions. The important question is: what will the regional changes of temperature and rainfall be? Studies with climate models and of climate observations indicate that regional anomalies will probably occur, but the magnitude and impact of these anomalies cannot be reliably predicted at present. The medium and long-term effects of such possible climatic changes are of such fundamental importance that they command serious attention.

The general realization of the finite nature of fossil fuel resources has caused a re-examination of the possibility of using those energy resources which are of a non-depleting nature, and are therefore considered renewable. These energy sources are becoming increasingly important both in developed and in developing countries. In the former strategies for the exploitation of such sources constitute a part of recent policies which aim at reducing the dependence on fossil fuel. In the developing countries, particularly those short of fossil fuel resources, the renewable sources of energy show promise of meeting some of the future energy needs, especially of rural areas. Without strong rural development programmes based on decentralized energy sources, urban migration will become torrential, exacerbating the already dire urban problems brought into focus at the United Nations Conference on Human Settlements in 1976. Similarly, the United Nations Conference on Desertification convened in 1977 emphasized the importance of using locally available renewable sources of energy to reduce excessive wood cutting in arid and semi-arid areas, which is one of the most important causes of desertification. The importance of renewable sources of energy has been recently brought into focus by the decision of the General Assembly of the United Nations to convene a United Nations Conference on New and Renewable Sources of Energy in Nairobi next year.

One of the tasks assigned to the Governing Council of the United Nations Environment Programme by the General Assembly of the United Nations in its resolution 2997 (XXVII) of 15 December 1972 is to:

"keep under review the world environmental situation in order to ensure that emerging environmental problems of wide international significance receive appropriate and adequate consideration."

In this respect, the United Nations Environment Programme has embarked on a number of in-depth reviews of the environmental aspects of production and use of all sources of energy. Three major studies have so far been carried out. The first one, dealing with the environmental impacts of production, transport, processing and use of fossil fuels, was reviewed by an International Panel of Experts that met in Warsaw in April, 1978. The second study, which deals with the environmental impacts of nuclear energy, was reviewed by an International Panel of Experts that met in Geneva in November, 1978, and revised at a second meeting held in Nairobi in April 1979. The third study, dealing with the environmental aspects of renewable sources of energy, was reviewed by an International Panel of Experts that met in Bangkok in January, 1980.

The present volume consolidates these three studies into one text, summarizing the state of knowledge at the end of the 1970s. I hope that this review will be found to give a balanced assessment of the environmental impacts of energy production and use as viewed by the United Nations organ responsible for the safety of the environment at the global level. It is also my sincere hope that the scientific community will pick up the many inadequacies in our knowledge of the environmental impacts of production and use of energy, and accelerate the efforts to find adequate solutions to the problems encountered or likely to be encountered.

Many scientists participated in the three studies that constitute this volume, either by contributing background papers and/or by participating at the review panels. To all of them, I would like to express my gratitude. My appreciation goes also to the United Nations bodies and to the scientific institutions that provided information and/or helped in the review process.

M.K. Tolba

Executive Director
United Nations Environment Programme

Nairobi, *October 1980*

Preface

At its fourth session, in 1976, the Governing Council of the United Nations Environment Programme requested the Executive Director to undertake the preparation of a comprehensive review of the environmental impacts of production and use of different sources of energy. I was then entrusted with planning and executing this Study. The plan was formulated early in 1977 and it was found most appropriate to divide the Study into three main parts: the first dealing with fossil fuels, the second with nuclear energy and third with renewable sources of energy. For each part a number of background papers was prepared by some experts and my task was to consolidate these papers, supplementing them with additional material gathered from the literature and from some United Nations Organizations and scientific Institutions, into drafts which were then reviewed by three main panels of experts. In the light of the comments of these panels and comments received from a number of reviewers, the drafts were then finalized and the reports released by UNEP in 1979 and 1980.

The three parts of the Study have been structured in such a way as to review the environmental impacts of each stage of the fuel cycle. This starts with the production of the raw material followed by its processing, transportation and use. Whenever possible the environmental impacts have been quantified on the basis of electricity production of 1000 MW (e)y to facilitate the comparison between the environmental impacts of the major sources of energy.

For the part on fossil fuels, background papers were prepared by Prof. L. Freiberg (Stockholm), Dr. B. St. John (Calgary), Dr. T. Tarnawa (Warsaw), Dr. M. Waldichuk (Vancouver) and Dr. J. Williams (Laxenburg). The following participated at the review panel that was convened in Warsaw, 17-21 April, 1978: Mr. M. Alwaer (National Oil Corporation, Tripoli), Prof. A. Beba (EGE University, Izmir), Prof. M. Berlin (University of Lund), Dr. C. Ducret (ECE, Geneva), Dr. L. Farges (IAEA, Vienna), Mr. M. Fila (IMCO, London), Dr. G. Foley (OECD, Paris), Prof. G. Goodman (Beijer Institute, Stockholm), Dr. L. Hamilton (Brookhaven National Lab., New York), Dr. W. N. Hurst

(Department of the Environment, Canberra), Dr. L. Feng (Research Institute of Petroleum, Peking), Dr. A. Martono (Electric Power Research Centre, Jakarta), Dr. Z. Nowak (Research Centre of Coal Processing, Warsaw), Dr. Y. Ogisu (National Research Institute for Pollution and Resources, Kawaguchi-Saitama, Japan), Dr. W. S. Osburn (Department of Energy, Washington), Dr. A. Podniesinski (Research Institute on Environmental Development, Warsaw), Dr. A. Pradinaud (Ministère de la Qualité de la Vie, Paris), Dr. L. Reed (Department of the Environment, London), Dr. P. Ruyabhorn (National Energy Administration, Bangkok), Dr. B. St. John (Petro-Canada, Calgary), Dr. A. Sauer (Ruhrkhole AG, Essen), Dr. K. W. Sedlacek (IPIECA, London), Dr. T. Tarnawa-Tomaszkiewicz (Ministry of Environmental Protection, Warsaw), Dr. A. Titkov (IAEA, Vienna), Dr. B. Turyn (General Bureau Energy Studies, Warsaw), Dr. G. S. van der Vlies (Shell International Research, The Hague), Dr. M. Waldichuk (Pacific Environment Institute, Vancouver), Dr. Wang Po-yung (Research Institute of Petroleum, Peking) and Dr. G. Woznjuk (Department of the Environment, Ministry of Coal, Moscow).

For the part on nuclear energy background papers were prepared by Dr. J. M. Costello (Australia), Dr. A. J. Gonzalez (Buenos Aires), Dr. A. K. Biswas (Laxenburg), Dr. J. R. Beattie (Culcheth), Dr. H. Howells (Risley), Dr. F. Kenneth Hare (Toronto), and Dr. Y. Sousselier (Paris). The following participated at the review panel that was convened in Geneva, 20-24 November, 1978: Dr. P. Albrecht (World Council of Churches, Geneva), Dr. Y. Ahmed (IAEA, Vienna), Dr. M. N. Aybers (Technical University of Istanbul), Dr. J. R. Beattie (UK Atomic Energy Authority, Culcheth, Warrington), Dr. D. Benninson (UNSCEAR, Vienna), Dr. A. Bishop (ECE, Geneva), Dr. A. K. Biswas (IIASA, Laxenburg), Dr. K. H. Buob (Institut Fédéral des Recherches en matiere de réacteurs, Würenlingen, Switzerland), Dr. M. Carriera Vasquez (Junta de Energia Nuclear, Madrid), Dr. J. H. Chesshire (University of Sussex), Dr. J. Costello (Australian Atomic Energy Commission, Sutherland, Australia), Dr. C. Ducret (ECE, Geneva), Dr. L. Farges (IAEA, Vienna), Dr. A. K. Ganguly (Bhabha Atomic Research Centre, Bombay), Dr. J. de Givry (ILO, Geneva), Dr. A. Gonzalez (Comision Nacional de Energia Atomica, Buenos Aires), Dr. G. Gorrie (Department of Environment, Canberra City, Australia), Dr. L. Hamilton (Brookhaven National Lab., New York), Dr. H. Howells (British Nuclear Fuels Ltd., Risely, Warrington, UK), Dr. R Ichikawa (National Institute of Radiological Science, Chiba-Shi, Japan), Dr. L. Ilyin (Institute of Biophysics, Moscow), Dr. H. P. Jammet (Institut Curie, Paris), Dr. Z. Jaworovski (Central Lab. for Radiological Protection, Warsaw), Dr. E. Komarov (WHO, Geneva), Dr. W. L. Lenneman (IAEA, Vienna), Dr. B. Lindell (National Institute of Radiation Protection, Stockholm), Dr. G. Ozolins (WHO, Geneva), Dr. Y. Sousselier (Commissariat à l'Energie Atomique, Fontenay-aux-Roses, France), Dr. P. Strohl (OECD, Paris), Dr. L. Teh-Ping (Institute of Energy, Peking), Dr. B. Wachholz (US Department of Energy, Washington, D.C.), Dr. B. Wallenberg (OECD, Paris) and Dr. J. L. Weeks (White Shell Nuclear Research Establishment, Pinawa, Manitoba, Canada). From these experts the following attended a second panel convened in Nairobi in the period 2-6 April, 1979: Dr. Y. Ahmed, Dr. J. R. Beattie, Dr. K. H. Buob, Dr. M. Carriera Vasquez, Dr. J. Costello, Dr. L. Hamilton, Dr. H. Howells, Dr. H. P. Jammet, Dr. Z. Jaworovski and Dr. J. L. Weeks. In addition Dr. H. T. Daw and Dr. C. H. Millar, both of IAEA, Vienna attended the meeting.

As for the part on renewable sources of energy, background papers were prepared by Dr. H. J. Allison (Oklahoma), Brace Research Institute (Canada), Dr. A. J. Ellis (New Zealand), Dr. H. Korniewicz and Dr. A. Podniesinski (Warsaw), Dr. M. Saleh (Cairo), Dr. J. W. Tatom (Smyrna, Georgia, USA) and Dr. E. N. Terrado (Manila). The following participated at the review panel convened in Bangkok in the period 7-11 January, 1980: Dr. J. Ahmed (IAEA, Vienna), Dr. J. Allison (National Science Foundation, Washington, D.C.), Dr. A. Arismunandar (Ministry of Mines and Energy, Jakarta), Dr. S. Arungu-Olende (UN, New York), Dr. A. K. Biswas (Biswas and Associates, Oxford), Dr. T. Cheickne (National Solar Energy Lab., Bamako, Mali). Dr. S. I. Dolgov (GKNT, Moscow), Dr. A. Ellis (DSIR, Wellington, New Zealand), Dr. Y. S. El. Mahgary (ESCAP, Bangkok), Dr. L. Hamilton (Brookhaven National Lab. New York), Dr. S. Hurry (University of Mauritius), Dr. P. Kambhu (ESCAP, Bangkok), Dr. J. Martinod (World Bank, Washington, D.C.), Dr. J. McDivitt (UNESCO, Paris), Dr. R. N. Morse (Victorian Solar Energy Research Committee, Melbourne), Dr. K. Openshaw (University of Dar es Salaam, Tanzania), Dr. J. Phéline (Commissariat à l'Energie Atomique, Paris), Dr. P. Ruyabhorn (National Energy Administration, Bangkok), Dr. W. Shearer (UNU, Tokyo), Dr. B. Sørensen (Niels Bohr Institute, Copenhagen), Mr. H. R. Srinivasan (Khadi and Village Industries, Bombay), Mr. H. S. Subasinghe (Ceylon Electricity Board, Colombo), Dr. J. Tatom (Smyrna, Georgia, USA), Mr. L. Wiltshire (UN, New York) and Dr. G. Woldegiorgis (National Energy Commission, Addis Ababa).

The Study benefited enormously from the contributions and comments of the previously mentioned experts. Dr. Mostafa Kamal Tolba, despite his many commitments, as the Executive Director of UNEP, followed closely the progress of the Study and provided most valuable advice and support. A great deal of credit goes also to Mr. R. B. Stedman, the former Deputy Executive Director; Mr. P. S. Thacher, Deputy Executive Director and Mr. S. Evteev, Assistant Executive Director who provided all possible assistance and encouragement.

Finally, several colleagues at UNEP provided useful comments and/or information that was of great help in drafting the parts of the Study. Mrs. D. Orrill, Mrs. Y. Da Costa, Miss E. Itaka and Mrs. A. Fernandez were responsible for typing the different manuscripts and greatly helped in the preparations for the panel meetings; their efforts are gratefully acknowledged.

Essam El-Hinnawi
Study Director
Chairman, Energy Task Force
United Nations Environment Programme

Nairobi, *October 1980*

ABOUT THE AUTHOR

Dr. Essam El-Hinnawi is Research Professor of Environmental Geochemistry at the National Research Centre, Cairo. He was awarded the State Prize for Science in 1967 for his achievements in geochemistry. He worked for a number of years as visiting research professor in Italy, and the Federal Republic of Germany. Besides his work at the National Research Centre in Cairo, he served as Secretary General of the Environmental Research Council, co-ordinating environmental research programmes in Egypt. He acted also on several advisory committees for policy planning at the Academy of Scientific Research and Technology. Since 1976, Dr. El-Hinnawi has been on leave of absence from the NRC, working with the United Nations Environment Programme as Chief of the State of Environment Unit and as Chairman of the Energy Task Force. Dr. El-Hinnawi has published more than 120 papers in mineralogy, environmental geochemistry, and natural resources. He supervised more than 25 M.SC. and Ph.D theses in these fields, and is the author of three books and editor of a recent one on Nuclear Energy and the Environment.

Introduction

ENERGY is one of the most important pre-requisites of life. Without energy our entire civilization — transport, industrial manufacturing, commercial activity and food production — would come to a standstill. Since prehistoric times human society has been consuming an increasing amount of energy. The modern industrial era brought with it a marked increase in consumption of energy and changes in energy sources from wood and coal to predominantly oil and natural gas. The relationship of economic prosperity to energy consumption has become an essential element of energy policy, for it couples the latter to economic policy and the general national welfare. Although there are differences in the amounts of energy required in different countries to achieve a given level of gross national product (GNP), there is nevertheless a rather consistent relationship between GNP and energy consumption. Recent development concepts aiming at the satisfaction of basic human requirements relates energy consumption to basic needs, production structure and styles of development rather than to GNP alone.

The rising global demand for energy has been met to an increasing extent by the use of fossil fuels, especially oil, which were cheap and plentiful. Recently, most of the countries have realized that non-renewable sources of energy are finite in extent and that diversification of energy sources is a must for future development. Concern for future energy supplies is reflected in the programmes of many national governments and in the efforts being made by a number of international organizations to assess global energy resources and possible rates of supply and substitution. Energy policies are nowadays influenced by several factors: energy resources, population growth, level and nature of socio-economic activity, the relative costs of energy, the adequacy and reliability of supply, the availability of technology and supporting infrastructure, the success of energy conservation programmes and concern about the environmental and safety aspects of production and use of energy.

World energy consumption increased almost 600 per cent between 1900 and 1965 and is projected to increase another 450 per cent between 1965 and the year 2000. Most of the world's consumption of energy from fossil fuels throughout all history has taken place during the past 40 years. Estimates of future demand and supply vary considerably, depending on the assumptions made about resource availabilities, economic growth, pricing policies, the responsiveness of energy demand and supply to changes in prices and incomes, and political and environmental factors (see, for example, WAES, 1977; World Energy Conference, 1978; World Bank, 1979).

Industrialized countries dominate the commercial energy market, accounting for about 85 per cent of world consumption; the developing countries account for only about 15 per cent (World Bank, 1979). The energy problem over the next two decades should be seen as one of transition, in which countries need to ensure the rational, non-wasteful use of the non-renewable sources of energy and to diversify their sources of energy establishing an appropriate and environmental realistic "energy mix" to meet their incremental needs.

ENVIRONMENTAL IMPACTS

Consideration of the Declaration and Principles of the United Nations Conference on the Human Environment convened in Stockholm in 1972 shows that the environment — defined as the whole outer physical and biological system in which man and other organisms live — is a whole, albeit a complicated one with many interacting components. The wise management of that environment depends upon an understanding of those components: of its rocks, minerals and waters, of its soils and their present and potential vegetation, of its animal life and of its climate. Good management avoids pollution, erosion, and the wastage of resources by irreversible damage. To prevent such types of environmental degradation is even more challenging and certainly more efficient than to redress them after they had occurred.

Of the many potential environmental impacts associated with any particular energy technology, some would be substantial and others small, some important and others of little consequence, some of short duration and others with long term effects, some might be adverse and others beneficial and they might occur in different geographic areas and might affect different communities in different ways. A distinction should be made between the assessment of the nature, scale and geographic distribution of the impact, and the evaluation which is concerned with its value or importance. For many environmental changes which are identified as impacts, the state of knowledge and the technology will often only permit a qualitative assessment. Only in a few cases is it possible to evaluate an impact quantitatively. Decisions must ultimately be made on the basis of a combination of cost/benefit analysis, other quantified inputs and qualitative information.

Discussion of the environmental impacts of various energy strategies has, in the past, tended to focus more attention on short-term aspects, such as occupational and public health and direct impacts on the physical environment, than on the long-term socio-economic and environmental consequences. However, there is now a growing disposition to analyse these long-term impacts which may range from those for which substantial data exist and around which there is a fair degree of certainty as to the risks involved, to those which are rather speculative in nature and for which very little data are available.

The biosphere consists of different organisms, plants and animal life supported by a number of physical characteristics such as topography, soils, climate, water supply and drainage. For a given development, these physical characteristics and hence the biosphere may be affected. Whenever pollutants are released, the analysis of the environmental impact of these pollutants requires the knowledge of the:

(a) quantity and types of pollutants released;
(b) dispersion of these pollutants in the environment;
(c) ecological pathways followed by the pollutants;
(d) relationships between the pollutants and the damage to man and his environment;
(e) the extent of the damage including its cost (where it is possible to make this assessment).

The total impact of some pollutants may depend on positive or negative synergistic effects. Although standards have been formulated for "acceptable" levels of several pollutants, it is prudent to assume that for exposure to many pollutants there is no threshold and that effects can occur at very low exposures. A further feature is that many pollutants may remain in the environment (and accessible to the food chain) long after the action releasing them has been discontinued. Attempts to assess the long-term impacts of these pollutants, although difficult, should be made taking into consideration the different pathways, biogeochemical cycles and fate of these substances in the environment.

The assessment of environmental impacts resulting from the different stages of any fuel cycle* is important in relation to policy-making and decisions about energy options or "mixes" to be developed. The conservation of the natural environment is essential to the maintenance and regulation of the food, air and water cycles on which human life depends and to socio-economic development. The most important impacts of any energy technology on the human environment are probably the impacts on health and safety and on social well-being.

An important part of the overall risk assessment process for the various energy options is the attempt to quantify the total harm to man and his "life-support-systems" caused by energy pollutants, e.g. radionuclides, potentially

---

*In this book, the "fuel cycle" refers to the entire programme from the extraction of raw material, through transportation, processing, storage, use of the fuel, to the management of wastes produced in all steps of the cycle.

toxic heavy metals, polynuclear aromatic hydrocarbons (PAH), nitrogen oxides (NO_x), sulphur oxides (SO_x) ... etc. These act directly (via inhalation, ingestion) or indirectly (via biogeochemical cycles and the human food-chain) to affect human health, agriculture or resource productivity, or to cause nuisance or other disamenities (e.g. corrosion). Although the amount of this risk is nowadays being worked out separately for each energy option on a comparative basis, traditionally, environmental health authorities have tried to assess the amount of harm to people resulting from direct or indirect exposure to various doses of a pollutant whatever its origins, i.e. in a more general way taking into account all its sources together.

Using experimental animals as substitutes for man, toxic effects (particularly acute effects) have frequently been found to follow a sigmoid curve with dose where, below a certain level of pollutant virtually no effect is found. This type of dose-response relationship is often confirmed by occupational health statistics (i.e. at relatively high pollutant levels). As a consequence, attempts have usually been made to derive safety standards and "clean-up" procedures for pollutants in the various environmental media of air, water, soil, food etc., so that they ultimately deliver a dose to man below a "safe" level, thus avoiding harm altogether. However, apart from acute and sub-acute effects caused by relatively high pollutant doses, there are long term effects on living organisms (including man) usually associated with smaller doses, e.g. chronic illness, carcinogenic, mutagenic or teratogenic effects. All these may appear much later on in the life-cycle or even be delayed until later generations. Chronic, and particularly "late" effects (genotoxic effects) are more and more coming to be regarded as being linearly related to low doses or repeated incremental doses (see for example, Ehrenberg and Löfroth, 1979).

The following chapters describe the impacts of production and use of different sources of energy on ecosystems (especially land, water and air). Whenever quantitative data are available, the impacts have been calculated on the basis of electricity production of 1000 MW(e)y (1GW(e)y), for easy comparison.

References

Ehrenberg, L. and G. Löfroth (1979): On the Assessment of Genetic and Carcinogenic Effects; in Goodman, G. T. and W. D. Rowe (Ed.) *"Energy Risk Management"*; Academic Press, London.

WAES (1977): *Workshop Alternative Energy Strategies.* Edit: C. Wilson; McGraw Hill, New York.

World Bank (1979): *World Development Report,* 1979. Washington, D.C.

World Energy Conference (1978): *World Energy Resources 1985-2020.* IPC Sci. Techn. Press.

Part I

FOSSIL FUELS

CHAPTER 1

Coal

Coal has long been used as an energy source and it occupied a prominent position amongst fossil fuels at the beginning of the present century. The importance of coal declined after the extensive discoveries of oil, which has, until recently, been cheap fuel. Nowadays, special emphasis is being made on expanding the production and use of coal to reduce the dependence on the fast-depleting oil resources.

Table 1 gives the annual production of coal in the World in 1975-1977 showing only a slight increase in total production; the average per capita consumption of coal in 1976 was about 2600 kg. The total world resources of hard coal and brown coal have been estimated at $10,310 \times 10^9$ tonne coal equivalent (tce); the reserves that are currently technically and economically recoverable are estimated to be about 640×10^9 tce. i.e. 6.3% of known resources*. It is expected that this amount will double by the year 2020 (Grenon, 1978). Table 2 gives the geographical distribution of coal resources and the amounts of technically and economically recoverable deposits.

Table 1 World Production of Coal (in million metric tons)[**]

	1975		1976		1977	
	H.C.	B.C.	H.C.	B.C.	H.C.	B.C.
Africa	75.4	0.1	81.2	0.1	85.5	0.1
N. America	595.0	21.5	603.5	27.9	631.4	32.8
S. America	7.9	0.1	9.4	0.1	9.8	0.1
Asia	652.6	20.1	673.4	23.2	698.3	26.1
Europe	475.9	630.3	482.0	655.0	486.0	700.0
Oceania	62.2	30.1	70.1	31.1	73.3	33.2
U.S.S.R.	484.7	160.2	494.4	192.8	500.0	206.0
World	2353.7	862.4	2413.7	930.8	2475.1	998.3
Total coal		3216.1		3344.5		3473.4

[**]H.C. : Hard coal with thermal value greater than 23210 kJ/kg;
  B.C. : Brown coal and lignite with thermal value less than 23210 kJ/kg.
  After : World Energy Supplies 1972 – 1976, United Nations (1978)

The expansion in coal production is faced by a number of obstacles, the most important of which are: the availability of qualified miners and engineers; the construction of a suitable infrastructure and of adequate transportation facilities; various environmental problems, which need to be solved, both in production and in consumption; the fact, that at present potential markets for coal are not yet sufficiently being developed in many parts of the world, since other sources of energy are still being offered at competitive prices; the lack of

*Recent estimates by World Energy Conference (1980) are: $11,062 \times 10^9$ tce for geological resources and 687×10^9 tce for technically and economically recoverable reserves.

Table 2 World Coal Resources (in 10⁶tce)*

	Geological Resources		Technical and Economically recoverable	
	H.C.	B.C.	H.C.	B.C.
N. America	1,286,225	1,399,525	121,938	65,031
S. America	25,106	9,263	4,901	5,860
Asia	5,494,025	887,127	219,226	29,591
Europe	535,664	53,741	95,010	33,752
Africa	172,714	190	34,033	90
Oceania	213,890	49,034	18,164	9,333
Total World	7,727,624	2,398,880	490,272	143,657

*After Grenon (1978).

interest on the part of the potential investors to commit themselves to the development of coal; the long lead times required for opening up new mines, establishing the necessary infrastructure, transportation facilities, etc.

ENVIRONMENTAL IMPACTS OF COAL MINING

Coal can be mined, according to its geological setting by a variety of methods, the most common of which are underground mining (room and pillar or long-wall techniques) and strip mining (area or contour). Recovery factors of 100 per cent are not possible in practice; at present the average recovery factors in most coal producing countries are around 35 per cent in underground room and pillar mining, 60-80 per cent in underground mining of long-wall faces and 80-90 per cent in surface mining (ECE, 1976). The amount of coal to be mined depends on the energy content of the coal, which varies with its type and its use — whether for domestic purposes, industry (steel, cement...etc), electricity generation, or a combination of these.

Occupational Hazards:

Coal mining has a number of occupational hazards. Besides accidents in underground mines (fires, explosions, land subsidence, etc) and in surface mining, coal-workers are exposed to pollutants causing respiratory ailments and to noise.

The accident fatality rate in underground coal mining has been estimated to be 0.4 deaths per 10⁶ tonnes of mined coal, on the basis of U.S.A. statistics for the period 1971 to 1975. In large modern mines, the rate is likely to decrease by 50%, i.e. 0.2 deaths per 10^6 tonnes of mined coal (Hamilton, 1977; Morris *et al.*, 1979). In surface mining, the accident fatality rate has been estimated to be 0.1 deaths per 10^6 tonnes of coal. The rate of accidental non-fatal disabling injuries in underground mining is estimated to be 30 per 10^6 tonnes of mined coal, while in surface mining it is estimated to be 5.5 per 10^6 tonnes of coal (Morris *et al.*, 1979). Table 3 gives the estimated number of fatal and non-fatal accidents resulting from coal mining per 1 GW(e)y. It should be noted that statistics from other countries might show different accident rates and the estimated number of fatal and non-fatal accidents per 1 GW(e)y could be higher or lower than these figures.

Table 3. Estimated fatal and non-fatal accidents from coal mining (per 1 GW(e)y)*

	Underground Mining	Surface Mining
Fatal accidents	1.20 (0.6)**	0.3
Non-fatal accidents	90 (45)**	16.5

* – Assuming that 3×10^6 tonne coal (energy content 2.74×10^7 kJ/tonne are required for the production of 1 GW(e)y from a power plant with 38% thermal efficiency.

** – Values between brackets are estimates for large modern mines.

Occupationally induced mortality and morbidity (due to respiratory diseases, especially black-lung disease known as coal worker's pneumoconiosis, CWP) in coal mines is subject to great uncertainty. It was estimated (New York Acad. Sci. 1971) that 25% of coal miners get black-lung disease from inhalation of coal dust. However, in many countries, stringent control measures have considerably reduced the incidence of the disease. Rae (1971) estimated that CWP might be virtually eliminated by reducing dust concentration in underground mines to 2 mg/m^3. However, this may not be possible for techno-economic reasons or in some coal mines, especially in developing countries.

In a recent study in the U.S.A., Rockette (1977) estimated the occupational disease induced-mortality in underground coal mining at 0.07 deaths per 10^6 tonne. Morris *et al.* (1979) considered this figure as the best estimate, with a range of zero to 0.47. The chronic respiratory disease incidence has been estimated to be about 12 times that of deaths (Morris *et al.* 1979). Using these estimates, the calculated number of occupational deaths per 1 GW(e)y would be 0.21 (range between 0 and 1.41). The calculated number of cases of chronic respiratory disease is 2.52 per 1 GW(e)y.

Some studies have been made on the relationship between trace elements in coal and the severity of CWP (Sweet *et al.* 1974; Carlberg 1971). No relationship has been found between the severity of CWP and the average concentrations of chromium, copper, iron, manganese, nickel, titantium and/or zinc. However, vanadium was found to be more positively correlated with CWP than either beryllium or magnesium. A 1975 survey of the presence of radon-222 and radon-220 daughters in 123 operating mines in the U.S.A. concluded that there was no apparent occupational health hazard from inhalation of radon-222 or radon-220 daughters (Lee *et al.* 1975).

On the other hand, the incidence of CWP in surface coal mines is much lower than in underground mines. Fairman *et al.* (1977) found that 4% of 1438 surface miners has some evidence of CWP, but only 7 miners (0.5%) had X-ray films interpreted as CWP. Most of these miners had previously worked in underground mines for prolonged periods.

Another occupational hazard in coal mining is the exposure to noise produced by mining equipment. A survey made in 1976 in North-Rhine Westfalia (Federal Republic of Germany) showed that about 61% of workers in underground mines were exposed to a maximum noise level of 90 dB(A) which is considered non-injurious according to present regulations in the FRG; about 31% were working at noise levels of 91 – 100 dB(A) and about 8% of the workers were exposed to 101 dB(A). Recently, several technologies have been applied leading to marked reduction in the level of noise of some of the mining equipment (Hurck, 1978).

Impact of Coal Mining on Land:

Surface coal mining (strip mining) has a more potential impact on land than underground mining. The amount of disrupted land varies from one place to another, depending on the geological occurrence of the coal-bearing formation and its characteristics. It has been estimated in Poland, for example, that 6–9 hectares of land are disrupted in the production of one million tonne of lignite (ECE, 1976).

Where surface mining is carried out in densely populated areas (for example, in the Rhineland, Federal Republic of Germany) it has a direct effect on human settlements and the total infrastructure in the area. Construction of new settlements, roads, etc., is necessary if mining operations are to move towards older inhabited areas. Reclamation of strip-mined areas has been successfully achieved in some countries. In the Rhine area, for example, huge wheel excavators selectively strip off and save the top layer of loess (an extremely fertile type of loam), and remove the remaining sand, gravel and clay overburden to expose the coal beds. Simultaneously, mammoth spreader machines fill the overburden back into mined-out pits while bulldozers level it out in preparation for applying the top layer of loess (Nephew, 1972). Fields of grain and hay are already thriving on land that was restored less than five years ago.

The degree of difficulty and the time required for the land to be restored depends to a large extent on the grading of the land after mining, soil restoration, fertilizer addition, the amount and type of reseeding, and the amount of rainfall and water available in the area. The water needs for reclamation vary from 1500–12000 m³ per hectare per year. The amount and type of treatment to obtain satisfactory results will vary from place to place as will the costs. It is possible, in spite of the great variability, to generalize on the range of costs that may be involved in restoring strip mined areas. The complete restoration conducted in the German Rhineland brown coal region is estimated at $7500 to $11,000 per hectare, and about $10,000 per hectare for the UK; costs estimated for the U.S.A. vary between $250 and $10,000 per hectare depending on the degree of reclamation used. Generally speaking, about 4% of the coal production costs are required for reclamation of land (OECD, 1978). In some areas, coal mine spoil has been used for brick manufacture, cement and light-weight aggregate production (Glover, 1978).

Impact of Coal Mining on Water:

Coal-mining water demands are modest, and include water for dust control, fire protection and coal washing. Average water use in coal mining varies from 63 to 120 litres per metric ton in underground mining and about 17 litres per tonne for surface mining. An additional 33 litres per tonne in both methods are required for waste disposal (UNESCO, 1979). Underground mining has little direct impact on surface streams but may have profound long-term effects on ground water resources. By its nature a mine tends to drain a large area, and if soluble minerals are present in the coal and associated rocks, these minerals may enter the streams draining a mining region and cause severe degradation of the water quality. This effect is important both where mines are drained by pumping and in mountainous regions where drainage is by gravity. Contamination caused by both deep and surface mining has substantially altered the water quality of some 17,000 km of streams in Appalachia, U.S.A. Acid drainage seriously pollutes about 10,000 km of streams, reducing or eliminating aquatic life (Nephew, 1972).

Different control measures are being taken in different countries to minimize the detrimental effects of acid mine drainage. Such measures include drainage control in the mine area, proper disposal of sulphur-bearing materials to ensure that pyritic refuse does not come in contact with water, sealing up abandoned mines to prevent water from entering the sulphur-bearing soil and chemical treatment of acid mine drainage. The latter consists essentially of treatment with various alkaline neutralizing agents, including limestone, lime, and caustic soda. Such treatment is designed primarily to treat solution acidity and remove those ions whose solubilities are sensitive in the acid to neutral range. New methods of treating acid mine drainage utilizing ion-exchange technology, reverse osmosis, and flash distillation reduce acidity and a substantial part of the dissolved ions. In some countries, stringent control measures are being implemented to avoid the hazards of acid mine drainage.

ENVIRONMENTAL IMPACTS OF COAL PROCESSING

Coal cleaning is accomplished by physically separating refuse and sulphur-containing pyritic material from coal. Chemically-bound organic sulphur is not removed. These physical benefication techniques are capable of removing up to 40–50% of the sulphur and 65–75% of the ash contained in the raw coal (Ferrell, 1978). The amount of pollution generated in the process of coal cleaning depends upon the amount of coal treated, the chemical and physical properties of the coal, and the top size to which the coal must be prepared.

Most air pollution arising from coal washing is the result of drying of the coal in a stream of hot combustion gases that are produced by the burning of coal. The air pollutants that are created are fine coal dust, ash, NO_x and SO_x, resulting from the combustion of the coal used to supply the hot gases for coal drying. In addition, dust is created during the crushing and grinding of the coal to the size required. Air pollution can also arise from burning of refuse piles at coal preparation plants which have caught fire as the result of spontaneous combustion. This can usually be controlled by preventing oxygen from reaching the carbonaceous material by reshaping the pile to avoid steepy sides and by covering the pile with soil and vegetation.

Coal washing requires varying amounts of water depending upon process and type of coal (UNESCO, 1979). The liquid effluent produced is called black-water and contains suspended coal fines. Black-water is usually sent to a tailings pond where the solids are allowed to settle and the clear water re-circulated. Re-circulation and treatment of wash water are integral parts of the operation of modern coal cleaning plants. Closed water circuits have grown in popularity because they eliminate discharge to streams, reduce makeup water, and allow for improved recovery of fine coal.

Solid wastes from coal cleaning and processing consist of coarse and fine refuse and constitute a potential source of acid mine drainage. The finely divided and well-exposed state of coal refuse enables acid-producing and weathering processes to work more effectively. Pollution from refuse piles can be controlled by proper reclamation to prevent seepage or by diverting any seepage from the piles to settling ponds where suspended solids and other pollutants can be retained and treated.

The occupational accidents in coal processing have been estimated at 0.02 deaths per 10^6 tonne of coal input to the plant and 1.3 disabling injuries per 10^6 (Morris *et al.*, 1979). Accordingly, 0.08 deaths and 5.2 dissabling injuries may be incurred per 1 GW(e)y.

ENVIRONMENTAL IMPACTS OF COAL TRANSPORTATION AND STORAGE

Inland coal transportation is generally by railway or inland waterway over long distances, by conveyor belt or road over shorter distances, and in the future will be increasingly by slurry pipelines. The principal environmental impact of surface transportation of coal is the fugitive coal dust, unless measures are taken to eliminate or substantially reduce its occurrence. A surface pipeline might create a barrier to animals and farm equipment, thereby creating possibly greater secondary impacts than railways.

The principal pollutant from loading/unloading, and transportation, and storage of coal is fugitive coal dust. It is estimated that about 0.1 per cent of coal are lost during transit and the loading/unloading operations (OECD, 1978). The amount of dust generated from the open storage piles varies widely depending on such factors as climate, topography, and characteristics of the stored coal. Dust from coal handling and open coal storage can be suppressed by different methods, for example by spraying, storing in enclosed bins or silos, etc. Water used in coal slurry pipelines is contaminated with suspended and dissolved solids and may cause water pollution problems if discharged without proper treatment.

The storage of any coal can present problems of spontaneous combustion from reactions between the coal and atmospheric oxygen at ambient temperatures. This is especially true in case of lower rank coals with high proportion of volatiles. Special precautions are generally taken to prevent such spontaneous combustion.

The accidental fatality rate in railway transport (based on statistics for rail freight transport in the U.S.A., see Morris *et al.*, 1979) has been estimated at 0.01 per 10^6 tonne of coal transported over a haul distance of 500 km, while in waterway transport by barge, it is estimated at 0.04 death per 10^6 tonne of coal. For pipeline transport, accidental death is 0.005 per 10^6 tonne of coal over a 500 km distance. According to these estimates, the calculated fatalities from coal transportation per 1 GW(e)y would be:

Railway transport of coal	0.03
Waterway transport of coal	0.12
Pipeline transport of coal	0.01

It should be emphasized that these values differ markedly with distance of transportation and are lower for shorter distances. Some countries, for example, the U.S.S.R., have introduced the concept of "energy parks" or complexes where the plants are located near the coal mining and cleaning plants. This makes it possible to reduce the environmental impacts of transportation of coal and to better manage the wastes resulting from all plants. The U.S.A. and other countries are considering this concept. In such cases, however, the electricity generated from these centralized complexes has to be transmitted through long transmission lines to centres of consumption.

ENVIRONMENTAL IMPACTS OF COAL UTILIZATION

Traditional Coal Combustion:

Coal is used directly at home, in industry, agriculture or for electricity generation. The percentage of coal used in the different sectors varies considerably from one country to another.

Direct combustion of coal has a number of environmental impacts due to the emission of air pollutants (sulphur oxides, SO_x; nitrogen oxides, NO_x; carbon monoxide, CO; carbon dioxide, CO_2; particulates; organic compounds; trace metals; radionuclides; etc.) and the creation of an ash waste which requires disposal. These problems are particularly evident in the power industry where large quantities of coal are burned.

When coal is burned, sulphur in the coal is oxidized mainly to sulphur dioxide (SO_2) but very small amounts of sulphur trioxide are also produced. Some proportion of the sulphur always remains with the bottom ash as sulphate and a small amount is retained as sulphate in the fly ash. Generally about 5% of the sulphur in hard coal is retained in the bottom ash and about 95% is emitted as SO_x, although there can be a greater retention with lower combustion temperatures.

The quantity of fly ash or particulate emissions is determined by the ash content of coal, the temperature of combustion and the design of the boiler. Fly ash is generally formed by evaporation and subsequent condensation of ash matter, or by elutriation of solid or molten ash particles by the moving gases. The former mechanisms leads to the formation of very fine particulates, often less than one micron in diameter, which are difficult to remove from the flue gas. Higher combustion temperatures produce more fine particulates. Technology is well developed for the control of particulates above the domestic use level. Domestic control is often applied in urban areas by requiring the use of specially prepared smokeless solid fuel.

The formation of nitrogen oxides (NO_x) during combustion is a complex mechanism which depends upon temperature of combustion, design of the combustion chamber and availability of excess combustion air. During combustion both the nitrogen in the fuel and the nitrogen in the air can be converted to NO_x. In larger combustors, the temperature of combustion is higher and NO_x formation is greater. NO_x formation can be controlled to some extent by controlling the excess combustion air or by certain modifications to the combustor.

Table 4 gives an estimate of the airborne, liquid and solid wastes produced by a coal-fired power plant. The characteristics and impacts of the different pollutants on the environment are discussed in detail in Chapter 4.

Coke Production:

Coking is the process of heating coal in an atmosphere of low oxygen content, i.e., destructive distillation. During this process organic compounds in the coal break down to yield gases (coal gas) and a residue of nonvolatile nature (coke). The gases are drawn off to a collecting main and are subjected to various treatment to separate ammonia, coke-oven gas, tar, phenol, light oil and pyridine. The collected coke-oven gas, being high in CO and hydrocarbon content, is used as a fuel in the coking process and in the steelmaking process. Hydrogen and carbon disulphide present in the gas are emitted as sulphur dioxide during combustion. Desulphurization of coke-oven gas prior to combustion is carried out in a number of countries, notably in FRG and Japan.

Table 4.   Effluents from a coal-fired power plant (per 1 GW(e)y). in tonne*

Airborne Effluents	
Particulates	3×10^3
Sulphur oxides	11×10^4
Nitrogen oxides	2.7×10^4
Carbon monoxide	2×10^3
Hydrocarbons	400
Liquid Effluents	
Organic material	66.2
Sulphuric acid	82.5
Chloride	26.3
Phosphate	41.7
Boron	331
Suspended solids	497
Solid Wastes	
Bottom ash + recovered fly ash	3.6×10^5

*Calculation from data given by Wilson and Jones (1974). Assuming power plant burns 3×10^6 tonne coal: sulphur contents of coal 2%; energy content of coal 2.74×10^7 kJ per tonne; thermal efficiency of power plant 38%; fly ash removal efficiency 99%; no flue gas desulphurization.

Water and tar vapour in the gases emitted from the coking process condense as the gases are cooled. The water is then separated from the tar in continuous decanters. This water, called ammonia liquor, is by volume the second largest product from coking. It contains phenol which, even when present in very minute quantities, constitutes a real threat to water purity. The total amount of phenol-carrying water discharged may be of the order of 130 to 180 litres per tonne of coal. Up to 98–99 per cent of the phenols can be removed by existing liquid extraction processes (OECD, 1978).

Coal Conversion Processes:

The concept of coal conversion is deceptively simple. It involves primarily two basic steps: the cracking of heavy hydrocarbons into lighter ones and the enrichment of the resultant molecules with hydrogen. Unlike the concept, the application of coal conversion is not simple. It involves the handling of enormous amounts of a highly variable material (upwards of 25,000 tonnes of coal per day) often at high temperatures and pressures; it requires containment and control of both highly corrosive process materials and those that pose a possible health threat; and it calls for treatment and disposal of a voluminous solid waste and a possibly hazardous liquid or gaseous waste (Braunstein *et al.*, 1977). Nonetheless, coal conversion is a viable technology.

Coal gasification:

Low-Btu and high-Btu gases are the products of prime interest in coal gasification. Low-Btu gas (heating value of 100–500 Btu/scf or 3700–18500 kJ/m^3) is usable as fuel feedstock or for power generation in combined gas-steam turbine power cycles. High-Btu gas (heating value of 900–1000 Btu/scf or 33300–37000 kJ/m^3) can be a substitute for natural gas and would find wide use for heating fuel and industrial feedstock. The basic difference between low-Btu and high-Btu processes is that air is used instead of oxygen in the gasifier to produce low-Btu gas. Also, the shift conversion and methanation operations are not used in low-Btu gas production. Most gasification technologies under development today incorporate the concept of hydrogasification in which the incoming coal is initially reacted with a hydrogen-rich gas to form directly substantial amounts of methane.

The rates and degrees of conversion of the various reactions that take place in gasification are functions of temperature, pressure, gas composition, and the nature of the coal being gasified. Aside from the various operating configurations used, most gasification systems can be categorized according to the following methods of contracting gas and liquid streams: suspension or entrained-bed reactor; fixed — or moving-bed reactor; fluidized-bed reactor, and molten-bath reactor (OECD, 1977; Braunstein *et al.*, 1977).

Although coal gasification was carried out on a commercial scale in the 1930s and 1940s, the technology was abandoned when a cheaper natural gas and oil became abundant. At present, revival of interest in coal gasification, especially in industrial countries, is based on the expected need to find a substitute to restrict oil imports. Several studies are underway to develop technologies for coal gasification with reasonable costs to increase the conversion efficiencies.

In-situ gasification of coal, which converts a coal seam into combustible gases, produces a gas rich in carbon monoxide, carbon dioxide, hydrogen, hydrocarbon gases, and other gases. Depending on the gas injected (air, oxygen, or oxygen and steam), the heating value of the product gas can range from as low as 50 Btu/scf when air alone is injected to as high as 280 Btu/scf when an intermittent air-stream sequence is used (Braunstein *et al.*, 1977). The idea of underground gasification is attractive. The method avoids mining and waste disposal and adds unaccessible, unworkable, or uneconomical coal deposits to the energy pool. Unfortunately, the gas has a low heating value and would either have to be used at the site or upgraded for transportation.

Coal Liquefaction:

Synthetic liquid fuels have a prime potential use in the firing of industrial and electric boilers and gas turbines. Advantages of coal liquefaction are that the entire range of possible products — including fuel oil, gasoline, jet fuel, and diesel oil — can be produced from coal by varying the type of catalysts and other operating conditions. This flexibility is very desirable from a process standpoint.

Industrial quantities of liquid fuels, suitable as a petroleum refinery feedstock, were produced in Germany in the 1930s–1940s by two different processes. In one (Bergius), the coal was dissolved in recycled oil and reacted in the presence of a catalyst with hydrogen (produced by the gasification of coal) at pressures of 700 atm. and temperatures of 450°C. The resulting liquid product was filtered, part of the oil recycled with fresh coal, and the balance of the primary oil product hydrogenated in a gaseous phase to make synthetic petroleum products. In the other process (Fischer-Tropsch), all of the coal was first gasified to a mixture of carbon monoxide and hydrogen. After purification and adjustment of the carbon monoxide-to-hydrogen ratio, the gas was reacted over a catalyst at moderate pressures (30–35 atm.) and temperatures (235°C) to produce liquid products. By the selection of different carbon monoxide-to-hydrogen ratios, catalysts, pressures, and temperatures, a wide range of liquid products can be made — from straight chain hydrocarbons to alcohols. A commercial plant using the Fischer-Tropsch process is in operation today in South Africa. South Africa domestic coals are gasified in Lurgi generators, the resulting gas purified, and then liquefied by two different catalytic processes. The plant uses 6000 tonne of coal per day and produces a wide range of liquid products — gasoline, diesel and other fuel oils, waxes, alcohols, and ketones. The plant size is 8,000–10,000 barrels of oil per day (OECD, 1977). Many of the new processes now being investigated to convert coal to oil are designed to produce a clean liquid product than can be used as a utility or industrial boiler fuel to raise steam or for use as a heat source for industrial processes. The objective is to have a product that is low in sulphur and ash and can be stored, transported and burned as a liquid.

Environmental Impacts of Coal Conversion:

Table 5 summarizes the estimated resource requirements and effluents for coal gasification and liquefaction. It should be noted that modern plants are designed for zero aqueous discharge, thus either avoiding water pollution concerns or reducing water intake requirements.

Conversion processes are dually oriented: (1) to convert plentiful coal into scarce liquid and gaseous fuel and (2) to remove or treat, during processing, environmentally unacceptable or health-endangering compounds. As a result, coal conversion methodology is concerned as much with processes for handling the by-products as with the products themselves. Knowledge of wastes and emissions from coal conversion process is still incomplete (Braunstein et al., 1977; Talty, 1978). However, it is known that wastes will be generated during each main process stage and that these wastes are largely controllable or convertible to environmentally acceptable forms. A more serious and less

Table 5.   Estimated resource requirements and effluents from coal gasification
and liquefaction*

	Low-Btu Gasification	High-Btu Gasification	Coal Liquefaction
Resource requirements			
Coal (tonne)	4.6×10^6	5.7×10^6	4.6×10^6
Land use (hectare)	1531	672	1562
Water (litre/y)	2.4×10^9	19.6×10^9	10×10^9
Effluents (air-borne), tonne/y			
Particulates	0.75	820	529
SO_x	1960	9061	1706
NO_x	982	6771	7409
CO	28	356	291
Hydrocarbons	28	108	2268
NH_3	40	49	–
Solid wastes (tonne/y)	53×10^4	57×10^4	62×10^4

*Calculation after date given by EPA (1978), on the basis of energy requirements to produce 1 GW(e)y; i.e. 8.3×10^{13} kJ/y. The energy content of coal assumed to be 2.74×10^7 kJ/tonne.

tractable problem arises from possible fugitive emissions produced by inadequate containment of process streams (leaky valves) or incomplete treatment of wastes. High-temperature process streams may contain carcinogenic organic compounds similar to those implicated in an increased incidence of skin and lung cancer among coke-oven workers.

Because coal is a dirty fuel, early technologies that used coal without sulphur-emission control contributed heavily to environmental pollution. Coal conversion, however, is specifically mandated to produce clean fuel from coal; thus, a wide variety of techniques for removing sulphur are integrated into coal conversion processes. Sulphur can be removed during coal pretreatment of the conversion process and from product and waste streams. Because organic sulphur is difficult to remove during coal pretreatment and because about 90% of the organic sulphur is converted to hydrogen sulphide in both gasification and liquefaction, this form of sulphur is estimated to be the major sulphur contaminant in gas streams. Most of the sulphur in coal is expected to be recovered from conversion streams as elemental sulphur, although the sulphur that reaches the stack gas as sulphur oxides and is collected by limestone scrubbing will end up as sulphate. Neither form of sulphur is particularly obnoxious environmentally, although the quantity to be disposed of is considerable: A 25,000-ton-per-day plant using a coal containing 4% sulphur will produce 1000 tons of sulphur daily. As opposed to throwaway disposal, new uses for elemental sulphur are being considered. Production of sulphur-based paving materials and plastics or use of sulphur in thermal and acoustical insulation are considered alternatives to landfill or minefill disposal (Braunstein *et al.,* 1977).

The fates of trace elements during the coal conversion process are not yet completely known; however, the majority of elements are expected to remain in the solid by-products such as ash, chars, and filter cake. Nontheless, material balance studies are required to trace the pathway of hazardous elements during conversion and to determine the concentrations and forms of the elements in gas streams, condensates, and waste streams. The release of conversion process wastewater into natural waters also constitutes a possible source of environmental pollution. The aqueous streams, which arise when process gases are condensed or scrubbed, may contain components of the product gas — carbon monoxide, carbon dioxide, hydrogen, hydrogen cyanide, and methane — along with contaminants such as sulphur and nitrogen compounds, ash (dust and particulates), phenols, emulsified tar, and oil. Fugitive tar collected in wastewater contains a wide variety of organic compounds, and wastewater particulates may collect and absorb polycyclic aromatic hydrocarbons (PAH). Of the water-soluble organics, phenols appear to be important, along with alcohols such as catechol, resorcinol, and their methylated derivatives. Basic components such as pyridines, quinolines, and indoles have also been identified in product wastewater. The environmental impacts of these pollutants arising from coal conversion processes have still to be clearly defined.

CHAPTER 2

Oil and Natural Gas

OIL is undoubtedly the most versatile fossil fuel and source of energy. The distillation of crude oils produces a variety of refined products, from straight-run petrol, middle distillates (such as kerosene, heating oils, diesel oils and jet fuels), wide-cut gas oil (waxes, lubricating oils and starting material for the production of gasoline) to residual oils, usually asphaltic in nature. Modern transportation systems, the petrochemical industry and several energy-requiring systems cannot operate without these products. Indeed, oil has recently played a major role in the socio-economic structure of the world, not shared by any other source of energy.

Table 6 gives the production and consumption of oil in the world. Over ten years (1966–1976), the production and consumption has nearly doubled. The annual change in 1976 over 1966 is about +5.6% (British Petroleum, 1977).

Table 6.   World oil production and consumption (10^6 tonne)*

	Production			Consumption	
	1975	1976	1977	1975	1976
N. America	557.4	538.6	529.3	849.0	908.3
Latin America	228.1	233.0	236.0	174.0	186.4
W. Europe	30.6	45.0	68.0	664.0	706.4
Middle East	972.7	1100.1	1084.7	66.8	73.3
Africa	245.9	285.4	298.0	51.3	54.2
Asia	89.1	103.1	111.3	356.2	377.9
Oceania	19.9	20.5	22.7	35.3	36.3
USSR	485.0	515.0	538.0	362.0	380.0
Eastern Europe	20.0	20.0	20.7	86.0	90.0
China	65.0	75.0	93.3	55.6	66.0
Total World	2713.7	2935.7	3002.0	2700.6	2878.8

* After World Energy Supplies: 1972–1976, (United Nations, 1978); (British Petroleum, 1977).

The estimated recoverable amounts of oil (proved reserves and estimated additional recoverable resources) are given in Table 7. It should be noted that these estimates do not represent the potential of oil reserves in the world since exploration activities, particularly in the offshore and outer continental shelf areas, will probably lead to the discovery of huge oil reserves, hitherto unknown.

Although the oil production has been steadily increasing in the world, some studies (e.g. WAES, 1977) predict that the production will level by 1985 and then decline towards the end of the present century. It is difficult at this stage to draw decisive conclusions in this respect since the oil production is controlled by several factors, technical as well as geo-political.

Natural gas (also that associated with oil) is becoming an attractive source of energy, both to reduce dependence on oil and from the environmental point of

Table 7. Estimated Oil Reserves in the World (in 10^6 tonne)*

	Proved recoverable Reserves (1 Jan. 1979)	Estimated Additional Recoverable Resources
Africa	8040 (9%)	34000 (16%)
North America	4480 (5%)	24000 (11%)
Latin America	7770 (9%)	12000 (6%)
Far East/Pacific	2390 (3%)	12000 (6%)
Middle East	51050 (57%)	52000 (24%)
Western Europe	2710 (3%)	10000 (5%)
USSR, China, Eastern Europe	12700 (14%)	64000 (30%)
Antarctic	—	4000 (2%)
Total	89140 (100%)	212000 (100%)

* After WEC (Survey of Energy Resources, 1980).

view. Table 8 gives the world natural gas production and consumption for 1975–1977. Between 1966 and 1976, the production and consumption of natural gas has increased about 1.6 times (British Petroleum, 1977).

The estimated recoverable amounts of natural gas are given in Table 9. These values are, again, approximate values and do not represent the true natural gas potential of the world. Recent exploration operations have led to the discovery of several fields of natural gas and further exploration efforts will lead to the discovery of more fields.

Table 8. World Production and Consumption of Natural Gas (in $10^9 m^3$)*

	Production			Consumption	
	1975	1976	1977	1975	1976
N. America	622.5	627.9	641.1	644.0	654.1
Latin America	40.3	41.1	41.1	45.7	47.3
W. Europe	167.4	175.2	172.9	177.9	195.2
Middle East	41.1	41.9	43.4	31.2	33.6
Africa	11.6	16.3	16.3	6.8	7.6
Asia	22.5	26.4	27.1	21.7	25.8
Oceania	5.4	7.0	8.5	5.7	7.0
USSR	265.9	295.3	320.0	274.2	294.8
Eastern Europe	48.1	52.7	55.0	58.4	62.7
China	3.9	3.9	4.7	5.5	6.1
Total World	1228.7	1288.4	1331.0	1271.0	1334.4

* After World Energy Supplies: 1972–1976, (United Nations, 1978); (British Petroleum, 1977).

Table 9.   Estimated Natural Gas Reserves in the World (in $10^9 m^3$)*

	Proved recoverable Reserves (1 Jan. 1979)	Estimated Additional Recoverable Resources
Africa	7300	26000
North America	7500	42000
Latin America	4700	10000
Far East/Pacific	3300	10000
Middle East	20500	30000
Western Europe	3900	6000
USSR, China, Eastern Europe	26900	64000
Antarctic	—	4000
Total	74100	192000

* After WEC (Survey of Energy Resources, 1980).

ENVIRONMENTAL IMPACTS OF OIL AND NATURAL GAS PRODUCTION

Exploration and production of oil and natural gas, whether carried out on land or offshore, have a number of environmental impacts. Accidents and equipment failures can cause harm to workers and to the environment. Fires, explosions and accidental oil spills are the most common accidents. Modern drilling methods and equipment incorporate a number of safety measures which have been devised particularly to prevent accidents and if they happen, to minimise their effects (van Eek, 1977).

Offshore production of oil has markedly increased since the 1950s (see Table 10) and extensive offshore exploration is taking place in more than 100 countries at present. To date more than 18000 oil wells have been drilled off the U.S.A. coast alone (Westaway, 1977).

Table 10.   World Offshore Oil Production*

Year	Offshore Oil Production (10^6 tonne)	% of Total Oil Production
1955	0.70	0.1%
1960	9.35	0.8%
1965	30.87	1.9%
1970	365.55	16.0%
1975	474.49	17.8%
1978	513.47	17.2%

*After World Energy Supplies 1950–1974 (United Nations, 1976) and World Energy Supplies 1973–1978 (United Nations, 1979)

Offshore oil and gas exploration and production is a major industrial development. The support it needs in terms of installations and trained personnel is large and is provided from coastal areas as near to the oil fields as possible. When these areas are not industrialized, labour and housing shortages may become acute, and considerable pressure is put on social and other community services. The siting of the necessary facilities, e.g. pipeline terminals, platform building sites, etc. leads to some problems in land management as it competes with other possible uses for the coastal zone.

The fatality rate of accidents in onshore and offshore operations has been estimated at 0.02 per 10^6 tonne of oil produced and 0.18 per $10^9 m^3$ of natural gas produced (EPA, 1978). The rate of injuries has been estimated at 1.72 per 10^6 tonne of oil and 16.5 per $10^9 m^3$ of gas. The number of fatalities and injuries which might occur by oil and natural gas production to produce 1 GW(e)y is given in Table 11.

Table 11. Fatalities and Injuries from Accidents during Oil and Natural Gas Production (per 1 GW(e)y).

	Fatalities	Injuries
Oil*	0.50	43.0
Natural Gas**	0.40	36.3

* Assuming that 2.5×10^7 tonne of oil are needed as input to refinery producing 2×10^6 tonne of residual oil for the production of 1 GW(e)y.

** Assuming that $2.2 \times 10^9 m^3$ of natural gas are required to generate the same amount of electricity.

Despite careful treatment of effluents discharged and stringent controls to minimize the number of accidental oil spillage, offshore and terminal operations will result in some discharge to surrounding waters. Although some field studies (for example, in Lake Maracaibo, Venezuela, and Timbalier Bay, Gulf of Mexico) showed that there is no ecologicial damage in these areas due to offshore operations (Westaway, 1977), further studies are necessary to determine the long-term effects of hydrocarbons and trace elements on the marine ecosystem. Tidal marshes, coastal wetlands, river swamps and sheltered bays are sensitive eco-zones supporting a variety of organisms at all stages of development, and low levels of hydrocarbons may have some local long-term effects on these organisms.

Although accidental spills from offshore operations are normally of relatively small volume, some large spills have occurred. Blowouts and natural hazards (hurricanes, storms, etc.) are the main causes of such accidents. In the period between 1953 to 1972, 43 accidents occurred during operations on the outer continental shelf in the U.S.A. About 290,000 to 1 million barrels of oil were spilled and the accidents caused 56 deaths and 108 injuries among workers (OCS, 1974). Technology now available is sufficient to ensure a safe conduct of operations for depths down to 200 m and for climatic conditions similar or less severe than those of the North Sea. However, there is a definite trend towards working at greater depths and in even harsher environments. For this reason, new technologies are being continuously developed and there may be a possibility that higher environmental risks may occur when applying these new technologies (OECD, 1977).

In the production of oil (offshore or onshore) large amounts of brine (about $8 \times 10^7 m^3$) are associated with the oil in an emulsion form. The brine is separated by the aid of chemical agents or electrostatic separation. The disposal of the separated brine is normally carried out by injection into the earth. Small amounts may however be disposed of in the near-by marine waters or in other surface waters, in which case, it will seriously affect the water quality of the receiving water body and consequently the aquatic ecosystem.

If H_2S content of the oil is high, sulphur emissions will constitute a potential air pollutant at the production site. The amount of emitted sulphur depends on the technology used for storing oil at the site before transportation, the sulphur content and the temperature of the crude during storage. In some cases, especially in gas production, desulphurization is necessary to satisfy the pipeline specifications on sulphur content. The recovery of sulphur depends, however, on techno-economic factors which include the availability of markets for sulphur. In small fields, the H_2S is vented into the atmosphere or flared. In large fields, however, Clause plants are used to recover sulphur from 85–95% of the H_2S; the rest being emitted to the atmosphere.

ENVIRONMENTAL IMPACTS OF TRANSPORTATION OF OIL AND NATURAL GAS

Marine Oil Transportation:

Marine transportation is the most important means of transporting oil from oil-producing to oil-consuming areas. Table 12 shows the growth in the transport of oil by sea over the period 1960 to 1975, a period of rapid growth and one which has seen a major change in the size and technological development of tankers. During this period, the total oil transportation by sea more than tripled (the finished products nearly doubled while the crude movement has increased by a factor of more than 4).

To carry this oil, the world's fleet of tankers has also grown. In 1954 the world tanker fleet consisted of just under 3500 ships totalling 37 million

Table 12. World Movement of Oil by Sea (in million metric tonnes)*

	Crude	Products	Total
1960	305	144	449
1965	567	180	747
1970	1,033	230	1,263
1973	1,404	291	1,695
1974	1,387	269	1,656
1975	1,273	235	1,508
1977			1,700

* Source: Walder (1977); IMCO (1978)

deadweight tons (dwt). By 1977 there were nearly 7000 tankers in operation, totalling approximately 340 million dwt (IMCO, 1978). One feature of this expansion has been the great increase in the size of ships. The largest tanker in service in 1954 was around 30,000 dwt; today there are several ships of more than 500,000 dwt (ships over 150,000 dwt are referred to as Very Large Crude Carriers, VLCC; those over 350,000 dwt as Ultra-Large Crude Carriers, ULCC).

Although the world tanker fleet generally has a good safety record, the fact that ships today are so much larger than they were 20 years ago means that the consequences of an accident are potentially much greater, as was shown by the Torrey Canyon incident in 1967 and the even greater Amoco Cadiz disaster in 1978. Experience has shown that such incidents can have serious effects upon the environment and marine life, damaging such important resources as fisheries and tourism for long periods.

Tanker accidents can happen almost anywhere. While oil is imported chiefly into the industrialized countries of Europe, Japan and North America, tanker routes pass close to the coasts of many other countries. Winds and current can move oil slicks large distances in a relatively short time, and the consequences of a major spillage can be even greater in developing parts of the world, simply because there are fewer resources for dealing with them. Another danger to the oceans (and to coastal states) comes from the cleaning operations carried out by crews of oil tankers. The tanks in which the oil is carried are normally cleaned while the ship is returning to the loading port. In modern ships the normal procedure is to use special machines which blast jets of high-pressure water on to the tank sides removing the oily residues which are left after the oil has been unloaded. This procedure results in a mixture of oil and water forming at the bottom of the cargo tanks. Some of the tanks are also filled with water on the return voyage, to make the ship low enough in the water for it to be properly manoeuvrable. The water used in this way also becomes contaminated with oily residues. In either case, the mixture of oil and water which results has to be disposed of before the tanker can take on a fresh cargo of oil, and in the past the normal practice was to pump this directly into the sea. For many years it has been recognized that the amount of oil being pumped into the sea is too great for the ocean to absorb; and a variety of methods have been introduced in an attempt to eliminate the problem of what is normally termed "operational pollution". The "load-on-top" (LOT) system and more recently the washing of tanks with crude oil have greatly reduced this type of operational pollution.

The Inter-Governmental Maritime Consultative Organization (IMCO) has played an important role in promoting safety at sea and reducing marine pollution from ships. IMCO has introduced a wide variety of international Conventions, Codes and Recommendations which have helped greatly in achieving these objectives (see IMCO, 1978).

Table 13 gives the amount of oil spilled in the ocean due to tanker accidents, in the period 1970–1978.

Table 13.   Oil spilled into Ocean (tonne)*

Year	Oil Spilled
1970	109,450
1971	109,125
1972	58,500
1973	52,200
1974	54,900
1975	110,700
1976	135,000
1977	121,050
1978	237,600**

* After IMCO (1979)
** Amoco Cadiz disaster contributed 220,000 tonne (Hann, 1979).

This amount of oil constitutes about 35% of all oil discharges in the sea (see Table 14).

The ecological impact of oil transportation often arises more from the development and maintenance of loading and unloading facilities on shore than from the transportation medium itself. Terminals are often constructed in or near ecologically sensitive coastal areas, such as estuaries. The installation may alienate valuable nursery grounds for fisheries. Whether the terminals service a large petroleum refinery or a pipeline for conveyance of oil to distant places, the risk of an oil spill always exists in the transfer of oil from the tanker to the shore facility. Even in the absence of any acute, large-scale oil spills, there is often the long-term chronic effect of continuing small spills and leaks. Thus one might conclude that the interface between different media such as land and water, where

Table 14. Sources of Oil in the Oceans (after EPA, 1978)

Source	Estimated Contribution (Barrels/y)	%
Production and Transport Tankers Dry Docking Terminal Operations Bilges Accidents	16,000,000	34.9
Direct Sources Coastal Refineries Municipal Waste Industrial Waste Off-shore oil production	6,500,000	14.3
Indirect Sources Rivers and urban Runoff Atmospheric fallout	14,000,000 4,500,000	31.2 9.8
Natural Sources Seepage	4,500,000	9.8
TOTAL	45,500,000*	100.0

*Equivalent to about 6,232,874 tonnes.

the mode of transport must be changed and the substance must be transferred to accommodate this change, is the most vulnerable to transportation impact. In this respect, increasing attention has been paid recently to the selection of sites for deep-water oil ports. An environmental risk index for the siting of deep-water oil ports has been developed recently for the east coast of Canada (Canada, Department of Fisheries and Environment, 1976). A similar activity is currently underway for site selection on Canada's west coast, where there is a need to have a terminal that would accept crude petroleum conveyed by tanker from Valdez, Alaska, and transmit it through a pipeline to areas of high demand in midwestern and southern U.S.A.

Impacts of Oil on Marine Environment:

Petroleum is composed of a large number of components ranging from the gaseous methane to the solid paraffins. The large majority of these compounds are hydrocarbons which are immiscible with water and are chemically inactive compounds. Lower alkanes (paraffins) are rapidly lost by evaporation and are the most readily biodegradable group. Aromatics are the most soluble, can be photodegraded and are the most toxic components of the oil.

The composition of crude oil depends on its origin. There are heavy crudes (high content of large hydrocarbon molecules), and light crudes (high content of small hydrocarbon molecules); they may contain much sulphur (Kuwait 2.4%) or be "sweet" (North Sea with about 0.3%S). Refinery products, e.g. petrol, kerosene, diesel oil, residual oil, have more homogeneous compositions and some of them are far more toxic than the crude from which they were derived.

Spilled petroleum, following release to the ocean, spreads over water. The extent of spreading depends upon the nature of the material and the prevailing wind and current systems. At this time, physical, chemical and biological processes begin altering the composition of the oil which, because of its immiscibility with water, is initially present as a separate phase. Although the rates and degrees at which changes take place in the environment are not known in detail, the general directions can be predicted with some degree of certainty (Goldberg, 1976). The greater the rate and extent of spreading, the greater the rate of evaporation which will eliminate about 50% of the hydrocarbons. The low molecular weight hydrocarbons will dissolve in the water column; the rate of solution depends on wind, sea agitation and water temperature. Emulsions (chocolate mousse) form as semi-solid lumps containing up to 80% water. Biological, chemical and photo-oxidation processes tend to increase the solubility of some compounds.

There is a wide variety of marine ecosystems which could be affected by marine transportation of oil. Some ecosystems are more sensitive than others to the environmental changes that arise from marine transportation activities and facilities development. Clearly, the many physical processes in the sea leading to dilution and dispersion of pollutants, along with renewal of water, vary from place to place, depending on such factors as geographical configuration, exposure to oceanic currents, winds, tides and runoff. Perhaps the most critical consideration in terms of ecological impact of marine transportation is the effect on biological productivity. In this respect, coastal ecosystems are by far the most vulnerable, because they are generally the most productive and because the impact of an oil spill, for example, can be most severe at the interface of water and land.

The latitudinal effect must be recognized when dealing with pollution or environmental problems of marine transportation. The higher metabolic rates associated with the higher temperatures of tropical and sub-tropical regions means that the restorative capability of ecosystems in such areas is higher than that of cold-climate ecosystems. By the same reasoning, arctic ecosystems heal slowly when injury is inflicted on them by pollution or other man-made activities. Oil spills and their adverse effects in the arctic can endure for a long time, particularly if the oil seeps under the ice or mixes with ice and snow. Bacterial degradation of oil slows down considerably at these low temperatures.

Coastal zones are generally more sensitive to oil pollution than the open ocean. The major world fisheries are located in the coastal zone; 90% of the world fisheries are produced in waters of the continental shelf which constitute only 10% of the ocean area (Waldichuk, 1977). It is here that not only the major groundfish stocks are located, but stocks of certain pelagic species are fished and crustacean and molluscan shellfish are harvested. The coastal waters are spawning and nursery areas for most species of demersal fish, such as cod and halibut, and for such pelagic species as herring. In some coastal waters, herring may deposit their spawn on intertidal vegetation and other substrate, rendering such areas expremely vulnerable to oil pollution. Anadromous species, which comprise an important component of commercial fisheries in some parts of the world, must pass through the coastal zone in their migration to riverine spawning grounds as adults and out to the high seas as juveniles.

Invertebrate species and some pelagic fishes, which use the intertidal zone for spawning and rearing of larvae and juveniles would be particularly affected by an oil spill impinging on the shoreline. If intertidal animals are not suffocated by a layer of oil covering the inshore substrate, it can be almost certain that their flavour will be impaired and acceptability on the market downgraded. It is usually the practice of regulatory agencies to close fisheries in areas thus affected, until it can be assured that the quality of fisheries products is acceptable for human consumption. Shorelines adjacent to oil transportation routes are always threatened by oil spills, because they are often at the mercy of onshore winds and tidal currents that can quickly bring an oil slick onshore.

Perhaps the most vulnerable of coastal ecosystems is the estuary. All kinds of physico-chemical and geochemical reactions take place there when freshwater mixes with sea water. The net result of such interaction is a unique type of ecological system. The estuary is often considered to be the most productive of coastal habitats. Estuaries provide nursery grounds for juvenile salmonids prior to these young fish continuing their journey out to sea. Certain micro-invertebrates serve as food for these juvenile fish, and any adverse effects on the food chain involving these organisms can have an impact ultimately on the commercial and recreational fisheries.

Although the normal surface flow in an estuary is seaward, because of freshwater input from the river, other physical characteristics of an estuary tend to present a situation where a pollutant like an oil slick can pose a threat. Tides and winds can combine with the geographical configuration of the coastline, for example, to move an oil slick into an estuary. There is little question that a large oil spill in an estuary can lead to acute ecological effects with possible long-term consequences. The chronic effects of small spillages and leaks of oil from a loading/unloading facility into an estuary would have little dramatic effect like a large-scale spill, but could lead to rather insidious long-term degradation of the estuarine ecosystem.

The character of the coastal land, where terminals are located, has a bearing on the environmental impact of a marine transportation system. The alteration of an estuary, in construction of a port, in general, and its back-up facilities, may have a profound effect on the ecology and use of the estuary by fish and other aquatic organisms. Dredging and filling can destroy benthic habitats for fishes and invertebrates and marshes for avian wildlife. Diversion of water flow in either a river as it enters the sea, or in coastal tidal currents, can modify various physical and chemical characteristics of the waters in the estuary and in the nearshore marine regime.

The marine environmental effects of oil pollution are regularly reviewed in various fora (Hoult, 1969; Cowell, 1971; NAS, 1973, 1975; Peters, 1974, Wolfe, 1977; GESAMP, 1977; API, 1977; BIO, 1977). Many collective and individual studies and reviews have been reported (GESAMP, 1977; Vernberg, *et al.*, 1977; Wolfe, 1977; Malins, 1977; Cowell, 1976) on the subject. A number of factors must be taken into consideration when assessing the impact of an oil spill. Oil is far from being a unique chemical compound with specific physical and chemical characteristics. Because of the inherent properties of oil which cause it to float on the sea surface, oil tends to become deposited in the intertidal zone once an oil spill reaches the shoreline. This has many undesirable consequences, one of which is that the aesthetic and recreational value of a beach is at least temporarily impaired. This effect can be counteracted by the application of suitable surface treatment agents for the protection of shorelines from oil spills. Also, sinking agents have been used in some oil spill cleanups to deposit the oil on the sea bottom before it reaches shore. In locations, where ecologically sensitive areas are threatened, emulsifying agents are used to disperse the oil in the water.

Oil that coats shorelines may persist for a long time, depending on the characteristics of the hydrocarbons. Crude oil and bunker C form tarry-like residues that may remain imbedded in beach sands or encrusted on rocks for months or even years. Weathering combined with the scouring action of waves eventually removes these residues. The lighter refined oils will usually evaporate without leaving the same type of persistent black residues. Depending on whether a beach is intensively washed by wave action or not, one beach may be more rapidly cleansed than another of hydrocarbon materials. Tarry residues may persist for a considerable time if oil is deposited out of the reach of waves until bacterial decomposition and other weathering processes break them up. Tarballs are found along shipping lanes in many parts of the world oceans.

The lethal effects of oil (refined oil is more toxic than crude oil) depend on its concentration and on the length of time an organism was exposed to it. Experiments with oysters showed that whereas they accumulate more hydrocarbons in their tissues with time of exposure to oil, the oysters, once they were placed in clean flowing seawater, purged themselves rapidly of the accumulated hydrocarbons. After about 2 weeks in clean water, only negligible hydrocarbon levels were found in the oyster tissues, and there was no evidence of pathological damage to the oyster as a consequence to the oil (Mertens and Allred, 1977). Experiments with shrimp, clams and some other marine organisms produced similar results. Planktonic eggs and larvae may be particularly exposed and sensitive to oil pollution especially to light refined oils. Kerosene-based dispersants with high aromatic content are especially dangerous in this respect and this type of dispersant has, therefore, been banned in many countries.

Subtidal and intertidal benthic organisms have suffered particularly heavy losses from discharges of light and heavy oils and their subsequent treatment. The long-term ecological effects of oils and dispersants on these communities are the most extensively studied and documented. Complete recovery may take several years and, therefore, the effects of repeated oilings are particularly important. In the intertidal zone, mortality of grazing invertebrates such as sea urchins may result in an explosion of the population of attached green algae, which, in turn, affects other parts of the ecosystem. This type of secondary effect, is an aspect that has to be considered when damage due to oil is evaluated. A special case is offered by coral island ecosystems. If the corals are killed, the natural protection of the island from erosion will be lost. Further, because of isolation, most of the organisms in the coral island ecosystem will not be replaced. This would considerably prolong the duration of the effects even if the oil itself disappears or is removed.

The introduction of hydrocarbons to a population of marine bacteria selects those capable of utilizing this food source at the expense, at least initially, of the rest of the population. Evidence that low concentrations of crude oil and oil components inhibit bacterial chemotaxis has been questioned. It is noted that the number of micro-organisms able to use oil hydrocarbons as a source of carbon and energy increases from clean to oil polluted marine areas. This can be observed in inshore waters as well as in the open oceans.

Susceptibility of micro-algae appears to vary enormously. In addition, effects of an exposure may take a long time to become apparent and may even be the result of a comparatively short exposure. Laboratory studies have indicated that very low concentrations of oil may stimulate primary production by phytoplankton, but that higher concentrations lead to reduction of carbon fixation, and finally to mortality. Other studies indicated that macro-algae, if coated with oil, may be mechanically stripped from their substrate (Waldichuk, 1977). Some lower forms are resistant to oil pollution and thrive in polluted environments. Low concentrations of oil have been shown to depress the growth of red-algae sporelings. Experiments appear to show that higher levels of oil inhibit biosynthesis of nucleic acids and their polymerization of macro-algae. Oil penetrates the higher forms of plant life, blocking intercellular spaces, increasing respiration and decreasing transpiration rates and affecting flowering and reproduction. It should, however, be borne in mind that these findings are the results of laboratory tests which may differ significantly from the actual behaviour of plants and animals in their natural environment.

Some littoral or salt marsh plants may tolerate repeated light oilings but heavy fouling often leads to mortality. These effects may take several years to appear. Mortality of some zooplankton species, including pelagic fish larvae, occurs from oil slicks at sea, with unknown ecological significance.

The combination of the variety of sublethal responses of marine organisms to crude and refined oils and their ecological implications are not fully understood. Effects of oil should be studied at the ecosystem level, rather than by single species bioassay. Attention should be paid to chronic and sublethal effects using, for example, histopathological techniques. Genetic alterations and other effects on single species should not be ignored.

Birds are particularly vulnerable to oil spills, whether they occur at sea or close to shore. Heavy mortality due to oil pollution of species with low annual breeding rates, such as auks, may have serious consequences. Populations of birds from discrete breeding colonies may be similarly seriously reduced or destroyed. When their inner feathers are coated with oil, insulation is lost, and a bird may die of exposure in any season. A total bird population is small compared to aquatic populations and runs a higher risk of extinction by whatever cause.

When an oil spill occurs, several options exist to deal with it:
(a)   leave it,
(b)   contain and remove it,
(c)   disperse it,
(d)   burn it, or
(e)   a combination of these.

Selection of the most suitable technique, or techniques, depends on the circumstances of each spill, involving consideration of the scale of the spill, its location, the prevailing and predicted weather, the proximity of sensitive areas and the availability of oil spill clearance equipment and manpower. The decision-making processes to clean-up each spill should include the ecological interests (fisheries, wildlife, etc.), the commercial interests (tourism, etc.), the legal interests (plant and oil owner), the financial interests (insurance broker etc.) plus representatives from the State and authorities affected, together with technical assistance and advice of specialists trained in clean-up activities. It is important that the decision-making should be professional and flexible.

The cleaning-up operations can be complicated by sea and meterological conditions. For example, because of rough sea conditions, the crude oil spilled by Amoco Cadiz in 1978 (off Brittany coast, France) was quickly emulsified with sea water as it leaked out of tanks. The resulting chocolate mousse, containing up to 80% water, became more and more viscous, resulting in pollution four times the initial volume of crude oil (Bocard et al., 1979). The costs of cleaning-up of oil spills varies from one place to another, but on the average it is estimated to be $1000/barrel. (EPA, 1978).

Pipeline Transportation:

Oil spills can also occur during the transportation of oil by pipelines. Records of such land spills show that the most important cause is damage to the pipelines. Pipelines which are not buried in the ground can cause disturbances to wildlife and to some extent to land use in certain areas. The Trans Alaska Pipeline, for example, was examined for its possible impact on the permafrost, migrating wildlife, seismic exposure, etc. The route and design of the line was selected with a view towards minimising its overall impact on the environment.

Transportation of Natural Gas:

Natural gas consists of the short-chain hydrocarbon compounds (e.g. methane, ethane, propane and butane) and is usually transported from the well head to treatment facilities and ultimately to the consumer by pipelines. Until recently these pipelines followed overland routes; the growing production of natural gas in offshore areas necessitated, however, the commissioning of submarine pipelines, e.g. in the North Sea. In liquid state, natural gas (LNG) is transported by special tankers.

The natural gas is liquefied cryogenically at a temperature of $-162^\circ C$ (Tanner, 1977). Before liquefaction, the gas must be dried, heavier hydrocarbons fractions removed and pre-treated to remove such components as H_2S, CO_2 and organic sulphur compounds that would solidify during liquefaction. Traces of mercury are sometimes present in natural gas which have to be removed because they may cause corrosion. The main environmental effects of a natural gas liquefaction plant are: (1) discharge of heat to the atmosphere or to fresh or marine waters, depending on the type of cooling system used; (2) occasional emission or flaring of excess components of gas.

The liquefied natural gas (LNG) is transported by LNG tankers. The capacity of these tankers has greatly increased in the last 15 years as a result of the increase in production and consumption of natural gas. (Table 15).

Table 15. Capacity of World LNG Tankers*

	1960	1970	1975
Number of ships	1	14	36
Fleet capacity (1,000m³)	5	1500	5000

*After Rob (1975).

Liquefied natural gas consists mainly of methane and there is a possibility of an explosive evaporation if the LNG comes in contact with water. There is also the tendency of stratification if two LNG different in composition are mixed, for example, in storage tanks. In such cases, an increase of pressure inside the storage tank might occur and lead to an explosion if safety measures are not adequately taken.

If liquefied natural gas is spilled, it boils rapidly. Although the vapour is not toxic, it may in high concentration cause asphyxiation by excluding oxygen. Moreover, the low temperature of the material may result in frostbite for anyone in the immediate vicinity of a spill. Both of these potential hazards are too localized to be of concern outside the boundaries of the liquefaction facility. The major hazard of liquefied natural gas is, however, fire and adequate safety measures are normally taken to prevent such accidents.

Little is known about the biological effects of LNG when spilled in the sea. Methane, the principal constituent of natural gas, is relatively non-toxic to freshwater species. There is no reason to suspect that this gas would be more toxic to marine organisms. Toxicity of natural gas would probably be associated with such sulphur-containing impurities as hydrogen sulphide and mercaptans. However, these would be normally removed from the gas prior to liquefaction.

ENVIRONMENTAL IMPACTS OF OIL PROCESSING (REFINING)

Refineries are large industrial installations with air and water emissions, large water requirements for processing and cooling — unless air cooling is used extensively — and safety problems due to the risk of explosions and fires. A 300,000 bbl/day refinery requires about 520 hectare of land for direct use and an additional 520 hectare as exclusion area. The latter is the area that surrounds the plant site where no resident population is allowed for safety-related reasons (OECD, 1978).

The principal types of refinery airborne emissions and their potential sources are give in Table 16.

Table 16. Potential Sources of Emissions from Oil Refining*

Type of Emission	Potential Source
Sulphur oxides	Boilers, catalyst regenerators, decoking operations, flares, heaters, incinerators, treaters, acid sludge disposal.
Particulate matter	Boilers, catalyst regenerators, coking operations, heaters, incinerators.
Hydrocarbons	Air blowing, barometric condensers, blowdown systems, boilers, catalyst regenerators, compressors, cooling towers, decoking operations, flares, heaters, incinerators, loading facilities, processing vessels, pumps, sampling operations, tanks, turnaround operations, vacum jets, waste-effluent-handling equipment.
Nitrogen oxides	Boiler catalyst regenerators compressor engines, flares.
Carbon monoxide	Catalyst regenerators, coking operations, gas generators, flares boilers, and process heaters.

*After: Mallat (1977).

Small amounts of carbon monoxide are also emitted, the sources of which include: catalyst regenerators, coking operations, gas generators, flares boilers, and process heaters. Of these sources, catalyst regenerators from catalytic cracking units and fluid bed cokers generate continuously significant amounts of carbon monoxide. Because catalytic cracking units could be significant point sources of carbon monoxide, emission standards in many countries require catalytic cracking units, to reduce CO emission, usually by secondary combustion. Table 17 gives an estimate of airborne and liquid effluents from an oil refinery processing 2.5×10^7 tonne crude oil (equivalent to that needed to produce 1 GW(e)y).

Table 17. Emissions from Oil Refinery (tonne)*

Airborne Effluents	
SO$_x$	21,000
Organic compounds	23,000
NO$_x$	18,000
CO	4,300
Ammonia	2,230
Liquid effluents (1.4×10^8 tonne waste water) containing:	
Chlorides	24,000
Grease	600
Ammonia nitrogen	600
Phosphate	3
Suspended solids	2,000
Dissolved solids	100,000
Trace metals (Cr, Pb, Zn, Cu)	22

*Calculated after Wilson and Jones (1974) and Korte (1977), normalized to production of 1 GW(e)y.

Refineries require large amounts of water, mainly for heat removal and in various process operations. Each process operation has different water usages associated with it, and the characteristics of the wastewaters produced differ considerably. In addition to cooling water and process wastewaters, ballast water, storm water runoff, and sanitary wastes also contribute to the total waste load that refineries must treat before discharge.

The most significant pollutants present in these various wastewaters are oil and grease, phenols, ammonia, suspended and dissolved solids, sulphides, and chromium. In addition, some wastewaters are highly alkaline while others are acidic. As a result of widley differing water usages and processes in refineries, the quantity and quality of wastewaters varies considerably from refinery to refinery. These wastes, however, are readily treatable with a combination of in-plant controls and treatment techniques and end-of-pipe treatment. The last consists mainly of primary separation of oil and solids and neutralization, followed by biological treatment using activated sludge systems, aerated lagoons, or oxidation ponds. The pollutants named above are largely removed; the effluents discharged from most refineries contain them in only low concentrations (Mertens and Allred, 1977).

While growing amounts of information on the biological effects of oil spills on the aquatic environment are becoming available, little has been written on the biological effects of refinery effluents. The toxicity of individual contaminants present in these effleunts, however, is fairly well documented. The effects, if any, of the contaminants present in refinery effluents on the aquatic environment can be determined by bioassay techniques. Unfortunately, current bioassay procedures used to monitor the effects of refinery effluents on water quality and aquatic life are cumbersome and expensive in terms of the required time, facilities, technicians, and test organisms. Consequently, only a fraction of all refinery effluents can be monitored for toxicity (Mertens and Allred, 1977).

Odour can be a potential nuisance in and around a refinery. The principal malodorous compounds existing in crude oil or formed during its processing into products are hydrogen sulphide and mercaptans. Should any of these escape from a refinery there is a risk of smells in the neighbourhood. Ethyl mercaptan has a perceptible smell when present in a concentration of only one part per thousand million; thus even a very small loss can create an unpleasant smell downwind under certain meteorological conditions.

Accidental spills at refineries can occur from stored crude oil or refined products. In 1974, a huge oil tank in the Mizushima refinery on Seto Inland Sea, Japan, developed an 8 m rupture and lost about 50,000 barrels of oil to the sea. The spill affected the area considerably and extensive cleaning efforts were undertaken to restore the area. The total cost of the damage has been estimated at $160 million (Nicol, 1976; Hiyama, 1979).

ENVIRONMENTAL IMPACTS OF OIL AND NATURAL GAS UTILIZATION

Use of Oil Products:

A variety of oil products is used as energy sources; gasoline, gasoil, kerosene and residual oil are the most common. Mobile sources contribute currently more to the total emission of carbon monoxide and hydrocarbons than stationary sources (Table 18). It is estimated that in the U.S.A. the motor vehicle contributes almost 65% to all anthropogenic carbon monoxide and about 46% to all anthropogenic hydrocarbons emitted to the atmosphere. Vehicle exhaust emissions are, therefore, of major public concern. The complete combustion of gasoline under ideal conditions should result in exhaust gases composed of water vapour and carbon dioxide. However under actual driving conditions the exhaust gases also contain unburnt hydrocarbons and carbon monoxide. The amount of these latter substances may vary considerably not only in accordance with the driving conditions, but also among different cars and different engine designs.

Table 18.   Emission Sources in Urban Areas*

	Pollutant per cent of total		
	HC	CO	NO_x
Automobiles	50 – 60	77 – 87	40 – 50
Trucks, buses etc.	5 – 10	8 – 10	8 – 13
Stationary sources	25 – 45	3 – 15	37 – 52

*After EPA (1974).

The fate and effect of lead additives of motor gasoline have been extensively discussed in recent years. Other additives used in motor gasoline are ethylene dibromide (or chloride) to reduce combustion chamber deposits and traces of other substances, e.g. boron compounds, to improve carburettor cleanliness, as dye or for similar purposes.

Regulations are now in force in some countries, to eliminate or reduce the use of antiknock additives and to reduce the emission of unburnt hydrocarbons, carbon monoxide and in some cases also oxides of nitrogen. It is assumed that by 1985 the emission of carbon monoxide and hydrocarbons will be reduced by about 75 per cent of the 1972 level. However, the reduction of the total emission into the atmosphere may be less since it is expected that from 1972 to 1985 the world car population will increase from 208 million units to 373 million units, i.e. almost double (UNEP Motor Vehicle Seminar, 1977). This would result in a reduction of only 50% of the total emission of these pollutants.

Spills of refined products might occur during transportation of the products from storage facilities to gas stations, etc. or during loading/unloading operations. However, safety precautions should prevent such spills and consequently health hazards and possible fires.

The use of residual oil in power stations gives rise to a number of effluents. These are summarized in Table 19. The environmental impacts of these effluents are discussed in Chapter 4.

Use of Natural Gas:

Natural gas can be used for domestic, industrial or power production purposes. Table 20 gives an estimate of effluents from a power plant operated with natural gas.

(residual)

Table 19.   Effluents from Oil-fired Power Stations (tonne/1 GW(e)y)*

Airborne Effluents	
SO_x	37,000
NO_x	24,800
CO	710
Hydrocarbons	470
Aldehydes	240
Particulates	1,200
Liquid Effluents	
Suspended solids	497
H_2SO_4	83
Chlorides	26
Phosphates	42
Boron	331
Chromates	2
Organic compounds	66
Solid waste	
(collected fly ash)	9,190

* Calculated after Wilson and Jones (1974). The power plant uses $2x10^6$ tonne residual fuel (1% S, 0.5% ash); thermal efficiency 38%; Fly ash recovery 99%.

Table 20.   Effluents from a Natural Gas-fired Power Plant (tonne/1 GW(e)y)*

Airborne Effluents	
SO_x	20.4
NO_x	$2x10^4$
Hydrocarbons	34
Other Organic Compounds	238
Particulates	510
Liquid Effluents	
Suspended solids	497
Organics	66
H_2SO_4	83
Chlorides	26
Phosphates	42
Boron	331
Chromates	2

* Calculated after Wilson and Jones (1974). The power plant uses $2.2x10^9 m^3$ of natural gas with energy value of 37,000 kJ/m^3, thermal efficiency 38%.

CHAPTER 3

Oil Shale and Tar Sands

OIL SHALE

OIL shale is a sedimentary rock containing hydrocarbons that can yield oil when retorted. The development of a commercial shale oil industry depends on the grade of the shale, the technology to produce the fuel, the ability to overcome institutional and environmental problems, and economic incentives that make investment worthwhile. Generally speaking, high grade oil shale contains more than 100 litres of oil per tonne of shale.

The recovery of oil from shale is accomplished by retorting, for which two generic methods have been developed: (a) on the surface in a retort vessel and (b) underground (in situ). Both surface and in situ retorting will produce a heavy oil that will be upgraded and converted into a transportable refinery feedstock. The ultimate products of refining can range from a low-grade boiler fuel to gasoline.

In the case of surface retorting of oil shale, mining is undertaken by underground or surface methods according to the geological occurrence of the shale. The amount of mined shale depends on the hydrocarbon content. On the average, about 50 million tonne of shale would be required annually for a plant producing 100,000 barrels of oil per day (Dickson et al., 1976). The mined shale is crushed to retortable size, heated in a retort to extract the oil and then disposed of the spent shale.

In situ retorting involves underground retorting with partial mining. To retort oil shale, the formation is fractured and ignited. The shale is heated to 400°C or higher to bring about pyrolysis of the kerogen (the organic portion of the oil shale) and production of shale oil and gas. Combustion is sustained by pumping compressed air, and the produced gases and liquids are forced or pumped horizontally through the fracture system to production wells which surround the site.

The development of a major shale oil industry will have a number of environmental impacts on the land, air, water, and natural resources of the region. The environmental disruption associated with oil shale mining is typical of that of any large surface or underground mining operation, except that size of the operation will mean that the scale of the disruption will be much greater.

In the case of surface retorting, the disposal of spent shale will create land disturbances of large magnitude, accumulation of toxic substances in vegetation, and contamination of ground and surface water from runoffs. Even under the best reclamation strategies, the naturally occurring ecosystems of the canyons in which the spent shale may be deposited will be completely covered and destroyed. The goal of reclamation is to establish a new ecosystem on the spent shale piles, which can be self-sustaining long after human involvement has ended. This goal involves stabilization of the pile against erosion and sliding, establishment of a suitable plant cover, and ultimately the generation of a plant succession system similar to other systems in the area. Some of the spent shale can be returned to the mine. This is most readily accomplished if surface mining is employed, since it can be done in conjunction with the return of overburden to the mined-out areas. A part of the spent shale can also be returned to the mine if underground mining is employed. In either case, disposal problems will remain since the volume of shale expands under retorting (10 to 30%, depending on the retorting process used) and not all the spent shale can also be returned to the mine. Furthermore, temporary disposal sites will still be required since several years of mine development are needed before backfill operations can begin. In general, the prospects for acheiving a long-term stable ecosystem on massive spent shale piles have not been fully assessed and it remains one of the major problems of oil shale development.

The oil shale industry will exert a potential impact on local water resources due to the demand for clean process water, the need for removal of process effluent discharges and mine dewatering. Mining operations, especially underground mining in arid and semi-arid regions, will cause long term hydrological disturbances in the region. The mining of shale horizons that separate fresh water aquifers from saline aquifers in the Mahogany Zone, U.S.A., for example, will lead to contamination of the fresh water with saline water; the fresh water aquifer recharges the streams of the region (Rattien and Eaton, 1976). The demand for water by the oil shale industry has not been accurately assessed. Estimates range from 100 to 200 million m³/year for a one million bbl/day operation. This large amount of water will have to be obtained in some cases by diversion of water courses locally available, which will cause environmental disruptions in the region.

During the retorting and upgrading of oil shale, waste water is generated. Such water contains phenols, hydrogen sulphide, ammonia and trace chemicals, and should be treated and re-cycled. A potential source of water pollution is leaching or runoff from the spent shale disposal pile into local aquifers. Except in catastrophic failure of the pile or flash flooding, catchment dams will probably be sufficient to retain any runoff water. The potential for water contamination due to leaching depends on several factors, such as the degree of compaction of the spent shale, and has yet to be fully assessed.

A number of air pollutants are encountered in oil shale processing. Table 21 summarizes the emissions produced from surface retorting and from in situ extraction of oil.

Table 21. Air pollutants from oil shale processing operations: 50,000 bbl of oil/day. (After Rattien and Eaton, 1976)

	Quantity (tonne/y)	
	Surface Retorting*	In situ Extraction
Particulates	320–3245	1576
Sulphur oxides	955–5835	8406
Nitrogen oxides	609–6412	2256
Carbon monoxide	291–3634	72
Hydrocarbons	1386–4041	971

* The range covers different technologies for surface retorting.

The question of the fate of toxic trace elements in oil shale conversion processes has received considerable attention due to the potential for highly toxic metals such as mercury, lead, beryllium, arsenic, cadmium, selenium and fluorine, to enter air, water, or soil and ultimately to create health hazards. However, the fate of these elements during oil shale retorting has not yet been clearly defined.

TAR SANDS

The term "tar sands" has been defined as sand having a "highly viscous crude hydrocarbon material not recoverable in its natural state through a well by ordinary production methods". Tar sands are more appropriately referred to as bituminous sands or oil sands as "tar" is defined as a substance resulting from the destructive distillation of organic matter. Major tar sands deposits occur in Albania, Canada, Malagasy Republic, Romania, Trinidad, U.S.A., U.S.S.R. and Venezuela, (Phizackerley and Scott, 1967). Estimates of world reserves of tar sand bitumen vary widely from conservative estimates of 915.2 billion barrels to the high estimate of 2,100 billion barrels of oil in place (Demaison, 1976).

Table 22 gives some of the reserve estimates of tar sands. Most deposits in countries other than those listed in this table are poorly recorded and are considered to have in-place bitumen amounting to less than 1 million barrels.

Table 22. Estimated reserves of major tar sands deposits of the world

Country	Reserve Estimate (Million Barrels oil)
Venezuela	200,000 (a) − 1,050,000 (b)
Canada	710,800 (a) − 895,000 (c)
U.S.S.R.	144,000 (b)
U.S.A.	27,000 (b) − 30,000 (d)
Malagasy Republic	1,750 (a)
Albania	317 (a)
Trinidad	60 (a)
Romania	25 (a)

(a) Phizackerley and Scott (1967).
(b) Demaison (1976).
(c) Demaison, Ibid., Govier (1973), Walters (1974),
    Phizackerley and Scott, Ibid., Jardine (1974).
(d) Frazier (1976).

The enormous deposits of the Orinoco Heavy oil belt (Venezuela) and the Alberta Tar Sands (Canada) account for over 90% of the world reserves of tar sand bitumen. Deposits in the U.S.S.R. and the U.S.A. account for nearly 8% of world reserves.

The first commercial scale tar sand plant, went into operation in 1967 in Alberta, Canada, and has produced an annual average of 45,000 bbl/day since then. The second Canadian plant, began production of synthetic crude in 1978. By 1985, production is expected to reach 125,000 bbl/day. In 1972, the U.S.S.R. started operating a commercial "thermal mining" plant to produce high viscosity crude from the Harega oil field, about 1,400 km northwest of Moscow. These operations are the only "commercial" size tar sand plants currently in operation. Small scale local exploitation of tar sand has been common in many localities throughout the world for many years, mainly for asphalt recovery or other non-fuel uses. At present, there are numerous pilot-scale operations being developed around the world to test and develop tar sand production technology or to enhance heavy oil recovery.

Tar sand deposits can be mined either from surface or deeper deposits depending on the geological setting of the sand-bearing formations. Surface mining technologies are generally similar to those used for coal. The tar sand is mined and transported to a processing plant where the bitument is extracted and the sand discharged. The extraction system is based on the hot water extraction process, cold water separation or anhydrous solvent separation process. Deeper deposits are generally uneconomic to mine by underground technologies, and the separation of bitumen from such deposits may be accomplished by in-situ extraction processes. These involve the application of heat, solvents, emulsification, bacterial action or thermal cracking. Two advantages of in-situ over surface mining methods are that the former would eliminate the need for handling and processing vast tonnages of bitumen-bearing materials and for disposing of the resultant spent sand tailings. One disadvantage, from a resource utilization point of view, is that recovery efficiency will probably be no greater than about 50% compared to about 90% bitumen extraction from mined tar sand.

The environmental effects of large-scale tar sands development could be widespread and severe, if insufficient planning and financing accompanies such development. Although many aspects of oil sand excavation are comparable to coal mining, the basic differences are the spent tailings sand and sludge material. Table 23 presents types of terrestrial environmental disturbance that could result from a surface tar mining operation. Of the types indicated, those that actually occur, as well as their duration and severity, will depend on details of mining and material handling methods employed; size of the operation; all elements of the pre-mining environment; and the degree to which environmental protection measures are employed during planning, mining, and surface restoration phases of the operation. Generally, factors determining the degree to which the area of a tar sand mine can be restored to a pre-mined status should not differ from those of a surface coal mine in the same area with one significant exception; namely, reclamation and revegetation of tailings sand. The most abundant solid waste material produced by extraction is the mineral matter sent as tailings to the

Table 23. Environmental impacts of tar sands mining (After Frazier et al., 1976)

Operation or Source of Impact	Potential Impact								
	Increased Landslide Risk	Destruction of Existing Vegetation	Alteration of Habitats	Topographic Changes	Drainage Diversion	Increased Noise	Solid Waste Generated	Changes of Ground Water Physical	Changes of Ground Water Chemical
Site Preparation		X	X	X	X	X			
Surface Clearing (cleared area)		X	X	X	X	X	X	X	X
Stripping		X	X	X			X		
Tar Sand Extracting (mined area)				X		X		X	X
Haul Road Transportation (construction)		X	X	X	X	X			
Tailings Disposal						X	X		
Bitumen in Tailings or Low Grade Tar Sand Waste		X				X			X
Solubles or Water-borne Particles in Overburden			X					X	X
New Surface									
Increases in Surface Slope From Waste Disposal	X								
Rehandling of Material: Backfilling, Grading and Recontouring						X			

settling pond. For a tar sand facility producing 10,000 bbl/day of synthetic crude oil product, about 23,211 tonne/day of this mineral matter will be rejected in this manner. Depending upon the requirements of an operation e.g., government regulations, extraction process, and method of transportation of tailings sand, methods of disposing of tailings sand include: (1) temporary disposal in pond areas until the sand can be disposed of permanently in mined out areas without interfering with mining operations; (2) permanent disposal behind dams near a mine or processing plant. Revegetation of tailings sand has been the subject of considerable research. The total direct land disturbance of a major tar sand mining operation can be very significant. It has been estimated that more than 200 km^2 will be disturbed over a period of 25 years to supply a plant producing 120,000 bbl of oil per day (ECE, 1978).

In-situ production of tar sands offers much less potential for surface impacts than surface mining. Some vegetation would normally need to be cleared and the ground graded to accommodate equipment installations, but only relatively small areas would be affected. Surface operations can be conducted so as to avoid or minimize interference with concurrent land usages such as livestock grazing, or agricultural activities. Produced sand and drill cuttings would constitute the major solid wastes associated with in-situ production. The produced sand should not be significant if gravel packing or other sand filtration methods are applied in the well bore. The sand that would be produced could be separated from the bitumen and probably discharged into a pit that would later be back-filled. The cuttings produced from the drilling operation could also be disposed off by procedures commonly employed in the oil production industry — burying in a pit and back-filling at the completion of the drilling operations. Disposal of drilling muds should pose little to no potential for environmental impacts because these muds are customarily collected and reused.

The mining operation of a tar sand plant produces water problems closely similar to those of a conventional coal mining operation. Ground water present in the tar sand must usually be removed by wells and discharged prior to mining operations and this water may be saline and/or toxic. Surface mining operations that disturb existing, or expose, new surfaces, increase availability of soluble and suspendable constituents for aqueous transport. Duration and severity of impacts on water quality resulting from these operations depend on terrain, climate, details of mining method and environmental protection practices, geochemistry of overburden, methods of transporting ore and/or tailings sand, surface drainage patterns, and the hydrogeology.

Excluding possible effects of chemical differences between tar sands and coal, the types of impacts on water quality of a surface tar sand mine would be essentially similar to those of a surface coal mine in the same area. The organic phases of tar sand bitumen are more similar to petroleum than to coal. Hence, exposed tar sand surfaces and unrecovered bitumen in spent processed sand, if exposed to the physical action of runoff, will be a potential non-point source of organic loading (primarily alkane- or paraffin-type hydrocarbons of heavy molecules with lesser amounts of aromatics) that is not associated with a surface coal mine operation.

Sulphur in tar sand bitumen is present as organic compounds which are oil soluble and are not generally leachable. Also, trace metals in the bitumen are present as oil soluble organic compounds, e.g. porphyrins and salts of organic acids. Although a majority of these compounds are insoluble in water, some could hydrolyze and the metallic ions become soluble in water. If the hot water process is used, some components of the bitumen, including metal compounds, aliphatic and aromatic organic acids, would be expected to react with the extraction water and dissolve (hydrolyze). The effect of this is to make the tailings pond water toxic to aquatic life, and hence unsuitable for discharge into natural water bodies. Further, some bitumen remains in the tailings and a thin slick of bitumen and lighter hydrocarbons form on the surface of the tailings pond. This slick can pose a hazard to water-associated birds that alight on the pond through bitumen-fouling of feathers. The currently accepted solution to the problem of tailings water contamination is to design "zero-discharge" plants, with the tailings water being impounded and recycled through the extraction plant to the maximum extent possible. Loss of water from a tailings pond on a "zero-discharge" plant can come from evaporation, from inclusion in deposited sludge on the bottom of the pond, and from seepage through dykes or into natural ground water below the pond.

Air emissions from mining and extraction operations of tar sands are quite similar to those from surface mining of coal except that as tar sands are exposed, volatiles in the bitumen are an additional emission source. Materials handling operations can produce dust, particularly if overburden or rehandled waste is dry and unconsolidated. Haul roads are a source of dust when transporting ore and spent sand. Water or hydroscopic materials can be used to reduce dust generated from this source. A problem that has been identified in the Athabasca area is the production of "ice fogs" from the large tailings ponds. These "ice fogs" are generated in very cold environments ($-30^{0}C$ and below) when exposed water surfaces are available, or when water vapour is otherwise released. The problem of ice fogs is not unique to tar sands operations, but accompanies many very cold climate operations.

Gaseous by-products would constitute the major atmospheric discharge associated with in-situ combustion. The volumes of gases produced with the bitumen could vary considerably depending on the deposit conditions, the type of production mechanism, and the characteristics of the bitumen in place. The produced gas would usually contain varying amounts of sulphur compounds (SO_x, H_2S, ... etc.) nitrogen, carbon dioxide, carbon monoxide, oxygen and hydrocarbons. A review of 24 different wet and dry in-situ combustion recovery projects on heavy oil reservoirs showed the following average range of concentrations of the produced gases (Farouq, 1972):

Oxygen	2.5–3.5%
Carbon dioxide	10–17%

Carbon monoxide	0–2%
Hydrogen sulphide	0–2%
Methane	0–2%
Nitrogen	the balance

The gases produced in conjunction with in-situ tar sand production will probably be very lean and lack sufficient heating value to justify collection and processing for marketing. They might be either vented from the water knockout and separation tanks to the atmosphere, or combusted further to completely oxidize the carbon monoxide, hydrogen sulphide, and methane. Other air emissions associated with in-situ recovery projects are the exhaust discharges from diesel or gasoline-powered equipment and dust generated from vehicles travelling the access roads.

Upgrading operations at a tar sand plant are the primary source of atmospheric emission with potentially detrimental effects on the environment. The emission that has provoked the most active debate in Canada is sulphur dioxide, but particulate materials, water vapour, hydrocarbons, vanadium and nickel and other substances potentially emitted from a bitumen upgrading plant are also of concern.

CHAPTER 4

Environmental Impacts of Pollutants From Fossil Fuel Combustion

1. SULPHUR OXIDES

SULPHUR is a relatively abundant element which plays an essential part in the environmental cycle. On land, it is found in rocks as sulphide and sulphate and in the oceans it is present predominantly as dissolved sulphate. In the atmosphere, however, the principal sulphur compounds are hydrogen sulphide, sulphur dioxide and sulphate aerosols and mists.

Sulphur compounds are not accumulating in the atmosphere. A cycle operates whereby sulphur is continuously transported between the different phases; and there is a balance between the release of sulphur into the atmosphere and its return to the Earth's surface, although over the last 100 years or so, the increasing amounts of atmospheric sulphur generated by man may have shifted the balance point. This is shown by recent changes in the sulphur content of polar ice, which had previously remained constant over the centuries (Koide and Goldberg, 1971).

Over the past several years estimates of natural and man-made (anthropogenic) sulphur emissions have been made for global sulphur budget calculations. Table 24 lists some of these estimates. On a global basis, fossil fuel combustion accounts for 75–85% of man-made sulphur emissions, and industrial processes such as refining and smelting account for the remainder.

TABLE 24. Emissions of sulphur into the atmosphere (10^6 tonne S(/y)

	Natural Emissions			Man-made
	Volcanoes	Sea-Spray	Biogenic	
Junge (1963)	–	45	230	40 (12.7)*
Eriksson (1963)	–	45	280	40 (11.0)
Robinson and Robbins (1972	–	44	98	70 (33.0)
Kellogg et al. (1972)	1	44	89	50 (27.3)
Friend (1973)	2	44	106	65 (30.0)
Bolin and Charlson (1976)	3	44	31	65 (45.5)
Hallberg (1976)	3	44	37	65 (43.6)
Granat (1976)	3	44	32	65 (45.1)
Garland (1977)	–	44	152	70 (26.3)
Davey (1978)	10	44	86	60 (30.0)
Cullis and Hirschler (1979)	5	44	97	94 (39.2)

* Figures between brackets are percentage of man-made emissions.

These estimates have large uncertainties, particularly in the case of natural emissions, where the magnitude of biological decay emissions is adjusted to make budgets balance. However, they show the magnitude of the contribution of man-made emissions. About 75–80% of man-made sulphur is deposited on land (Grant *et al.,* 1976). The Northern Hemisphere accounts for about 93% of the global man-made SO_2 emissions and has approximately $\frac{2}{3}$ of the Earth's land area. Thus most of the global man-made sulphur is deposited back onto the continental Northern Hemisphere, whereas a relatively small amount finds its way to land in the Southern Hemisphere (MARC, 1978).

Though they are not well understood, the mechanisms by which SO_2 is oxidized to sulphates are important because they determine the rate of formation and, to some extent the final form of sulphate. Atmospheric SO_2 may be oxidized to SO_3 and converted to sulphuric acid aerosol, or it may form sulphite ions that are then oxidized to sulphate. Subsequent to the oxidation, sulphuric acid or sulphate may interact with other materials to form other sulphate compounds. The most important sulphate formation mechanisms identified to date include: direct photo-oxidation, indirect photo-oxidation, air oxidation in liquid droplets, catalyzed oxidation in liquid droplets and catalyzed oxidation on particulates. The conversion of SO_2 to sulphate aerosol in power plant plumes is slow in the early part of the plume; that is, close to the point of emission. As ambient air mixes with the plume, the rate of conversion increases. Thus tall stacks reduce ground-level concentrations of SO_2 but increase sulphate aerosol formation by reducing sulphur losses of SO_2 and by increasing the atmospheric residence time, which results in increased SO_2 to sulphate conversion (Wilson *et al.,* 1977) in the atmosphere and a consequent spread of sulphates over a greater area.

Effects of sulphur oxides:

Although more quantitative data on dose-response relationships are needed, sufficient evidence exists to conclude that atmospheres polluted with oxides of sulphur directly and indirectly attack and damage a wide range of material. Much of this damage is due to the conversion of sulphur oxides to highly reactive sulphuric acid, and/or sulphates.

Atmosphere containing sulphur compounds can corrode several materials (for example, overhead powerline hardware, steel structures, etc.). It attacks and damages a variety of building materials — limestone, marble, mortar — as well as statuary and similar works, causing their physical deterioration (Kucera, 1976).

Although low doses of sulphur dioxide may have beneficial effects on plants, because sulphur is one of the major plant nutrients, an increased uptake of SO_2 will cause more and more changes in the system, first reversible, later irreversible, until a breakdown of the system occurs. Plant species and varieties vary in sensitivity to sulphur dioxide as a result of the interaction of environmental and genetic factors that influence plant response. Temperature, humidity, light, other pollutants and the stage of plant growth all interact in affecting this sensitivity. Sulphur dioxide absorbed by plants may produce two types of visible leaf injury, acute and chronic. Acute injury, which is associated with high concentrations over relatively short intervals, usually results in drying of the injured tissues to a dark brown colour. Chronic injury, which results from lower concentrations over a number of days or weeks, leads to a gradual yellowing (i.e. chlorosis, in which the chlorophyll-making mechanism is impeded). Different varieties of plants vary in their susceptibility to sulphur dioxide injury (Glass, 1978).

The interaction between soils and air-borne oxides of sulphur has received relatively little attention. The two major effects of SO_2 pollution of the soil are a decrease of pH and an increase in sulphate (Halstead and Rennie, 1977). A change of these two variables may then affect the structural and microbial characteristics of the soil. In some systems where the soils are sulphur deficient or where structure is improved by the addition of sulphur, increased productivity may result. However, these circumstances are relatively rare. When the rate of addition or the amount added is excessive, changes in the composition, structure, and function may result. Any effect of SO_x on vegetation will indirectly affect the soil.

Sulphur compounds are removed from the atmosphere by two processes (Glass, 1978): (a) dry deposition including the absorption of SO_2 on exposed surfaces and the sedimentation and impaction of particulates and (b) wet deposition in which sulphur compounds are frequently deposited as acid precipitation. Dry deposition is a continuous process depending mainly on the concentration of sulphur oxides near the ground, the yearly amounts deposited per unit are generally decreasing with increasing distance from the source. Wet deposition is much more variable being dependent both on the rainfall pattern and on the burden of sulphur compounds within the mixing layer. It can be substantial in areas exposed to precipitation from air which has passed large emission sources. In cold climates air pollutants deposited during the winter usually accumulate in the snow pack. When this melts, much of the pollutant load is released in concentrated form with the first melt water. This may lead to sudden increases of acidity in the watercourses and also to some extent in the soil.

Fresh water bodies in many areas of eastern North America and northern Europe, that today lie in and adjacent to the areas where precipitation is most acid, are threathened by the continued deposition and further expansion of acid precipitation. Many of these bodies of fresh water are poorly buffered and vulnerable to acid inputs. These ecosystems appear destined to suffer greater acidification and loss of fish populations. Equally as serious as damage to fish are the less conspicuous effects of the acidification of fresh water, including changes occurring in communities of aquatic organisms such as microdecomposers, algae, aquatic macrophytes, zooplankton and zoobenthos.

The acidification of thousands of lakes and rivers in southern Norway and Sweden during the past two decades has been linked to acid from precipitation., (Dochinger and Seliga, 1976); Braekke, 1976; Wright and Gjessing, 1976; Almer *et al.,* 1974; Dickson, 1975). In turn, this increased acidity has resulted in the decline of various species of fish, particularly trout and salmon. The fish population in rivers and lakes in 20% of the area of southern Norway have been affected by increasing acidity. About 10,000 Swedish lakes are estimated to have been acidified to a pH below 6.0 and 5,000 lakes to a pH of less than 5.0 (Dickson, 1975). Along the west coast of Sweden, about 50% of the lakes have pH values of less than 6.0 and pH has decreased as much as 1.8 units since the 1930's. Fish populations have been correspondingly decimated or seriously affected (Dickson, 1975).

Other evidence indicates that not only are fish affected by acidification, but that a variety of other aquatic organisms in the food web are adversely altered (Dochinger and Seliga, 1976; Braekke, 1976; Hendrey et al., 1976). In general, algal communities in lakes with pH under 6.0 contain fewer species, with a shift toward more acid-tolerant forms. In particular, the chlorophyceae (green algae) are reduced in acid lakes. Some acid lakes and streams contain greater amounts of benthic moss (Sphagnum) and attached algae, and the growth of rooted plants is reduced. There is a tendency toward fewer species of aquatic invertebrates both in the water column and in sediments in acid lakes and streams. The rate of decomposition of organic matter is reduced, with bacteria becoming less dominant relative to fungi. Swedish workers have observed thick fungal felts over large areas of sediments in some acidified lakes. They concluded that decreased decomposition of organic matter on the bottom of lakes, coupled with greater abundance of submerged mosses and fungal mats, reduces nutrient cycling from the sediments. This in turn leads to depletion of nutrients and reduced productivity in acid lakes.

Acid precipitation also causes other changes in lake-water chemistry as well. Elevated concentrations of aluminium, manganese, zinc, cadmium, lead, copper, and nickel have frequently been observed in acidified lakes (Wright and Gjessing, 1976; Beamish, 1976; Dickson, 1975). The abnormally high concentrations are apparently due in part to direct deposition with precipitation as well as increased release (solubility) from the sediments in acidified lakes (Dickson, 1975; Galloway et al., 1976; Galloway and Likens, 1977). These heavy metals may represent a major physiological stress for some aquatic organisms.

In recent years concern has been expressed that forest growth may also be affected far away from emission sources, where the concentration of acid in air and precipitation is lower than where acute damage and visible symptoms occur. The rate of forest growth has declined in southern Scandinavia and in the northeastern U.S.A. between 1950 and 1970, but it is not possible to state unequivocally that this decline is caused by acid precipitation (Abrahamsen et al., 1976; Dochinger and Seliga, 1976; Tamm, 1976). Terrestrial ecosystems are very complex, with numerous living and non-living components. Since acid precipitation is only one of many environmental stresses, its impact may enhance, be enhanced by, or be swamped by other factors. Recent experiments indicate that acid precipitation can damage foliage; accelerate cuticular erosion; alter responses to associated pathogens affect the germination of conifer seeds and the establishment of seedlings; affect the availability of nitrogen in the soil, decrease soil respiration; and increase leaching of nutrient ions from the soil (Abrahamsen *et al.,* 1976; Malmer, 1976; Tamm, 1976). Although many of these factors might be expected to adversely affect tree growth, it has not yet been possible to demonstrate unambiguously decreased tree growth in the field. However, it is possible that acid damage might have been partly offset by the nutritional benefits gained from nitrogen compounds commonly occurring in acid precipitation. Changes already detected in soil processes may as yet be too small to affect plant growth.

Sulphate aerosols contribute to a large extent to the reduction in visibility. In regions prone to high relative humidities some phenomena may aggravate the sulphur dioxide—sulphate—visibility relationship (Glass, 1978). First, there is an increased conversion rate of sulphur dioxide into sulphate thus, more net sulphate may be formed due to the reduced opportunity for SO_2 removal by dry deposition. Second, the visibility reduction of a sulphate aerosol always increases as the relative humidity increases because the particles grow in size. Therefore, if sulphur dioxide emissions cannot be kept at the recommended ambient values, rural and urban sulphate levels can be expected to increase, with a further visibility reduction. This can have socio-economic and climatic implications. The first ranges from simple citizen dissatisfaction to decrease in revenue and property values in area of scenic attraction. Climatological effects include the reduction of solar radiation for photosynthesis, heating or cooling of the atmosphere resulting in changes in the length of growing seasons, and changing precipitation levels (Higgins, 1977).

One of the primary problems in determining the health effects of sulphur dioxide continues to be the development of an understanding of the manner in which this gas interacts with other substances in the atmosphere. Laboratory studies have demonstrated that the levels of sulphur dioxide found in the ambient air are innocuous until combined with other substances. Three categories of human disease appear to be influenced by atmospheres containing sulphur dioxide and associated pollutants. These are: altered pulmonary ventilation and increased prevalence of lower pulmonary disease in children; increased frequency of severity of asthmatic attacks; and increased prevalence of chronic respiratory disease. During so-called air pollution episodes an increased mortality has been observed in susceptible groups, e.g. persons with heart and lung diseases, and the aged. Moreover, high concentrations of sulphur dioxide and particulates have been associated with acute morbidity.

The results from investigations on exposed workers are to a certain extent contrary to the findings in epidemiological investigations of the general population. For example, in studies of workers exposed to sulphur dioxide, it is obvious that medical symptoms have been reported first when the exposure was considerably higher than during even acute air pollution episodes. However, the type of exposure has not been comparable with the exposure from ambient air. Furthermore, occupational groups consists of relatively young and healthy persons while the general population includes also the aged and the infirm. Among occupational groups there is a certain self-selection which means that those who object to the exposure change jobs. All of these factors make it impossible to use the negative results from investigations of exposed workers as a basis for making meaningful quantitative risk evaluation pertaining to the general population. It should also be pointed out that effects were observed at concentrations of about 3 mg/m^3 SO$_2$ in short-term experiments on human volunteers (WHO, 1979).

It is evident from animal experiments that high concentrations of sulphur dioxide can bring about pathological changes in the airways of a similar nature to those seen in patients with chronic bronchitis. At lower exposure levels, about 3 mg/m^3, changes in lung functions have been seen as well as a reduced lung clearance. Although animal experiments have provided evidence that synergistic action takes place between sulphur dioxide and particulates, the conclusion must nevertheless remain that the effects in animal experiments occur at a much higher level of SO$_2$ than in the case of human beings.

A World Health Organization Task Group made an evaluation of the lowest concentrations of sulphur dioxide and particles which could be expected to cause health effects. Acute effects can be expected after simultaneous exposure for 24 hours to 250–500 μg/m^3 of both sulphur dioxide and particles. Respiratory symptoms can also be expected as a result of long-term exposure to 100 μg/m^3 sulphur dioxide and particulates (WHO, 1979). Based on these evaluations, guidelines for the protection of health of the public were developed in terms of 24–h values (100–150 μg/m^3) and in terms of annual means (40–60 mg/m^3).

This evaluation is based primarily upon results from epidemiological investigations. This leads to difficulties in quantitative generalizations. It cannot be excluded that substances other than sulphur dioxide and particulates may have been of crucial significance. For example, in those investigations on which the evaluations are founded the concentrations of nitrogen dioxide were not measured. Nor is there information available on the concentration of sulphur compounds other than sulphur dioxide, e.g. sulphate particulates. When it comes to applying the results to the present exposure conditions it should also be stressed that the particulates primarily consisted of soot which is not the case today. All of these factors mean that it is not possible, with any substantial degree of certainty, to establish values for sulphur dioxide and particulates (alone or in combination with each other) which can be expected to give rise, or, not to give rise, to health effects.

Table 25. SO$_x$ emissions from fossil fuel fired power plants and the estimated health effects (per 1 GW(e)y).

	Coal	Oil	Natural Gas
Tonne SO$_x$ emitted	11x10^4	3.7x10^4	20
Excess deaths	4.5 (0–27)	2.3 (0–13)	0.001

Assuming stack height of 300 m, 3x10^6 people within 80 km around power plant. For other information regarding efficiency of power plants, sulphur content of fuel, see Chapter 1 and 2.

Morgan *et al.* (1978) estimated the health damage due to sulphates to vary from 0 to 12 deaths per 10^5 people per μ/m^3 sulphate with a median of 3.7. This value was applied by Morris *et al.* (1979) to estimate the health hazards from coal-fired power plants. Table 25 gives the amount of SO_x emitted from fossil fuel fired power plants with no stack gas desulphurization together with the calculated mortality due to sulphur oxides.

Control Technology:

Several technologies have been developed for flue gas desulphurization, namely, limestone slurry scrubbing, lime slurry scrubbing, manganese slurry scrubbing — regeneration to H_2SO_4, sodium solution scrubbing — SO_2 reduction to sulphur, and catalytic oxidation (Cheremisinoff, 1976; Yan, 1976). The selection of technology is essentially controlled by its cost. Of these flue gas desulphurization systems, lime/limestone wet scrubbing is now considered a commercial process, land use restrictions may prevent the application of lime/limestone systems in regions of high population density. Assuming a coal with a sulphur content of 3% and a scrubber efficiency of 90% the amount of waste water produced is about 6.1×10^6 tonne (containing 7×10^5 dissolved and suspended solids) per 1 GW(e)y. Such water requires a settling pond about 7.2 hectare (3 m deep); the recovered solids require 10 hectare (6 m deep) storage facility. Such scrubbers reduce considerably the emissions of SO_x and hence the associated health effects.

2. NITROGEN OXIDES AND PHOTOCHEMICAL OXIDANTS

Nitrogen oxides, primarily NO plus smaller quantities of NO_2, arise from a different source than other pollutants. Basically, nitrogen in the combustion air (plus, possibly, small quantities of nitrogen chemically contained in fossil fuels) combines with the oxygen in the air during the combustion process to produce NO. Later, most of the NO oxidizes further to form NO_2. Most NO_2 is formed outside the boiler, often at a considerable distance downwind from the plant stack.

The emission of nitrogen oxides varies greatly from one country to another (see Table 26). However, the major source of world-wide atmospheric NO_x is naturally produced, especially by biological processes. Man's contribution is nevertheless a cause for concern because the emissions are concentrated in the urban areas (NO_x concentrations in urban atmospheres are 10 to 100 times higher than those in rural atmospheres).

Table 26.   Emissions of Nitrogen Oxides as NO_2 (10^3 tonne/year) in some selected countries*

Source	Canada 1972	Fed. Rep Germany 1971	Italy 1972	The Nether- lands 1970	Norway 1970	United Kingdom 1970
Area 10^3 km^2	9976	248	301	36	324	244
Transportation	1092	414	314	108	46	360
Stationary Combustion	445	1296	432	140	19	1602
Industrial Process	63	41	118	35	15	–
Miscellaneous	63	–	–	–	–	–
Grand Total	1663	1750	864	284	80	1962

* Emissions of nitrogen oxides are expressed as NO_2. Most sources are emitting much more NO than NO_2. Oxidation of NO into NO_2 is a function of time and the photochemical processes. Thus NO/NO_2 are dependent on the distance between source and measuring site, time of day, and meterological parameters. The reason for expressing NO_x emissions as NO_2 is due to the fact that formally NO was measured together with NO_2 after oxidation into NO_2. Recently, however, NO concentrations may be measured as such. (To express the NO_2 emissions as NO multiply the values given in the table by a factor of about 0.6). After OECD (1976).

Effects of Nitrogen oxides:

Nitrogen dioxide is a toxic gas in industrial environments. It can, if inhaled in high concentrations, produce plumonary oedema. The concentrations found in ambient air rarely exceed 0.1 to 0.2 ppm and would not produce pulmonary oedema. However, there is evidence from animal experiments that concentrations not much higher than those occasionally found in ambient air can produce cellular alterations and structural changes resembling those seen in some human lung diseases. On the other hand, very little work has been done on the toxic effects of NO mainly because it used to be thought that it was oxidized immediately to NO_2. However, NO is a very active molecule, capable of forming complex compounds with haemoglobin as does carbon monoxide and more work is needed to study its true effects (OECD, 1978).

A limited number of epidemiological studies have been published in which an increased incidence of acute respiratory tract infections in school children has been associated with exposure to nitrogen oxides. A simultaneous exposure to other air pollutants took place, however, and it is not possible to identify a direct relationship between exposure to nitrogen dioxide alone and respiratory diseases. An evaluation of the health effects of nitrogen dioxide must, therefore, rely on experimental data as relevant epidemiological studies are not available. There is indication, however, that adverse effects occur after short-term exposure to about 1–2 mg/m^3 NO$_2$ (0.5–1 ppm). These effects include not only morphological changes and other toxicological effects observed at higher NO$_2$ levels, but also certain effects seen at lower levels. These include increased airways resistance, increased sensitivity to broncho-constrictors and enhanced susceptibility to respiratory infections. In healthy humans adverse effects have been seen after ten minutes of exposure to about 1.3–3.8 mg/m^3 (0.7–2 ppm). Some data suggest that exposure to about 0.2–0.4 mg/m^3 may give rise to adverse effects in asthmatics under conditions of a simultaneous exposure to broncho-constrictors (WHO, 1977). At present, there is no evidence that nitric oxide concentrations typically observed in the ambient air have a significant biological effects (WHO, 1977).

In 1972, WHO believed there was insufficient information upon which to base air quality guidelines in the absence of conclusive epidemiological data. The 1976 Task Group felt it was appropriate and prudent not to wait further but to employ toxicological and experimental data from animals and humans to derive guidelines for the protection of public health. The Task Group selected 0.5 ppm NO$_2$ as an estimate of the lowest observed effect level for short term exposures. In view of the uncertainty concerning the lowest adverse effects level and the high biological activity for NO$_2$ the Task Group concluded that a considerable safety factor was required.

Accepting a safety factor of 3–5, the minimum exposure levels consistent with the protection of public health would become 0.10–0.17 ppm maximum, one hour concentration not to be exceeded more than once per month (WHO, 1977). However, some experts also feel that these exposure levels might have to be lowered if there is biological evidence showing interaction between nitrogen dioxide and other pollutants present if some segments of the population appear to be highly sensitive to NO$_2$. Because of the lack of information on the effects of long-term exposure to nitrogen dioxide, only a short-term exposure has been proposed. A number of countries have established ambient air quality standards for NO$_2$. They cover a wide range from 0.01 to 0.1 ppm for 24 hour averages and from 0.02 to 0.45 ppm for 20 minutes to 1 hour averages (OECD, 1978).

Photochemical Oxidants:

Nitrogen oxide emissions are of particular concern because they are "starting" materials for atmospheric reactions which lead to the production of photochemical oxidants (photochemical smog). Oxidants are formed when nitric oxide reacts with hydrocarbon vapors in the presence of sunlight. Major components formed are ozone, nitrogen dioxide, peroxyacyl nitrate and peroxybenzoyl nitrate.

It was believed that photochemical oxidants give rise only to a local problem in the large urban areas because of either topography or population distribution. However, recent evidence from field studies conducted in Europe and Eastern

North America has established that photochemical pollutants and their precursors can be transported up to several hundred kilometres. This long range transport implies that emission control on a local scale may be grossly insufficient in Europe and Eastern North America (OECD, 1978).

Different effects on man and his environment, attributed to photochemical air pollution, have been noted in many parts of the world. The variations in the type of effect are due to a number of factors such as the variety of pollutants present, their respective concentrations, their cyclical occurrence, their interaction, etc. Few systematic research programmes on health effects, plant damage or other types of effects have been carried out during episodes of photochemical air pollution.

While a large number of compounds fall into the group known as photochemical oxidants, the one which is formed most abundantly in the ambient air, and about which most is known is ozone. It is difficult to obtain an accurate measurement of all the photochemical oxidant compounds in the air, therefore, most air monitoring stations only measure ozone which generally constitutes over 90% of the total photochemical oxidant in the air. Most of the health studies, have been based on exposure to ozone rather than to a combination of oxidants as would occur in the air. In fact, little is known about the toxicity of many specific photochemical oxidant compounds other than ozone and PAN (peroxyacetyl nitrate), but there is a strong suspicion that some, mainly aldehydes (formaldehyde, acrolein), other peroxyacyl nitrates, sulphates and nitrate aerosols, are dangerous to humans.

The primary toxic effects of oxidants are increased susceptibility to infectious pulmonary disease; pulmonary and systematic biochemical changes; eye, nose and throat irritation, nausea and headaches; impairment of pulmonary function; structural changes in lung tissue; and chromosomal alterations of white blood cells. From laboratory experiments on humans and animals, it appears that some of the effects definitely can be attributed to ozone, because ozone does not possess any lachrymatory properties Hazucha et al. 1973; Bates and Hazucha, 1973; Hazucha and Bates, 1975; Watanabe et al. 1973; Sato and Frank, 1974).

Carefully controlled laboratory experiments in the United States (EPA, 1976) have shown that ozone interferes with the normal function of the lung in healthy adults at levels above 0.25 ppm, when ozone is the only air pollutant present. Some of the health effects can, however, be expected to occur at ozone levels considerably below that observed in this laboratory test. This is because ozone is known to have synergistic properties, i.e., the health effects are much more pronounced when ozone and other common air pollutants e.g. SO_2, NO_2, are breathed simultaneously than when breathed alone.

To obtain more information and to help answer questions which cannot be answered through human experiments, scientists have tested a variety of animals under controlled conditions for long periods of time. These studies showed that, at levels and durations of exposure comparable to those experienced in many areas of the United States, serious health effects occur in animals. These include chromosome changes, permanent damage to the elastic recoil properties of lungs, decreased fertility, birth defects, and, possibly, lung cancer. A further observation from animal studies is that exposure for several hours to very low concentrations of ozone (0.08 ppm) reduces the body's capability to resist bacterial infections (OECD, 1978, WHO, 1978).

Consideration has to be given to the limitations of interpreting laboratory animal and human studies. However, these studies have proved both necessary and useful in understanding the health effects of ozone. There is a need for more systematic epidemiological studies to be undertaken so that more definite conclusions on the long-term effects of relatively low concentrations of oxidants can be reached. More controlled animal and human studies with ozone and combinations of pollutants would also prove valuable.

Photochemical oxidants can affect plants in a number of ways, e.g., necrosis, bronzing, silvering, etc. of leaves, and in reducing yield and output. This latter effect is of particular importance for commercial crops. Damage to commercial crops has been reported in a number of countries, mainly Canada, the Netherlands and the United States. As a result of this damage occuring over several years, farmers in certain areas such as Ontario have switched to less sensitive species or to other crops. The commercial crops affected by ozone include white beans, tomatoes, tobacco, lettuce and spinach (OECD, 1978).

An increasing number of countries are using indicator plants to monitor for oxidants as this method is cheap and very useful in situations that do not justify the establishment of a complex and costly monitoring network. In this respect, Canada has reported on several studies concerning indicator plants that show leaf injury to sensitive species of tobacco plants to be consistent with ozone concentrations obtained by instrumental methods (OECD, 1978).

In north-western Europe, north-eastern United States, Los Angeles, Tokyo and Sydney, visibility reduction is an important effect of photochemical pollution episodes. Photochemical reactions lead to the oxidation of sulphur dioxide and nitrogen dioxide to aerosol species which cause the dense summer hazes regularly observed over hundreds of kilometres. Since one of the main removal mechanisms of these aerosol particles is by precipitation, these species may travel quite long distances during photochemical episodes. The aerosol in the polluted photochemical air mass has been observed to reduce the intensity of the solar beam to one tenth of the incident value.

In 1972, the WHO published a tentative air quality criteria value for urban areas of 0.06 ppm ozone. In 1976, WHO convened a Task Group to produce an updated air quality criteria document. The no-effect value for ozone was found to be 0.1 to 0.2 ppm. The Task Group reached a consensus that a one hour mean ozone concentration of 0.05 to 0.1 ppm not to be exceeded more than once per month, should be a guideline for the protection of public health (WHO, 1978).

Control Technology:

Theoretically, it is possible to reduce NO_x concentrations in combustion gases by: (1) modifying the burners or firebox, or both; (2) decomposing nitric oxide and possibly nitrogen dioxide, back to the elements oxygen and nitrogen; or (3) scrubbing the effluent gases. Of the three possibilities, modifications of the combustion equipment have been shown to be the most effective and probably offer the most promise of further NO_x reduction at combustion sources. For stationary sources, the control principle has been based on reducing either the flame temperature or the availability of oxygen, both of which prevent NO formation. Similar principles of control are applicable to motor vehicles. Catalytic principles, which have been applied to reduce NO_x from chemical processes, may also be applicable to the control of NO_x in motor vehicle exhaust.

3. CARBON MONOXIDE

Carbon monoxide (CO) is a product of incomplete combustion of carbonaceous fuels and is formed whenever carbon-bearing materials burn, if the oxygen furnished is less than that required to form carbon dioxide. By far, the largest source of CO emission is motor vehicles. Industrial sources normally are remote from urban areas, and their emissions make no significant contribution to the high CO levels experienced in inner-city areas. Even in the immediate vicinity of major industrial sources, CO levels are generally well below the allowable ambient air quality standard. The background concentration of carbon monoxide present in "clear air" has been found to be 0.13–0.14 ppm in the northern hemisphere and 0.06 ppm in the southern hemisphere. The worldwide emissions of CO have been estimated to be more than 200 million tonnes annually. Several processes for removing carbon monoxide from the atmosphere are known. The oxidation of CO by OH seems to be the most important sink. Additionally, carbon monoxide is removed by natural processes such as the metabolic conversion of CO to CO_2 and methane by soil micro-organisms.

Table 27 gives the carbon monoxide emissions from different sources in the United States, and illustrates that transportation devices account for about 75% of the carbon monoxide emitted. Distinct seasonal variations in carbon monoxide emissions are known; these are primarily the result of both traffic and meteorological variables.

Table 27. Carbon Monoxide Emission Estimates by Source Category for the United States*

Source	CO (10^9 kg/year)	Percentage
Transportation	101.0	73.7
Fuel combusion in stationary sources	1.6	1.2
Industrial processes	10.8	7.9 .
Miscellaneous	23.7	17.2

* Source: UNEP/GC.61/Add. 1.

There is no evidence that the carbon monoxide discharged as a result of man's activities is of any global significance. Adverse effects are known to occur in urban areas (especially road tunnels and confined spaces with heavy traffic), where high levels of carbon monoxide can affect the oxygen-carrying capacity of the blood. When inhaled, carbon monoxide combines with haemoglobin, whose vital function is to transport oxygen. Since carbon monoxide has an affinity for haemoglobin some 240 times that of oxygen, the prime result of this reversible combination is to decrease the capacity of the blood to transport oxygen from the lung to the tissues. It should be noted, however, that carbon monoxide is naturally present in the blood; the normal background concentration of blood carboxyhaemoglobin (COHB) is about 0.5%, and this is attributed to endogenous sources such as catabolic processes.

4. PARTICULATES

Suspended particulate matter can result from natural and man-made sources. The former include volcanic activity, dust storms, forest fires, sea salt spray etc., while the latter include mainly emissions from fuel combustion and industrial processes. Robinson and Robbins (1968) estimated the annual global emission of particulate matter from natural and man-made sources to be of the order of 2600×10^6 tonnes, of which man-made sources constitute 296×10^6 tonnes (or about 11.4%). The burning of fuel (especially coal) for heating and for the generation of power has been one of the major contributors to the suspended particulate matter in urban air. Vehicular traffic also generates particulate compounds from the exhausts of petrol-engined vehicles, and black smoke from those of diesel vehicles.

The impact of particulates on biota depends on their physical and chemical state. The size of the particulates and the presence of absorbed chemicals are the most important factors that determine the detrimental effects of particulates. Trace elements are known to be preferentially concentrated on the smallest particles (Linton et al. 1976). Recently, more concern is being expressed about the potential toxicity of polycyclic aromatic hydrocarbons (PAH) which are released in coal combustion and can be condensed on fine particulates. Studies in Norway (Lunde et al., 1976) have shown significant quantities of these compounds on particulates collected from regional air samples over southern Norway. Since the PAH compounds are known to be active in metabolism (including enzyme induction) and generate a wide range of metabolites, research is beginning to focus on their importance as carcinogenic agents for all animals breathing fine particulates, and for aquatic organisms taking up these compounds from the surrounding water (Lech and Melancom, 1977). Current research indicates important effects induced in fish at dilutions of a few parts per billion, well below levels observed in the Norwegian rainwater samples, but the hazard for human use of the fish is still unknown. Since most particulates in the future will be in the fine aerosol size, this may or may not represent a reduction in ecological hazard, depending on activity of these organic compounds.

Particulates, are generally not considered harmful to vegetation (Jacobson and Hill, 1970; U.S.D.A. Forest Service, 1977). In numerous and varied situations, however, particulates have been implicated in subtle adverse vegetation effects (Lerman and Darley, 1975). Evidence is also available to indicate that soil is an important sink for particulates. Such particulates can cause changes in the metabolism of micro-organisms in the soil and can affect the nutrient cycle, etc. (Bond *et al.*, 1976; Lighthart and Bond, 1976). Another consequence of fine particulates is the possible reduction of photosynthesis by reducing light due to atmospheric contamination (Czaja, 1966; Darley, 1966; Treshow, 1975). Kellogg (1977) pointed out that particulates lead also to cooling of atmospheric temperature, but it is difficult to estimate the net climatic changes due to emission of particulates from fossil fuels combustion.

Different technologies are now available for the removal of particulates from flue gases (wet collection devices, electrostatic seperators, etc.), and it is estimated that wider use and further improvement of these technologies may reduce particulates to about one-third of their present level by the year 2000. Unfortunately, however, the present removal devices remove particles down to one micron in diameter only. Finer particles remain suspended and are apt to be carried far from the source and even into the upper layers of the atmosphere. These fine particles contribute relatively little to the total weight of particulate emissions. There is growing evidence that these fine particles, which can lodge in the deep recesses of the lung, are the ones most responsible for adverse health effects. Small particles can interact with sulphur dioxide in the air to create a much worse health hazard than can SO_2 or particulate pollution independently. There is also growing evidence that small particles tend to worsen the impact of other pollutants on the environment. For short-term exposures, no satisfactory, direct evidence relating concentrations of total suspended particulates to effects is available (WHO, 1979). Because of this, a guideline for short-term exposure levels can only be inferred. A very approximate 24–h guideline for suspended particulate matter in the order of 150–230 $\mu g/m^3$ has been recently proposed by WHO (WHO, 1979).

5. TRACE ELEMENTS

Trace elements potentially hazardous to human health and ecosystems are present in fossil fuels, especially in coal. The trace elements of concern are, among others, arsenic, cadmium, chromium, mercury, lead, manganese, vanadium, flourine and beryllium. Concentrations of these elements vary considerably among different types of fossil fuels. The calculated emissions of trace metals from coal and oil-fired power plants are given in table 28. It should be noted that these values will differ from one plant to another, depending specially on the composition of the fuel.

Table 28.   Trace Elements emitted from Power Plants (tonne per 1 GW(e)y)*

	Coal-fired Plant	Oil-fired Plant
Arsenic	0.04	0.08
Cadmium	0.01	0.02
Chromium	16.00	0.07
Lead	0.90	1.00
Manganese	3.50	0.13
Mercury	0.58	0.003
Nickel	0.36	20.00
Vanadium	0.44	70.00

*Calculated after Freiberg (1977). For fuel and power plant characteristics see Chapter 1 and 2. No desulphurization of flue gas.

Little is known about the fate of trace elements emitted from tall stacks. Both modelling experiments and systematic grid sampling around power plants have indicated that less than 10% of the emissions can be accounted for within 50 km of the plant, (Vaughn *et al.*, 1975).

It is difficult to make an evaluation of the health effects of exposure to trace metals resulting from coal or oil fired power plants. The fate of the trace elements in the environment, the pathways taken and the possibility of synergistic and/or catalytic effects (in relation to other pollutants) are among the main factors that complicate the determination of the health effects of trace elements. Along the pathways in aquatic and terrestrial environments, interactions of trace elements with the bisophere occur. Organisms, especially micro-organisms in aquatic environments, can absorb, concentrate and transform trace elements into more concentrated forms or more toxic compounds. Bio-transformation of trace elements is particularly important in determining effects on man and other organisms. Trace elements may enter food chains and undergo bioaccumulation in passing through higher forms of life.

Of the different trace elements, lead is of particular importance since organic lead compounds (such as tetraethyl lead or tetramethyl lead) are generally used as additives to gasoline and are emitted to about 90 per cent as inorganic lead compounds. Lead has been observed in the urban air, in rush hours in concentrations about 1 $\mu g/m^3$. About 30–40% of inhaled lead is absorbed. Absorbed lead is accumulated chiefly in the skeleton where over 90% of the total body burden is found. The remainder of the body burden of lead is distributed within the brain, liver, kidney and bone marrow. The concentrations in these organs are in general below 1 µg/g. The metabolism of lead is complex. In blood and soft tissues, it probably has a relatively short biological half-time, 15–20 days (Rabinowitz et al., 1974; Chamberlain et al., 1975) while in the bone it has a half-life of over 10 years. The deleterious effect of inorganic lead compounds on different organs is well documented. Lead can cause anaemia by interfering with the synthesis of haemoglobin in several ways. Lead can also damage both the central and the peripheral nervous systems (Seppäläinen et al., 1975).

6. RADIONUCLIDES

Studies have been made in the past few years of the amounts of naturally occurring radioactive substances emitted in the airborne effluents of coal-fired power plants (Eisenbud and Petrow, 1964, Terrill et al., 1976, Hull, 1971; Lave and Freeburg, 1973 and McBride et al., 1978).

It should be noted that on the average, a member of the world population receives a whole body dose of about 100 mrem/y from natural radiation (cosmic rays, radioactive substances in the ground, etc.) and about 50–80 mrem/y from man-made sources (medical sources, nuclear explosions, power production etc.), of which less than 1.0 mrem/y is due to radiation from coal and nuclear power industry. The global dose commitment from one year of coal-power production at the present global installed capacity of 1000 GW(e) corresponds to 0.02 days of natural radiation exposure. One year production of nuclear power at present global installed capacity of 111 GW(e) gives 0.83 days of natural radiation exposure (UNSCEAR, 1977 p. 16).

7. CARBON DIOXIDE

Carbon dioxide is the final oxidation product of carbonaceous fuels; it is also an abundant compound, intimately involved in the natural cycle and essential to the maintenance of life. It exists in the ambient air at a concentration of around 300 ppm, and it is only if man's activities increase this value so as to interfere adversely with natural processes that carbon dioxide can be considered a pollutant. Carbon dioxide is involved in continuous cycles of interchange between atmosphere and oceans, soil and rock layers, and the biosphere. Both land and marine plants withdraw and use carbon dioxide to create carbohydrate compounds. Animals consume plants, releasing carbon dioxide back to the atomsphere in the process of biological oxidation (respiration). Although the geochemical equilibrium keeps the atmosphere content of carbon dioxide fixed at a level of around 300 ppm, it has been reported that there has been an increase in atmospheric carbon dioxide concentration (Williams, 1978).

It is generally agreed that the carbon dioxide concentration has increased from about 300 ppmv in 1900 to about 330 ppmv in 1975 (UNEP, 1979). One reason for this increase must be the burning of fossil fuels, which has increased at a rate of slightly more than 4% per year during most of this century, and which presently adds about 5×10^{15}g carbon per year to the atmosphere. A second possible source of the increasing CO_2 is deforestation and destruction of soil organic matter, particularly in the tropics. There is much uncertainty in the magnitude of this source, but it is believed to be around $1-5 \times 10^{15}$g carbon per year at present (Hampicke, 1979; Bolin, 1979). Not all of the CO_2 added to the atmosphere by man's activities in any one year remains there and it is estimated that about 50% is taken up by other sinks. The major sink is the ocean but there has recently been considerable debate about whether the rate of CO_2 transfer into this sink is large enough to account for the difference between the additions of anthropogenic CO_2 to the atmosphere and the observed yearly increase in atmospheric CO_2 concentration. Other sinks have also been proposed, including reforestation or growth of the biota in temperate latitudes, peat accumulation and dissolved organic matter in the ocean. Despite uncertainty about the sources and sinks of CO_2, it seems clear that continued use of fossil fuels (especially coal) will lead to a continued increase in the atmospheric CO_2 concentration, at least during the next 30 to 50 years during which the biogeochemical conditions, which have controlled CO_2 transfers during the last few decades, are not expected to change.

The reason for concern about the increasing atmospheric CO_2 concentration is because of the so-called greenhouse effect of the gas, which means, all other factors being constant, that an increase in the atmospheric CO_2 level would lead to an increase in the earth's surface temperature. Results from climate models suggest that a doubling of the CO_2 concentration would give an increase in the globally average earth surface temperature of 1.5–3°C (Schneider, 1975; Bach, 1978). However, the important question is: what will the regional changes of temperature and rainfall be? Studies with climate models and of climate observations (UNEP, 1979) indicate that the regional anomalies would occur, but the sign, magnitude and location of these anomalies cannot be reliably predicted at present. Thus it is not possible to make a detailed assessment of the impact on agriculture, water resources and human society, but if a doubling of the atmospheric CO_2 concentration were to occur within the next 100 years, then the change in globally averaged surface temperature would be greater than natural changes which have occurred during the last few thousand years, and this must be expected to have an impact on natural ecosystems as well as human activities.

In view of the concern about the CO_2 problem, several measures have been suggested to limit the amount of CO_2 going into the atmosphere, to remove CO_2 from the atmosphere or from the stacks of power plants, or to generate a climate cooling to compensate for the CO_2- induced warming (Marchetti, 1979). However, most of these measures are very costly or require a scale of operation presently not undertaken with technologies.

With regard to the input of CO_2 from fossil fuels into the atmosphere, a continuation of the present increase of 4% per year in the CO_2 input will probably lead to significant increases in the atmospheric CO_2 concentration within the next 50 years. Energy conservation is an important factor in energy strategies that would lead to significant reductions in energy demand, a process which would be of great importance in controlling the CO_2 problem. In contrast, recent responses to the "energy crisis" in the form of expansion in coal utilization or the development of some synthetic fuels will add to the problem.

The major uncertainties with regard to CO_2 at the present time concern the role of the terrestrial biota in the carbon cycle and the regional climatic impacts of an increase in the atmospheric CO_2 concentration. Until these uncertainties are resolved, it is not possible to say that the combustion of fossil fuels should be limited or halted but it is likewise irresponsible to say at the present time that the use of fossil fuels should be increased. Accelerated research efforts should, therefore be directed to eliminating these uncertainties; in particular, a detailed assessment is required of the present rates of deforestation and soil destruction and associated fluxes of CO_2 into the atmosphere, further information is also needed regarding historic trends in these processes and regarding other changes in the terrestrial biota, which could also influence the atmospheric CO_2 concentration.

CHAPTER 5

Other Environmental Impacts of Power Production from Fossil Fuels

LIKE any other power station, the interaction between foşsil fuel fired power plants and their surroundings has received considerable attention. The impact that a power station may have on the environment depends to a large extent on its location with respect to the load centre, populated areas, open water, agricultural land etc.

The siting of thermal electric generation power plants has recently become more complex. As the sizes of generators, boilers, and associated equipment have grown the problems of site location have increased. The present-day concepts of large fossil fuel fired plants will be met with greater problems of air and water quality control, as the number of plants to be constructed increases. There are several parameters that govern the suitability of the site of a power plant: the geological conditions of the site; the meteorological parameters, the availability of areas for fuel storage and disposal of wastes; terminals for the transportation of fuels to the plant switchyard areas and location, access to transmission lines etc. The near-by availability of a source of cooling water is one of the most important prerequisites in the selection of a site. For fossil fuel fired power plants, land requirements may vary from plant to plant depending on the situation of the plant with regard to open water, population areas and meteorological conditions. The average size requirement for a 1000 MW(e) plant varies from 80–120 ha. Additional land is needed for the switchyard and transmission lines. For coal-fired plants some additional 10 ha may be needed for coal storage at plant.

THERMAL DISCHARGES

The heat produced by burning fossil fuels is extracted through suitable coolants, this in turn runs a turbine to operate the generator to produce electricity. At the exhaust of the turbine the steam is condensed to water to maximize the energy conversion and then is returned to the boiler to repeat the cycle. A large amount of heat is rejected in the condensing process, and the rejected heat is greater than the heat equivalent of the electric energy generated. The thermal efficiency of modern fossil fuel fired power plants is approximately 38–40% which means that about 60% of the heat energy generated has to be rejected to the environment, in the vicinity of the power station.

The bulk of the waste heat is transferred from the steam to the cooling water in the condensers. Water is commonly used as the absorbent because of its general abundance, low cost, high specific heat, and ability to dissipate heat in the evaporation process. The cooling water is extracted from some suitable source (river or lake etc.), passed through the condenser where its temperature is increased by about 7^0C, and returned to the source body of water. Eventually the warmed sink gives up this extra heat to the atmosphere. Such a system, which is referred to as "once-through" cooling, may cause unacceptable environmental changes. In such cases, it is necessary to eliminate thermal discharges to the water source by passing the heated cooling water through a separate "cooling pond" or "cooling tower" system and then return it to the condenser for re-use.

A cooling pond is a large, shallow body of water that achieves its cooling by natural evaporation. Warm water from the condenser is pumped into one end of the pond and cooler water is extracted from the other end. A source of make-up water is required to replace the water lost by evaporation (usually about 3–5% of the water throughput). Cooling ponds are relatively inexpensive but require a rather large land area. A 1000 MW(e) plant may require about 400–1000 ha. Sometimes the water is mechanically sprayed into the air to enhance the evaporative cooling, in which case it is called a "spray pond".

Cooling towers can be classified as *wet* or *dry*. Wet cooling towers achieve cooling by evaporation, and so, like cooling ponds, require a source of make-up water. Dry cooling towers are closed systems and achieve cooling by conduction and convection. In such cases there is no loss of water. For economic reasons, dry cooling towers are very seldom used for power plants. For wet cooling towers, an outside source of make-up is required to compensate for the water lost by evaporation (about 3%), as well as water lost during "blowdown". Blowdown is the continuous or periodic flushing of the cooling system to remove solids and chemicals which accumulate in the circulating cooling water. This flushing can become a water pollution problem unless special treatment is provided. These towers also create huge plumes of water vapour which can contribute to local fogging and icing problems.

A water body provides the environment for many species of organisms, and changes in its temperature, chemical composition and flow rate may affect the number and kind of such organisms. Thermal discharges affect the water-based ecosystem to various degrees (Hutchinson, 1976; Hochachka and Somero, 1973; Coutant, 1975; Biswas, 1974; Gibbons, 1976). Heat influences all biological activity, ranging from feeding habits and reproduction rates of fish via metabolism to changes in nutrient levels, photosynthesis, eutrophication and degradation rate of organic material. It is also necessary to distinguish between thermal effects in temperate and tropical or sub-tropical habitats. In temperate habitats, as water temperature starts to decrease at the end of the summer, the ecosystem activity is reduced as well. Addition of heated water from power plants may then maintain the level of ecosystem development. In contrast, in the tropics and sub-tropics, the water temperature would be high in summer, and additional thermal input could be detrimental. This could be especially important for water bodies that become shallow in the summer, and under such circumstances, the siting of power plants becomes an important consideration. In any case, it should be noted that whereas many studies are available on the effects of thermal discharges in temperate climates, information is more limited on tropical and sub-tropical climates, where the effects could be more pronounced. Seasonal changes could be critical to certain species at specific stages of their life cycle (Hutchinson, 1976). Although studies have been conducted on a variety of organisms, predictions as to the effects of temperature changes and maximum temperatures still cannot be made with certainty especially if the changes are gradual, occur for only a short time, or affect limited areas or volumes of large bodies of water.

Properly controlled thermal discharges have resulted in an increase in the ability of certain commercially valuable aquatic species to multiply; at the same time decreasing the time for the species to reach maturity. Experience has shown that at a number of plant locations the discharge of waste heat to a stream or reservoir has improved fishing in the vicinity of the discharge during the cooler months of the year. This increase in fish populations in heated discharge regions due to higher availability of nutrients has led to the apparent paradox that most fish kills have occurred not during the release of heated effluents but whenever a plant shutdown occurs suddenly due to load dumping or equipment failure. This phenomenon, known as "cold shock", results from the sudden cooling of the water plume to whose temperature the fish have become acclimatised, leaving them with a suddenly reduced body temperature, in a state of stupor, and suffering a loss of equilibrium (Coutant, 1977). Unless the temperature is raised again quickly, the fish may die in a matter of minutes or a few hours. The effect is most pronounced during the cold season when ambient water temperatures are low. It can be minimized either by lowering the temperature slowly to give the fish a chance to adjust to the new temperature or to seek other pastures, by diffusing the heated water well, or by combining the discharge mixing zones of several plants so that the shutting down of any one plant has a less drastic effect on the temperature of the receiving water.

Another major cause of fish kills is the entrainment of small fish in the intake water at velocities too high for them to escape, leading to death either by impingement on the trash screens or by thermal shock as they are carried through the condenser pipes. Thus, the use of water for cooling purposes at steam-electric plants may have effects on aquatic organisms other than those resulting from purely thermal discharges. The destructive effects of passing fish and their larvae or eggs through pumps and condensers may indicate the need for intake screens, preferably with travelling screens having little or no impingement velocity. The most visible environmental impact of some power plants has been due to fish kills resulting from improperly designed intake structures, leading to impingement and mutilation of passing fish on the trash screens. Most recent designs incorporate wide-intake, low-velocity systems with trash racks whose main function is to remove driftwood and other solids, with finer screens to remove algae and other entrained plants and animals.

Before entering the condenser system the water has to be purified, softened and demineralized, so that the water leaving the plant will be cleaner than the inlet stream. In addition, chemicals used intermittently for defouling the condensers could adversely affect fish and fish food organisms. Thermal discharges enhance the solubility of chemicals and the rate of biochemical reactors, and this may be significant in view of the wide range of plant chemicals, such as detergents, algicides, corrosion inhibitors, that may be contained in the plant effluents (IAEA, 1975).

Much has been written and experience acquired about the possible beneficial uses of thermal discharges (Belter, 1975; IAEA, 1975, 1977; Biswas and Cook, 1974; Lee and Sengupta, 1977) from power plants. Of all beneficial uses, the most promising appears to be the employment of waste heat for residential and industrial space heating in the winter and absorption-type air conditioning in the summer. Such waste heat utilization for district heating has been demonstrated successfully in several countries, for example in Sweden (Josefsson and Thunell, 1967), FRG, Finland, France and the U.S.A. Agriculture is also a potential user of waste heat. Irrigation with heated water could promote winter seed germination and growth and extend the growing season. Hot houses are used to grow tropical or sub-tropical crops in more temperate regions. However, a number of problems need to be solved before largescale use of heated water for irrigation could become common practice. Also, the effect of any plant shutdown on such uses of warm water has to be explored. An obvious advantage of raising the temperature of the receiving body of water is in the provision of ice-free shipping lanes and ports. However, the range of waterway affected may be limited to 16–30 km of ice-free water, and the adverse effects on the ecology would be most pronounced during the summer months. Another potential use of condenser discharge water is in aquaculture. Marine and freshwater organisms may be cultured and grown in channels and ponds fed with heated water. For example, it may be possible to grow commercially valuable oysters in the area where they cannot normally reproduce or survive due to low water temperatures. Culture experiments with shrimp, eel, white fish and other species have been carried out with thermal discharges in some countries, and it has been reported that growth rates measured for shrimp and snapper are much higher in the 7^0–8^0C warmer discharge water than at natural temperatures (Belter, 1975). In the U.S.A., catfish production operations are being carried out in conjunction with steam-electric power plant operations in Tennessee and Texas. Aquaculture is also practiced in France and other countries.

TRANSMISSION LINES

The environmental impact of high-voltage transmission lines* may be considered with regard to six aspects: aesthetic considerations, land requirements, communications, hazards, ozone, and habitat effects. Since underground lines at this time are limited to 250 kV and to short-range distribution systems, high-voltage transmission lines will remain part of our environmental scene for a considerable time. Efforts have been made to design

---

* High-voltage transmission lines are common to all electric power stations: nuclear, fossil-fueled or hydro-electric.

more attractive towers and to select cables permitting longer free distances between towers. Routing can take advantage of topographical features to reduce visibility, and use of higher transmission voltages will reduce the number of lines radiating from major power centres. Land requirements will vary with the number of lines and the height of towers, but the right-of-way for major transmission lines may be 30–120 m wide. In general, use of the land below the lines is restricted to pasturing or low-intensity farming, though in most cases the utility may be sole owner of the right-of-way, may keep it fenced and restricted. In any case, it is evident that the land demand for easements associated with any new power station that may be 60 kms or more for major load centres will greatly exceed the land area of the plant itself and of the exclusion area. Regarding communication problems, high tension lines may cause some interference with nearby radio and television reception and may introduce fluctuation on signal strengths on windy days or under icy conditions. Tall towers and multiple lines may pose a hazard to air traffic, particularly under conditions of poor visibility, and air traffic terminals need to be suitably located.

Ozone may be generated by any corona or electrical discharge in air or other oxygen medium. The quantities produced depend on the severity of the discharge and the quantity of oxygen in the affected volume. Corona discharges can increase as a result of abrasions, corrosion effects, foreign particles or sharp points on electrical conductors, or incorrect design that produces excessively high potential gradients. Experimental work has been conducted to determine the added ozone production around a 1000 kV high-voltage test line. The results showed an insignificant increase in ozone concentrations over that produced by sunlight. Consequently, current types of high tension lines are not expected to produce harmful ozone concentrations.

Power line right-of-way has great potential as a wildlife habitat. Shear clearing through heavily forested areas is not inconsistent with good forestry and wildlife management practices. A common management practice in large sections of unbroken forest land is to open the tract by means of small, evenly spaced clearings. The rationale for this practice is to provide diversity and food in the forest environment. Wildfires originally provided this type of habitat. Power line right-of-way creates long linear forest openings that are indefinitely maintained to prevent power outages. The sunlight penetrating the forest via the right-of-way stimulates understorey growth adjacent to the power line. Periodic line maintenance may perpetuate these beneficial wildlife habitat conditions.

Transmission lines can have adverse impacts when crossing wetland areas. There is some evidence to indicate that behavioural modifications may occur for waterfowls, which could result in the absence of birds covering an area within 1 km of the transmission lines. The swaying of the lines in the wind, their reflective properties, and the humming of the lines could explain the abnormal behaviour of the birds. Direct current transmission could also have effects on migratory birds using magnetic homing. Electric fields associated with a transmission line can produce a charge on animals or human beings, within the range of its influence (Janes, 1977). Questions have been raised on the resultant effects of this displacement of currents on biological systems, but this is normally well below the values generally accepted as "safe" levels. Comprehensive reviews of existing studies on biological effects of high voltage electric fields are available (Kaufman and Michaelson, 1974; Bridges, 1975, 1977). Kaufman and Michaelson (1974) conclude that "research to date has failed to provide convincing evidence that human exposure to stationary or low-frequency electric fields has any harmful biological effect". According to Bridges (1975), "although the great bulk of evidence suggests that there are no significant effects of electrical fields encountered under extra-high voltage lines, further research is needed".

AESTHETIC AND RECREATIONAL FACTORS

Power plants vary widely in the visual impression they make on the viewer. Depending on one's point of view their appearance may be judged as blending harmoniously with their surroundings or as an insult to an otherwise attractive countryside. Objectively, suitable architectural treatment utilizing colours and textures, reflections in cooling ponds and a distribution of masses can do much to enhance the plants' appearance. There has been some objection to the appearance of tall hyperbolic cooling towers. To the observer who does not cherish their stark graceful lines they may appear objectionably obtrusive. Attempts to camoflage them by painting them or surrounding them with a shroud rarely have been successful; where their appearance is considered unacceptable, alternative means of providing comparable cooling capacity may have to be explored. The aesthetic impact of high voltage transmission lines has also been criticized. At the voltages in common use, 230 kV and 500 kV, underground lines are not feasible now. The plant location and power line routing should not therefore impinge on areas valued for their historic or touristic significance.

References

Abrahamsen, G. *et al.* (1976): Effects of Acid Precipitation on Coniferous Forest. In: *F. H. Braekke, Editor, Research Report* FR–6, SNSF Project, NISK, Aas, Norway.

Almer, B. *et al.* (1974): *Effects of Acidification of Swedish Lakes.* Ambio, 3, 30.

Andren, A. W. *et al.* (1975): Selenium in Coal-Fired Stream Plant Emission. *Environ. Sci. Technol.* 9:856–858.

API (1977): Proc. 1977 Conference on Prevention and Control of Oil Pollution. Amer. Petrol. Institute, New York.

Bach, W. (1978): The Potential Consequences of Increasing CO_2 Levels in the Atmosphere. In *J. Williams: Carbon Dioxide, Climate and Society.* Pergamon Press, Oxford.

Bates, D. V. and M. Hazucha (1973): The Short-Term Effects of Ozone on the Human Lung. *Proc. Conf. Health Effects of Air Pollution.* Nat. Acad. Sci. Washington, D.C.

Beamish, R. J. (1976): Acidification of Lakes in Canada by Acid Precipitation and the Resulting Effects on Fishes. In: *Proc. The First Internat. Symp. on Acid Precipitation and the Forest Ecosystem.* L. S. Dochinger and T. A. Seling, eds. U.S.D.A. Forest Service Gen. Tech. Report NE–23.

Belter, W. G. (1975): Management of Waste Heat at Nuclear Power Stations; in *"Environmental Effects of Cooling Systems at Nuclear Power Plants"*, IAEA, Vienna.

BIO (1977): Oil/Environment 1977. Intern. Symposium on the Recovery of Oiled Northern Marine Environments. Bedford Institute of Oceanography, Dartmouth, Canada.

Biswas, A. K. (1974): Energy and the Environment. Planning and Finance Service, Environment Canada, Ottawa, Report No. 1.

Biswas, A. K. and B. Cook (1974): Beneficial Uses of Thermal Discharges; Planning and Finance Service, Environment Canada, Ottawa, Report No. 2.

Bocard, C. *et al.* (1979): Cleaning Products Used in Operations after the Amoco Cadiz Disaster. 1979 Oil Spill Conf. p. 163 American Petroleum Institute.

Bolin, B. (1979): Global Ecology and Man. Paper presented at the World Climate Conference, WMO, Geneva.

Bolin, B. and R. J. Charlson (1976): The role of the Tropospheric Sulphur Cycle in the Short-Wave Radioactive Climate of the Earth. *Ambio,* 5:47.

Bond, H. *et al.* (1976): Some Effects of Cadmium on Coniferous Soil and Letter Microcosms. *Soil Sci.* 121(5): 278–287.

Braekke, F. H. (1976): Impact of Acid Precipitation of Forest and Freshwater ecosystems in Norway. Research Report No. 6 Acid Precipitation on Forests and Fish. Aas, Norway. 111p.

Braunstein, H. M. *et al.* (1977): *Environmental, Health and Control Aspects of Coal Conversion,* an Information Overview. 2 Volumes ORNL/EIS–94, Oak Ridge.

Bridges, J. E. (1975): *Biologic Effects of High Voltage Electric Fields*. Electric Power Research Institute, Palo Alto. California Report 381–1.

Bridges, J. E. (1977): Environmental Considerations Concerning the Biological Effects of Power Frequency (50 or 60 Hz) Electric Fields. *Proc. IEEE* Paper F77–256–1.

British Petroleum (1977): *BP Statistical Review of the World Oil Industry 1976*.

Carlberg, J. R. *et al.* (1971): Total Dust, Coal, Free Silica and Trace Metal concentrations in bituminous Coal Miner's lungs. *J. Amer. Industr. Hyg. Assoc.* July, p. 43.

Chamberlain, A. C. *et al.* (1975): Uptake in Inhaled Lead from Motor Exhaust. *Postgrad. Med. J.* 51, 790.

Canada Dept. Fish. and Environ, (1976): see Fisheries and Environment Canada.

Cheremisinoff, P. N. and R. A. Young (1976): Control of Fine Particulate Air Pollutants, Equipment Update Report. *Pollution Engineering,* August 1976 p. 22.

Coutant, C. C. (1975): Temperature Selection by Fish, a Factor in Power Plant Impact Assessment. In *Environmental Effects of Cooling Systems at Nuclear Power Plants,* IAEA, Vienna p. 575.

Coutant, C.C. (1977): Cold Shock to Aquatic Organisms: Guidance for Power Plant Siting Design and Operation. *Nuclear Safety,* 18, 329.

Cowell, E. D. (1971): The Ecological Effects of Oil Pollution on Littoral Communities. *Proc. Sympos. Institute of Petrol.* London.

Cowell, E. D. (1976): Oil Pollution of the Sea. In *R. Johnston Marine Pollution*. Academic Press, London.

Cullis, C. F. and M. M. Hirschler (1979): Emission of Sulphur into the Atmosphere. International Symposium Sulphur Emissions and the Environment.

Czaja, A. T. (1966): The Effect of Dust, Especially Cement Dust, upon Plants. *Agnew. Bot.* 40: 106–120.

Darley, E. F. (1966): Studies on the Effect of Cement Kiln Dust on Vegetation. *J. Air Pollut. Contr. Assoc.* 16:L 145–150.

Davey, T. R. A. (1978): *Anthropogenic Balance for Australia 1976*. Australian Miner. Industr. Council Env. Workshop. Hobart. October 1978.

Demaison, G. T. (1976): Tar Sands and Supergiant Oil Fields. 51st Annual AAPG Meeting, San Francisco.

Dickson, W. (1975): Institute of Freshwater Research, Drottningholm, Sweden, Report No. 54, 8.

Dickson, E. *et al.* (1976): in: UNEP (1976); GC 61/Add. 1.

Dochinger, L. S. and T. A. Seliga (1976): Workshop Report on Acid Precipitation and the Forest Ecosystem. *USDA Forest Service, Gen. Tech. Rept.* NE–26 Northeast. For. Exp. Sta. Upper Darby, Pa. 18 pp.

ECE (1976): *Increased Energy Economy and Efficiency in the ECE Region* ECE, Geneva. United Nations.

ECE (1976): Symposium on Environmental Problems Resulting from Coal Industry Activities. Katowice, 18–22 October 1976, ECE, Geneva.

ECE (1978): *Environmental Impacts of Alternative Energy Technologies* ECE, Geneva.

Eisenbud, M. and H. G. Petrow (1964): Radioactivity in the Atmospheric Effluents of Power Plants that Use Fossil Fuels. *Science,* 144, 288.

EPA (1974): *Transportation Controls to Reduce Automobile Use and Improve* Air Quality in Cities. EPA–400–11–74–002, Washington D.C.

EPA (1976): in OECD; photochemical oxidants and their precursors in the atmosphere, OECD ENT (78/6), 1978.

EPA (1977): *A Summary of Accidents related to Non-Nuclear Energy.* EPA–600/9–77–012, Washington, D.C.

EPA (1978): *Energy/Environment Fact Book.* EPA–600/9–77–041, Washington, D.C.

EPA (1978): *Environmental Effects of Increased Coal Utilization;* Ecological Effects of Gaseous Emissions from Coal Combustion. EPA–600/7–78–108, Washington, D.C.

Eriksson, E. (1963): The Yearly Circulation of Sulphur in Nature. *J. Geophys. Res.* 68, 4001.

Fairman, R. P. *et al.* (1977): Respiratory Status of Surface Coal Mines in the U.S.A. *Arch. Environ. Health,* Sept/Oct.

Farouq, A. S. M. (1972): A Current Appraisal of in-situ Combustion Field Tests. *J. Petrol. Techn.* p. 477–486.

Ferrell, G. C. (1976): *Coal, Economics and the Environment.* IIAEA Professional Paper pp–78–2.

Fisheries and Environment Canada (1976): Potential Pacific Coast Oil Ports: A Comparative Environment at Risk Analysis.

Frazier, N. A. *et al.* (1976): *Production and Processing of US Tar Sands — An Environmental Assessment.* US EPA — 600/7–76–035 Washington, D.C.

Freiberg, L. and R. Cederlof (1977): Combustion Production of Fossil Fuels and their Possible Health Effects on the Swedish Population. Symposium at Karolinska Institute Stockholm March 1977.

Friend, J. P. (1973): The Global Sulphur Cycle. In *"Chemistry of the Lower Atmosphere"* (Ed. Rasool, S.) Plenum Press, New York.

Galloway, J. N. *et al.* (1976): Acid Precipitation in the Northeastern United States: pH and Acidity. *Science* 194: 722–724.

Galloway, J. N., and G. E. Likens (1977): *Atmospheric Enhancement of Metal Disposition in Adirondack Lake Sediments.* Research Project Tech. Resources Research, Department of the Interior, Washington, D.C. July 1977 40 pp.

Garland, J. A. (1977): The Dry Deposition of Sulphur Dioxide to Land and Water Surfaces. *Proc. Roy. Soc. Lond.* A 354, 245.

GESAMP (1977): *Impact of Oil on the Marine Environment.* Report No. 6, 250 pp. FAO Rome.

Gibbons, J. W. (1976): Thermal Alteration and the Enhancement of Species Population; in *"Thermal Ecology II"*, Edited by G. W. Esch and R. W. McFarlane, U.S. ERDA, Washington, D.C.

Glass, N. R. (1978): *Environmental Effects of Increased Coal Utilization.* EPA–600/7–78–108, Washington, D.C.

Glover, H. G. in: Goodman, G. T. and M. J. Chadwick (1978): *Environmental Management of Mineral Wastes,* Sijthoff and Noordhoff, The Netherlands.

Goldberg, E. D. (1976): *The Health of the Oceans.* The UNESCO PRESS, Paris.

Govier, G. W. (1973): *Alberta's Oil Sands in the Energy Supply Picture.* Canad. Soc. Petr. Geol. Symposium. Calgary. Alberta 5–9 Sept. Mem. 3 p. 35–49.

Granat, L. (1976): A Global Atmospheric Sulphur Budget. *SCOPE* No. 7 102.

Grenon, M. (1978): *Fossil Fuel Reserves and Resources*. IIASA Research Memo. 78–35.

Hallberg, R. O. (1976): A Global Sulphur Cycle Based on a Pre-Industrial Steady State of the Pedosphere. *SCOPE* No. 7 93.

Halstead, R. L., and P. J. Rennie. (1977): The Effects of Sulphur on Soils in Canada. In: *Sulphur and Its Inorganic Derivatives in the Canadian Environment*. National Research Council, Environmental Secretariat, Publication No. NRCC 15015. Ottawa.

Hamilton, L. D. (1977): Alternative Sources and Health, in. R. J. Budmitz, CRC Forum on Energy. CRC Forums, Cleveland, Ohio.

Hampicke, U. (1979): Net Transfer of carbon between the Land Biota and the Atmosphere. SCOPE, 13, 219. John Wiley, New York.

Hann (1979) See Oil Spill Conference 1979. American Petrol. Institute.

Hazucha, M. *et al.* (1973): Pulmonary Function in Man after Short-term Exposure to Ozone. *Arch-Envir. Health* 27, 183.

Hazucha, and Bates (1975): in OECD (1978).

Hendrey, G. R. *et al.* (1976): Acid Precipitation: Some Hydrogeological Changes. *Ambio* 5(5–6):224–227.

Higgins, I. T. T. (1977): *Airborne Particles*. National Acad. Sci. Washington, D.C.

Hiyama, Y (1979): Survey of the Effects of the Seto Inland Sea Oil Spill in 1974. 1979 Oil Spill Conference p. 699 American Petroleum Instit.

Hochachka, P. W. and G. N. Somero (1973): *Strategies of Biochemical Adoption*. Saunder Publ. Co. Philadelphia.

Hoult, D. P. (1969): Oil on the Sea, *Proc. Symposium Scientific Engineering Aspects of Oil Pollution of the Sea*. Plenum Press, New York.

Hull, A. P. (1971): Radiation in Perspective, Some Comparisons of the Environmental Risks from Nuclear and Fossil-Fueled Power Plants. *Nucl. Safety* 12, 185.

Hull, A. P. (1974): Comparing Effluent Releases from Nuclear and Fossil Fueled Power Plants. *Nucl. News* 17, 51.

Hurck, G. (1978): *Safety in Coal Mining, the Way Ahead*. Glückauf, 114 No. 3, p. 65.

Hutchison, V. H. (1976): Factors Influencing Thermal Tolerances of Individual Organisms in *"Thermal Ecology II"* Edited by G. W. Esch and R. W. McFarlane. U.S. ERDA, Washington, D.C.

IAEA (1975): *Environmental Effects of Cooling Systems at Nuclear Power Plants*, Vienna.

IAEA (1977): Urban District Heating Using Nuclear Heat. Proc. of a Symposium, Vienna, IAEA Doc. STI/PUB/461, Vienna.

IMCO (1978): The International Conference on Tanker Safety and Pollution Prevention – 1978. London.

IMCO (1979): *IMCO NEWS*, No. 1. IMCO, London.

Jacobson, J. S., and A. C. Hill. (1970): *Recognition of Air Pollution Injury to Vegetation:* A Pictorial Atlas. Air Pollut. Cont. Assoc. Pittsburg. Pa.

Janes, D. E. (1977): Background Information on High Voltage Fields. *Envir. Health Persp.* 20, 141.

Jardine, D. (1974): Cretaceous Oil Sands of Western Canada. Canad. Soc. Petrol. Geologists. Mem. 3, p. 50.

Jones, P. W. *et al.* (1976): Efficient Collection of Polycyclic Organic Compounds from Combustion Effluents. *Environ. Sci. Technol.* 10: 806–810.

Josefssen, L. and J. Thunell (1967): Nuclear District Heating, a Study for the town of Lund. in *Containment and Siting of Nuclear Power Plants*, IAEA, Vienna.

Junge, C. E. (1963): Sulphur in the Atmosphere. *J. Geophys. Res.* 68. 3975.

Kaufman, G. E. and S. M. Michaelson (1974): Critical Review of the Biological Effects of Electric and Magnetic Fields; in *"Biologic and Clinical Effects of Low Frequency Magnetic and Electric Fields*. Thomas Publ. Co. Springfield, Ill.

Kellogg, W. W. (1977): *Effect of Human Activities on Global Climate* WMO No. 486 Tech. Note. No. 156, WMO, Geneva.

Kellogg, W. W. *et al.* (1972): The Sulphur Cycle: *Science,* 175, 587.

Koide, M. and E. D. Goldberg (1971: Atmospheric sulphur and fossil fuel combustion. J. Geophys. Research 76, 6589.

Korte, F. (1977): Potential Impact of Petroleum Products on the Environment, Petrol. Industry Seminar, p. 475 UNEP, Nairobi.

Kucera, V. (1976): Effects of Sulphur Dioxide and Acid Precipitation on Metals and anti-Rust Painted Steel. *Ambio* 5: 243–248.

Lave, L. B. and L. C. Freeburg, (1973): Health Effects of Electricity Generation from Coal Oil and Nuclear Fuel. *Nucl. Safety,* 14, 409.

Lech, J. J. and M. Melancom (1977): Hazardous Chemicals in Fish: In *Documentation of Environmental Change Related to the Columbia Generating Station*. Tenth Semi-annual Report, pp. 132–142. IES Report 82. University of Wisconsin.

Lee, H. *et al.* (1975): *Potential Radioactive Pollutants resulting from Expanded Energy Programmes*. USEPA 68–03–2375.

Lee, S. S. and S. M. Sengupta (1977): Waste Heat Management and Utilization. Proceed. of a Conference, Miami Beach, 1976. Machan. Engin. Dept. Univ. Miami.

Leleuch, H. (1973): Economic Aspects of Offshore Hydrocarbon Exploration and Production. *Ocean Mag.* 1, 187.

Lerman, S. L. and E. F. Darley (1975): Particulates. In *Responses of Plants to Air Pollution*. J. B. Mudd and T. T. Kozlowski, eds. Academic Press, N.Y. pp. 141–158.

Lighthart, B. and H. Bond (1976): Design and Preliminary Results of Soil Litter Microcosms. *Int. J. Env. St.* 10: 51–58.

Linton, R. W. *et al.* (1976) Surface Predominance of Trace Elements in Airborne particles. *Science* 191:852–854.

Lunde *et al* (1976): in OECD (1978).

Malins, D. C. (1977): *Effects of Petroleum on Arctic and Subarctic Marine Environments and Organisms*. 2 Volumes. Academic Press, New York.

Mallatt, R. C. (1977): Refinery Emissions and Effluents Control in the U.S. Petroleum Industry. UNEP Petroleum Industry Seminar, p. 387. UNEP, Nairobi.

Malmer, N. (1976): Acod Precipitation: Chemical Changes in the Soil. *Ambio* 5:231–234.

Marc (1978): *Atmospheric Pathways of Sulphur Compounds,* by D. M. Whelpdale. Monitoring and Assessment Research Centre, London, Report No. 7.

Marchetti, C. (1979): Constructive solutions to the carbon dioxide problem. In *W. Bach et al.: Man's Impact on Climate*. Elsevier Publ. Co. Amsterdam.

McBride, J. P. *et al.* (1978): Radiological Impact of airborne effluents of coal-fired and nuclear power plants. *Nuclear safety,* 17, 497.

McCaslin, J. (1975): World offshore production, reserves listed by fields, *Oil and Gas* J. May 5, 226.

Mertens, E. W. and R. C. Allred (1977): *Impact of Oil Operations on the aquatic environment.* UNEP Petroleum Industry Seminar, p. 591 UNEP, Nairobi.

Morgan, M. G. *et al.* (1978): A probabilistic methodology for estimating air pollution health effects from coal-fired power plants. Energy Systems and Policy 2, 287–309.

Morris, S. C. *et al.* (1979): in *An Assessment of national Consequences of increased coal utilization* US Dept. of Energy TID–2945, Vol. 2 Washington, D.C.

NAS (1973): *Water quality criteria.* National Acad. Sci., Washington, D.C.

NAS (1975): *Petroleum in the marine environment:* National Acad. Sci., Washington D.C.

Nutusch, D. F. S. and J. R. Wallace (1974) Urban aerosol toxicity: The influence of particle size. *Science* 186: 695–699.

Nephew, E. A. (1972): Healing wounds. *Environment,* 14, 12.

New York Acad. of Sciences (1971): *The Sciences,* June–July.

Nicol, C. W. (1976): The Mizushima Oil Spill. Environment Canada Report EPS–8–EC–76–2.

OECD (1977): *Environmental Impacts from offshore Exploration and Production of oil and gas.* Paris.

OECD (1977): *Potential Environmental impacts from the production of synthetic fuels from coal.* Paris.

OECD (1977): *Siting of major energy facilities.* Paris.

OECD (1978): *Environmental Policies to promote Expansion of Coal Production, Transportation and utilization with minimum environmental impact.* ENV/EN/78.3, Paris.

OECD (1978): *Photochemical oxidants and their Precursors in the atmosphere.* ENV(78)6.

Outer Continental Shelf (OCS) (1974): *An Environment Assessment.* U.S. Gov. P.O. 5 Volumes, Washington, D.C.

Pellizzari, E.D. *et al.* (1975) Collection and analysis of trace organic vapor pollutants in ambient atmospheres. *Environ. Sci. Technol.* 9: 552–555.

Peters, A. F. (1974): *Impact of Offshore Oil Operations.* Applied Science Publishers, Barking, England.

Phizackerley, P. H. and L. O. Scott (1967): Major Tar Sand Deposits of the World. *Proc. 7th Petrol. Congress,* Vol. 3, 551.

Rabinowitz, M. *et al.* (1974): Studies of human lead metabolism by use of stable isotope tracers. *Environ. Health. Persp.* 7, 145.

Rae, S. (1971): Pneumoconiosis and coal dust exposure. *British Med. Bull.,* 27, 52.

Rattien, S. and D. Eaton (1976): The prospects and problems of an Emerging Energy Industry. In *J. M. Hollander and M. K. Simmons: Annual Review of Energy* Vol. 1 Ann. Reviews Inc. Palo Alto, California.

Robb, J. E. (1975): *Transportation of Energy.* 9th World Energy Conference, Detroit paper 5.

Robinson, E. and R. C. Robbins (1972): Emissions, concentrations and fate of gaseous atmospheric pollutants. In. *Air Pollution Control,* Part II. Ed. Strauss, W. John Wiley, New York.

Rockette, H. (1977): Mortality among coal miners covered by the UMWA Health and Retirement Funds. DHEW (141 OSH) Publ. 77–155 Washington, D.C.

Sato and Frank (1974): in OECD (1978).

Schneider, S. H. (1975): On the Carbon Dioxide Climate Confusion. *J. Atmos. Sci.* 32, 2060.

Seppäläinen, A. M. *et al.* (1975): Subclinical neuropathy at "safe" levels of lead exposure. *Arch. Environ. Health* 30, 180.

Sweet, D. V. *et al* (1974): The relationship of total dust, free silica and trace metal concentrations to the occupational respiratory disease of bituminous coal miners. *J. Amer. Indus. Hyg. Assoc.,* August.

Talty, J. T. (1978): Assessing coal conversion processes. *Environmental Sci. Techn.* 12, 890.

Tamm, C. O. (1976) Acid precipitation: Biological effects in soil and on forest vegetation. *Ambio* 5: 235–238.

Tanner, A. L. (1977): Gas processing and liquefaction techniques in *Proc. Seminar on Natural Gas,* Ottawa, 21–23 Febr. 1977 SCC and AREC.

Terrill, J. G. *et al.* (1967): Environmental aspects of nuclear and conventional power plants. *Ind. Med. Surg.* 36, 412.

Treshow, N. (1975) Interaction of air pollutants and plant disease. pp. 307–334 In: *Responses of Plants to Air Pollution.* J. B. Mudd and T. T. Kozlowski, eds. Academic Press, N.Y. 383 pp.

United Nations (1978): World Energy Supplies, 1972–1976. *Statistical Papers Series J. No. 21.* Sales No. E. 78. XVII.7 UN, New York.

UNEP (1976): *Review of the impact of production and use of energy on the environment.* GC 61/Add. 1.

UNEP (1977): Motor Vehicle Seminar, Industry Programme, Nairobi.

UNEP (1979): A systems study of Energy and Climate. Status Report by IIASA SR–79–2 B.

UNESCO (1979): *Hydrological problems arising from the development of energy.* Technical Paper Hydrology No. 17, Paris.

UNSCEAR (1977): *Sources and Effects of Ionizing Radiation.* United Nations Scientific Committee on the Effects of Atomic Radiation. United Nations, New York.

USDA, Forest Service (1977) File report of 1974 to 1976 surveys for oxidant injury in the Sequoia National Forest. Forest Pest Management, San Francisco, Region 5.

Van Eek, W. H. (1977): State of the Art of Environmental Conservation, Exploration and Production. UNEP Petroleum Industry Seminar p. 59. UNEP, Nairobi.

Van Hook, R. I., and W. D. Schults (1977) Effects of trace contaminants from coal combustion. Proc. Workshop Aug. 2–6, 1976, Knoxville, Tennessee. Energy Research and Development Administration, Washington, D.C. *ERDA* 77–64, 79 pp.

Vaughn, B. E. *et al.* (1975): Review of potential impact on health and environmental quality from metals entering the environment as a result of coal utilization. Battelle Memorial Institute, Richland, Washington.

Vernberg, F. J. *et als.* (1977): *Physiological Responses of Marine biota to pollutants.* Academic Press., New York.

WAES (1977): See under C. Wilson.

Walder, C. A. (1977): Environmental Aspects of the Transportation of oil by Sea. UNEP Seminar on Petroleum Industry. UNEP, 1977.

Waldichuk, M. (1977): Overview on Marine Environment. UNEP, Nairobi.

Walters, E. J. (1974): Review of the World's Major Oil Sand deposits. *Canada Soc. Petrol. Geologists. Mem.*3, p. 240.

Watanabe, S. *et al.* (1973): The cytotoxicity of ozone. Japan. *J. Publ. Health* 20, 554.

Westaway, M. T. (1977): Environmental impact of offshore development. UNEP Petroleum Industry Seminar, p. 121, UNEP, Nairobi.

Williams, J. (1978): *Carbon Dioxide, Climate and Society.* Pergamon Press, Oxford.

Wilson, C. (1977): *Energy: Global Prospects 1985–2000.* McGraw Hill, New York.

Wilson, R. and W. J. Jones (1974): *Energy, Ecology and the Environment.* Academic Press, New York.

Wilson, W. E. *et al.* (1977): *Sulphates in the atmosphere.* EPA–600/7–77–021, Washington, D.C.

Wolfe, D. A. (1977): *Fate and effects of Petroleum hydrocarbons in marine ecosystems and organisms.* Pergamon Press, New York.

World Bank (1979): *World Development Report,* 1979, Washington, D.C.

WEC (1980): Survey of Energy Resources, 11th World Energy Conference, Munich.

WHO (1977): *Environmental Health Criteria* 4. Oxides of nitrogen. Geneva.

WHO (1978): *Environmental Health Criteria* 7. Photochemical Oxidants. Geneva.

WHO (1979): *Environmental Health Criteria* 8. Sulphur Oxides and Suspended Particulate Matter. Geneva.

Wright, R. F., and E. T. Gjessing (1976), Acid precipitation: Changes in the chemical composition of lakes. *Ambio* 5(5–6): 219–223.

Yan, C. J. (1976): Evaluating Environmental Impacts of stack gas desalphurization Processes. *Environ. Sci. Techn.* 10, 54.

PART II
NUCLEAR ENERGY

CHAPTER 6

Radiation

THE nuclear power industry produces a number of environmental impacts similar to those which are caused by fossil fuels (especially by the coal fuel cycle) e.g. thermal pollution, land use, non-radioactive and radioactive emissions, occupational risks etc. However, the principal concern with releases from the nuclear fuel cycle operations has been with radioactivity and its effects on the biosphere, and specially on human health. Although the principal objective of radiation protection is the achievement and maintenance of appropriately safe conditions for activities involving human exposure, the level of safety required for the protection of all human individuals is thought likely to be adequate to protect other species, although not necessarily individual members of those species (ICRP, 1977). The ICRP therefore believes that if man is adequately protected then other living things are also likely to be sufficiently protected. However, ecological changes are sometimes unforeseen and there is a case for maintaining a watch on the ecology of areas which receive higher radiation doses (Royal Commission Environmental Protection, 1976). In this Chapter, a brief discussion of the sources of radiation, radiological pathways and radiation effects is given to facilitate the follow-up of the following chapters. For more details, the reader should refer to books on radiation protection, UNSCEAR (1977) and ICRP (1977).

SOURCES OF RADIATION

Radioactivity is a natural and spontaneous process by which the unstable atoms of an element emit or radiate the excess energy of their nuclei as particles or photons and change (or decay) to atoms of a different element or to a lower energy form of the original element. The second element may also be radioactive and will continue the process of disintegration in a cascade of different elements (or decay chain) until a non-radioactive, stable form is reached. Man has always been exposed to ionizing radiation from various natural sources. A distinguishing characteristic of this natural radiation is that it involves the entire population of the world and that it has been experienced at a relatively constant rate over a very long period of time. This radiation exposure arises from several sources; external sources, such as cosmic rays and radioactive substances in the ground and in building materials, and internal sources in the form of naturally occurring radioactive substances in the human body, particularly K–40. The extent of this natural exposure varies greatly from place to place, as well as locally. In some locations where high uranium or thorium content in rocks or soil exists (such as monazite sands, granites, phosphates, shales etc.), the population in those locations receive a comparatively higher population dose than does the world's population in general.

In addition to those natural sources of radiation, a number of technologically enhanced exposures to natural sources exist, for example, in high altitude flights, processing and using of some ores and materials such as phosphates and some building materials. Man-made ionizing radiation consists of radiation used for medical purposes, radiation produced from nuclear explosion, nuclear power production and radiation emitted from some consumer products (radio-luminous products, etc.).

On the average, a member of the world population receives a whole body dose of about 100 mrem/y* from natural radiation and about 50 to 80 mrem/y from man-made sources. The latter dose varies from one country to another and from person to person depending on the rate of use of medical X-rays, radiopharmaceuticals, nuclear power production etc. From the doses attributed to man-made sources, less than 1.0 mrem/y is due to radiation emitted from the coal and nuclear power industry. Table 29 summarizes the estimated global whole body dose commitments for different sources and practices. It is expressed as the duration of exposure of the world population to natural radiation which would cause the same dose commitment.

Table 29. Global Dose Commitments from Various Radiation Sources (after UNSCEAR, 1977.)

Source of exposure	Global Dose Commitment (days)**
One-year exposure to natural sources	365
One-year of commercial air travel	0.4
Use of one year's production of phosphate fertilizers at the present production rate	0.04
One-year global production of electric energy by coal-fired power plants at the present global installed capacity (1000 GW(e))	0.02
One-year exposure to radiation-emitting consumer products	3
One-year production of nuclear power at the present global installed capacity (111 GW(e))	0.83
One-year of nuclear explosions averaged over the period 1951–1976	30
One year's use of radiation in medical diagnosis	70

** The global dose commitment is expressed as the duration of exposure of the world population to natural radiation which would cause the same dose commitment. The occupational contribution is included.

The global dose commitment corresponds to 0.83 days of natural radiation exposure for one year of nuclear power production at the present installed capacity of 111 GW(e). Assuming that the present nuclear technology remains

* mrem/y: millirem per year. mrem = 1/1000 of a rem. For definition of rem, see p. 100.

the same, one year of energy production at the projected nuclear installed capacity in the year 2000 (1100–1700 GW(e)) would lead to a global dose commitment equivalent to 8.3–12.8 days of natural radiation exposure.

RADIOLOGICAL EXPOSURE PATHWAYS

The release of radioactivity into the environment can cause radiation exposure to the biosphere. This can be effected by a number of possible routes or pathways. These "environmental pathways" are illustrated in Fig. 1. Radiation exposure may be either external, internal or both. External radiation exposure occurs when the source of radiation is external to the body, for instance if the radioactivity has been deposited on the ground or passes overhead as a cloud. Internal exposure will occur if some of the radioactivity released gets inside the body. This may occur directly by inhaling some of the released radioactivity or by ingestion of activity through consumption of contaminated food. The latter may occur through a "food chain" which may be a single stage, such as direct contamination of leafy vegetables, or involve a series of stages leading to human food material. The nature of these food chains depends on ecological factors; radioactive materials, because of their chemical nature, may be concentrated by the flora or fauna exposed to the contaminant.

The process is highly site specific and the consequential exposure of man will also depend on food consumption rates and habits. It is for the latter reasons that ICRP introduced the concept of "critical group". For any proposed release of radioactivity into the environment it is usually found that there is one pathway whereby a group of people (the "critical group") may receive a radiation exposure higher than that received by the rest of the population. Adequate control of the exposure of members of this group will then ensure that individual members of the public are safeguarded. The size of the critical group is variable dependent on the particular circumstances and is related to the special dietary and working habits of the members of the group. These food chain pathways are often the most important environmental pathways leading to the intake of radioactivity by man and his subsequent radiation exposure.

RADIATION EFFECTS

The hazardous nature of ionizing radiation first became apparent in connection with exposures from man-made sources. The discovery of X-rays and the identification and separation of radioactive substances at the end of the last century brought, in addition to great benefits, unforeseen hazards. The need for

Fig. 1 Radiological Exposure Pathways

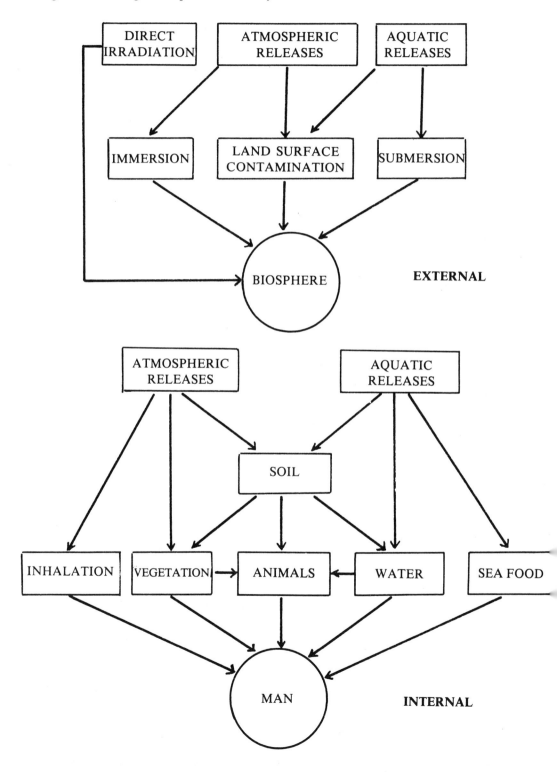

protection from radiation became obvious as observations of radiation injury in man and radio-biological experiments in animals increased the knowledge of the gross effects of radiation. The first international action to be taken in this respect was by the second International Congress of Radiology in 1928 through the establishment of an International Commission on Radiological Protection (ICRP) to provide guidance and formulate recommendations within the field of radiation protection. Because of the increase in radioactivity in our environment from man-made sources such as accelerators, nuclear reactors and artificially-produced radio-isotopes, the potential for radiation hazards and the problems of radiation protection have also increased. This has made it necessary for UNSCEAR, ICRP and similar bodies to keep under review the evaluation of radiation levels and effects and formulate recommendations for radiation protection.

The principal effects of exposure to radiation are a consequence of the alteration and destruction of cells in the irradiation issues. The observed effects of non-fatal doses are classified as *somatic* (affecting the individual exposed) and *genetic* (affecting the offspring of the exposed individual). The effects of radiation are also classified as *stochastic* and *non-stochastic* (ICRP). Stochastic effects are those for which the probability of an effect occurring, rather than its severity, is regarded as a function of dose, without threshold. Non-stochastic effects are those for which the severity of the effect varies with the dose, and for which a recommended exposure level may therefore occur.

At the dose range involved in radiation protection, genetic effects are regarded as being stochastic. Some somatic effects are stochastic; of these, carcinogenesis is considered to be the chief somatic risk of irradiation at low doses and therefore the main problem in radiation protection. The somatic effects which are non-stochastic are specific to particular tissues, as in the case of cataract of the lens, non-malignant damage to the skin, cell depletion in the bone-marrow causing haematological deficiencies, and gonadal cell damage leading to impairment of fertility.

The prevention of non-stochastic effects would be achieved by setting dose-equivalent limits at sufficiently low values so that no threshold dose would be reached, even following exposure for the whole of a lifetime or for the total period of working life. For stochastic effects, on the other hand, the situation is different. The mechanism of induction of malignancies whether by radiation or other agents, is not known. For an individual case it is impossible to establish a causative relation to a radiation exposure, and therefore the principal available information on radiation carcinogenesis in man stems from epidemiological studies of irradiated population groups with some confirmation from animal experiments. The induction of malignancies is well established after high doses approaching or exceeding 100 rad. However, the statistical evidence for most malignancies is still insufficient to infer the shape of dose-probability curve at low individual doses, which one anticipates in connection with the use of nuclear power. In particular, it is not known whether a threshold exists.

For radiation protection purposes it is generally regarded as conservative to assume that there is a linear relationship between dose and the probability of a stochastic effect, within the range of doses encountered in human exposures. A consequence of this prudent assumption is that doses can be regarded as additive, in the sense that equal dose increments equally increase the probability of an effect by an amount which is independent of the previously accumulated dose. It is also assumed that the relevant slope of the linear relationship is that of a straight line drawn from the points observed at high dose and passing through the origin. If the relationship were in fact to be sigmoid, as some animal experiments and theoretical considerations suggest, this linear assumption would be conservative because it would overestimate the effect of low doses.

The same considerations apply to the genetic effects, which are also stochastic, and are the result of point mutations and chromosomal effects in the germ cells of irradiated individuals. The severity of genetic effects ranges from insignificant to lethal; slight defects will tend to continue in the descendants for many generations, while a severe defect will be eliminated rapidly by the early death of the individual or the zygote carrying the defective gene. At present, there is no direct evidence of radiation-induced genetic effects in man.*

Some recent studies (Bross and Natarajan, 1972; Mancuso *et al.*, 1977; Najarian and Colton, 1978) have been interpreted by some to indicate that the commonly used risk estimates, which are based on the UNSCEAR and BEIR Committee Reports, underestimate the effect of low-dose radiation, i.e., that the linear theory is not conservative in estimating risk at low doses but rather underestimates it. Bross and Natarajan believe that they have identified subgroups in the population which are especially sensitive to radiation damage. Their belief derives from finding an association between some "indicators of susceptibility" (viral infections, bacterial infections, and allergy) shown by leukemic children from birth until diagnosis of leukemia. They concluded "the apparently harmful effects of antenatal irradiation are greatly increased in certain susceptible subgroups of children possessing the indicators associated with a slightly higher intrinsic risk of leukemia." However, Smith *et al.* (1973) pointed out that the incidence of these indicator diseases before the clinical onset of leukemia is the same in children who had received no irradiation *in utero* as in those who had (see also, Rothman, 1977).

Mancuso *et al.* (1977) and Kneale *et al.* (1978, 1979) claim to have demonstrated a radiation-induced excess of cancers in workers at the atomic energy plant at Hanford. Najarian and Colton (1978) claim a similar excess of leukemia and cancer in workers at the Portsmouth Naval Shipyard in New England. The data on which some of these claims have been based, and the analyses involved have been criticized by a number of authors (Mole, 1978); (Reissland and Dolphin, 1978; Marks *et al.*, 1978). In the Hanford workers, there

---

* Kochupillai *et al.* (1976) observed some radiation related genetic changes in Kerala, South India, where background radiation doses are between 1.5 and 3 rem/y (more than 10 times the normal exposure in the world). However, their findings have been questioned by Sundaram (see UNSCEAR, 1977, p. 432).

is the possibility of an association with the work experience for the cancer of the pancreas and multiple myeloma (Marks *et al.,* 1978), and if the cancer-doubling estimates of Mancuso *et al.* were correct, something other than radiation was the cause of the observed cancers. Similarly, the absence of any apparent latent period effect casts doubt on conclusions about the contribution of radiation to the high numbers of cancer deaths among the Portsmouth Naval Shipyard workers. Evans *et al.,* (1979) recently found an increase in chromosome damage (chromosome aberrations in peripheral blood lymphocytes) in dockyard workers in the U.K. with increasing exposure to low levels of radiation. However, they pointed out that the exposed population studied is small, subject to very low levels of radiation exposure for periods up to 10 years and is therefore unlikely to provide useful data on the incidence of malignant disease. Thus, although these claims of higher risks from low levels of radiation described by these authors have become the subject of considerable public debate, examination to date of the results does not support the findings of Bross and Natarajan; Mancuso *et al.* and Najarian and Colton.

The difficulties in establishing the effects of very low dose levels of a few millirem or less per year to the general population from the nuclear fuel cycle are two-fold. One arises from the fact that no statistically valid epidemiological data are available. This is due to the virtually unsurmountable difficulties imposed by the long-term nature of epidemiological work that would need to be conducted under controlled conditions on a very large population, and over several generations in case of genetic effects. Even epidemiology studies on comparable populations living exposed to different natural radiation levels have proved difficult and inconclusive. It also would be imperative to establish a unique cause and effect relationship, even if correlations between morbidity and radiation exposure were apparent, to rule out other causes such as economic conditions, nutritional factors, or improved diagnostic facilities due to better medical services. Another problem arises from the fact that, although it is easy to observe deaths and major malignancies such as may be observed at dose levels above 50–100 rad, it is more difficult to observe minor incidences of illness in a much larger population, such as might result from low-level exposures, and to distinguish them unambiguously from similar illnesses that could be caused by a host of other causes.

BASIC CONCEPTS

Risk and Detriment:

Risk can be defined as the number of undesirable events (for example, deaths) occurring in a given period of time. Risk can be expressed, quantitatively as the product of the magnitude of the consequences that result each time an event occurs and the frequency of that event. In equation form:

Risk	=	Frequency	X	Magnitude
(Consequences/year)		(events/year)		(Consequences/event)

For example, if two persons were killed, on the average in each train accident and if there were an average of ten accidents per year in a country, then the risk to the population in that country from that event is 20 deaths per year. In radiation protection, the word risk is used to describe the probability that a given individual will incur a deleterious effect as a result of a radiation dose. If several effects could result from a given radiation, the total risk, R, is expressed by the following equation:

$$R \simeq \sum_i P_i$$

i.e. the summation of individual risks P_i of suffering effects i.

In order to identify, and where possible to quantify, the deleterious effects of exposure to radiation, the ICRP (ICRP 1977) introduced the concept of "detriment". In general, the detriment in a population is defined as the mathematical "expectation" of the harm incurred from an exposure to radiation, taking into account not only the probability of each type of deleterious effect, but also the severity of the effect. For effects on health, if P_i, the risk of suffering the effect i, is small and the severity of the effect is expressed by a weighing factor g_i, then the detriment to health, G, in a group of N persons is given by:

$$G = N \sum_i P_i g_i$$

Several approaches are possible to give values to the weighting factors g_i. One possibility is the use of the reduction of life (in relation to normal life expectancy) or, in the case of occupational exposures, the working time lost. For radiation protection purposes, it can be assumed that the detriment is dominated by the induction of fatal malignancies and of severe genetic effects in the first two generations, assigning a weighting factor of 1 to all these effects.

Absorbed Dose and Dose Equivalent:

The "absorbed dose D" is defined as the mean energy imparted by ionizing radiation per unit mass at a certain point. The unit of absorbed dose in the International System (SI) is the *gray* (1 joule per kilogram) : $1 Gy = 1 J/kg$. In the present study use is made of the commonly used absorbed dose unit, the *rad*, where 1 rad = 0.01 Gy. The absorbed dose is insufficient by itself to determine the risk or the detriment resulting from irradiation because it is found that in biological systems, the degree of the deleterious effect produced by a given dose depends on many factors such as the type of radiation and how the radiation is distributed. This fact reduces the usefulness of the absorbed dose for radiation assessments, which require a quantity reasonably representative of the radiation detriment.

In radiation protection it has been found convenient to introduce a further quantity that correlated better with the more important deleterious effect of exposure to radiation, more particularly with the delayed stochastic effects. This quantity, called *dose equivalent,* H, is defined by the equation:

$$H = DQN$$

where D is the absorbed dose, Q is a quality factor and N is the product of all other modifying factors depending on irradiation conditions such as dose rate fractionation, etc. At present, the ICRP has assigned the value of 1 to the factor N. The quality factor, Q, is intended to allow for the effect on the detriment of the microscopic distribution of absorbed energy. The following average values of Q are recommended by ICRP (ICRP 1977) for both external and internal radiation:

	$\overline{Q}$
X-rays, gamma rays and electrons	1
Thermal neutrons	2.3
Other neutrons, protons and singly-charged particles of rest mass greater than one atomic mass unit of unknown energy	10
Alpha-particles and multiple-charged particles (and particles of unknown charge), of unknown energy	20

The special name for the unit of dose equivalent in the SI is the sievert (Sv):

$$1 \text{ Sv} = 1 \text{J/kg}$$

In the present study, use is made of the commonly used unit of dose equivalent, the *rem,* where 1 rem = 0.01 Sv. It should be noted that the dose equivalent is used in radiation protection under normal conditions and should not be used to assess the likely early consequences of severe accidental exposures in man.

Effective Dose Equivalent:

For stochastic effects, the concept of the "effective dose equivalent" has been introduced. This concept is based on the principle that at a given level of protection, the risk should be equal whether the whole body is irradiated uniformly or whether there is non-uniform irradiation. The effective dose equivalent H_E is given by:

$$H_E = \sum_T W_T H_T$$

Where W_T is a weighting factor representing the fraction of risk resulting from tissue T when the whole-body is irradiated uniformly and H_T is the dose equivalent in tissue T.

The values of W_T recommended by the ICRP are:

Tissue	W_T
Gonads	0.25
Breast	0.15
Red Bone Marrow	0.12
Lung	0.12
Thyroid	0.03
Bone Surfaces	0.03
Remainder	0.30

Regarding the remainder, a value of $W_T = 0.06$ is applied to each of the five organs receiving the highest dose equivalents; the exposure of the other remaining organs is neglected. The gastro-intestinal tract is treated as four separate organs (stomach, small intestine, upper large intestine and lower large intestine). The skin, lens of the eye, hands, forearms, feet and ankles are not included in the remainder. Regarding the skin, however, the assessment of detriment from exposure of population groups may require consideration of the small risk of fatal cancer resulting from exposure of the whole skin to soft beta radiation. In this case, a value of W_T of 0.01 may be applied to the mean dose equivalent over the entire surface of the skin.

Assessment of Radiation Health Impacts:

Assessment of radiation dose is only an intermediate step in the assessment of radiation risks and radiation detriment. For stochastic radiation effects (cancer and genetic effects) the assessment of *individual* risks makes use of the effective dose equivalent defined in the previous paragraph. The use of this quantity implies the assumption of a linear, non-threshold relationship between the risk of deleterious effects and radiation dose. Values of risk factors are given by ICRP (ICRP 1977) and in BEIR report for different tissues at risk. The average mortality risk factor for radiation-induced cancers (uniform whole body irradiation) is about 10^{-4}/rem for both sexes and all ages (ICRP 1977). The average risk factor for genetic effects, as expressed in the first two generations, would be substantially lower than this, when account is taken of the proportion of exposures that is likely to be genetically significant, and can be taken as about 4×10^{-5}/rem (ICRP 1977).

The estimate of the total radiation health impact of a given practice on an exposed *population* can be characterized by the mathematical expectation of the number of cases of deleterious stochastic effects resulting from the practice. In the case of nuclear power, it is convenient to normalize the impact for 1 GW(e)y.

The mathematical expectation of deleterious effects is proportional both to the average individual dose and to the number of individuals exposed. It is also proportional to the product of these two quantities, which is called the *collective dose*.

The collective dose, expressed in man rem, can also be viewed (although this is not formally correct) as the sum of the doses to all exposed individuals. In the assessment of the total radiation health impact of a practice, it is necessary to extend this summation to include all individuals exposed by the practice, irrespective of country limits or of regional distribution, i.e., it is necessary to make a global assessment. As long as only small increments to the background dose received from natural sources of radiation are considered, all individual dose contributions must be added. A certain average dose to 100 individuals would be expected to have the same total health consequences as 1% of that dose to 10,000 individuals, provided that all individual doses are small. On the other hand, for each individual, the smaller dose in the second case may mean that the added risk becomes insignificant in comparison with the overall risk to which the individual is exposed.

A unit practice which is limited in time, for example, the production of 1 GW(e)y, may result in environmental contamination that would deliver radiation exposures and consequently radiation harm over a long period after the end of the practice. In assessing the total radiation health impact of the practice, it is therefore necessary to include future radiation doses in the summation of doses to give the *collective dose commitment* of the unit of practice. The assessment of the future collective doses can be made through environmental pathway analysis of the type which has been used by UNSCEAR in assessing dose commitments from nuclear test explosions as well as from nuclear power production. The concept of collective dose commitment is illustrated in Fig. 2.

Fig. 2   The Concept of Collective Dose Commitment

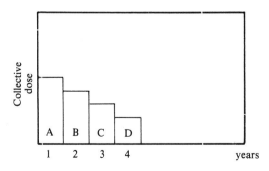

If the collective dose from a practice is A for the first year, B for the second year etc., the collective dose commitment for the practice = A+B+C+D. The collective dose commitment is used for two different purposes which should not be confused: (a) as a quantity which is proportional to the total radiation detriment (the total number of stochastic effects can be calculated by multiplying the collective dose commitment with the risk coefficients mentioned earlier) (b) as a tool in the assessment of the future individual doses and risks.

In the case of very long-lived radionuclides such as U–238 and I–129 where the radioactive substances would still exist even after millions of years, a situation may arise where an extremely low annual dose (and corresponding negligible individual risk) might prevail over periods of millions of years. Even in a situation where it would be extremely unlikely that any single case of injury would appear even over a period of, say, 10,000 years, the total life-time of the radionuclide might conceivably cause a definite number of injuries over millions of years. An even more extreme case would be that of non-radioactive carcinogenic substances of high chemical stability and persistence in the environment. For this reason, it is not sufficient to assess the collective dose commitment, with collective doses summed up over infinite time, without also indicating how the resulting harm might be distributed in time. If the implication of a certain collective dose commitment is a certain number of cancer deaths and a certain number of genetically disabled persons, it would be of interest to know whether this harm is expected to occur in the near future or if it will be distributed over an extremely long period of time. The reliability of assessments over periods of thousands or millions of years is quite small. The relevance of the assessment is also uncertain, since it is based on such assumptions as the existence and size of human populations in the far future and on the persistence of the validity of the dose-response relations into that future.

For the assessment of the maximum annual doses in the future, it is generally assumed that a given practice will cause a steady state situation. In that situation the introduction of new radioactive material into the environment is compensated by the decay or removal of the already accumulated material. Assuming that the affected population remains constant, in such a steady state situation, the future annual collective dose is equal to the collective dose commitment from that practice (Fig. 3).

Fig. 3 Future Annual Collective Doses

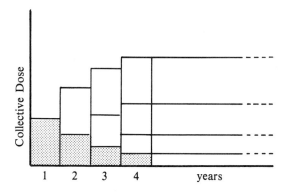

For long-lived radionuclides, the total period of the practice may not be long enough to permit the steady state situation to be reached. However, the summation of the annual collective doses over a period equal to the total duration of the practice period will give a sum called the *"incomplete collective dose commitment"* which is equal to the maximum future annual collective dose (at the end of the practice). In A practice duration of only three years will give an incomplete collective dose commitment of $A + B + C$, which is also the maximum future annual collective dose (after three years). The maximum per caput annual dose can be obtained by dividing $A + B + C$ by the size of the population involved.

In the assessment of the maximum future annual individual doses (and risks) from nuclear power production, it is relevant to compare the doses (and risks) and those to which the individual is exposed from other sources (for example, background natural radiation, other radiation sources etc.). It is also relevant to compare the detriment and benefit of the same practice or the detriments of alternative practices accruing the same benefits to society (for example, alternative sources for the production of 1 GW(e)y). In recommending a system of dose limitation, the ICRP adopted the following main features: (a) no practice shall be adopted unless its introduction produces a positive net benefit; (b) all exposures shall be kept as low as reasonably achievable, economic and social factors being taken into account; and (c) the dose equivalent to individuals shall not exceed certain recommended limits (for details of the system of dose limitation and recommendations, see ICRP 1977).

CHAPTER 7

The Nuclear Fuel Cycle

Nuclear Power Growth

NUCLEAR energy has been developed and used commercially for about two decades to meet a fraction of the world's electrical energy needs. The total installed nuclear generating capacity in the world in 1978 was 110.8 GW(e) from 227 power reactors operating in 20 countries (IAEA, 1979), constituting about 6% of the world electrical power capacity. In the past three years, several estimates of the nuclear power growth by the turn of the century have been made (Table 30). The estimates (countries with Centrally Planned Economies not included) demonstrate the wide range of possibilities which are wider after 1985.

Table 30. Comparison of Nuclear Power Growth Estimates (GW(e))

	1985	1990	1995	2000
OECD–NEA (1976)	479–530	—	—	2005–2480
Hanrahan et al. (1976)	358	620	—	1410
Krymm and Woite (1976)	350–400	—	—	1500–1800
Giraud (1976)	420	940	1500	2100
Messer (1977)	310–360	550–650	—	—
Williams (1977)	300–350	500–600	730–870	900–1100
Braatz and Dibbert (1977)	372–456	700–865	1110–1360	—
WAES (1977)	291–412	—	—	913–1772
Duret et al. (1978)	303	—	—	1543
OECD (1978)	278–368	504–700	750–1220	1000–1890

The values given by OECD (1978) represent two trends: the lowest figures are for "present trend" which takes cognizance of current patterns of energy utilization and supply, as well as present delays in the construction of new reactors, and generally assume a continuation of this trend; the higher values represent an "accelerated" trend which reflects the goals of ambitious nuclear power programmes, planned in response to the possible unavailability of conventional fuels. Recently, the IAEA (1979) estimated that the total nuclear power growth (for the whole world) will be in the range of 1100–1700 GW(e) by the year 2000 (see Table 31). For the purpose of this study, the lower estimates of IAEA which are similar to the "present trend" values of OECD given in Table 30 will be used (viz, 300 GW(e) for 1985, 475 GW(e) for 1990 and 1100 (GW(e) for the year 2000).

It should be noted that estimates of future nuclear energy production, extending over long periods of time, are unavoidably subject to large margins of uncertainty and can provide no more than very general indications, whose validity must be constantly subjected to critical revisions. The numerous energy forecasts (including those given above) rest on multiplicity of different

Table 31.   Estimates of World Installed Total Electric and
Nuclear Capacity (GW(e))

	1978	1980	1985	1990	2000
Total Electric	1900	2100	2700–3000	3300–3700	5500–6600
Total Nuclear	110	170	300–350	475–600	1100–1700
% Nuclear	5.8	8	11.7–12	14–16.2	20–26

After (IAEA, 1979).

assumptions and different aggregating procedures which make them roughly
indicative and subject to change. The main elements affecting these forecasts are
(a) world and regional scenarios of economic development; (b) correlation of
economic growth and energy consumption; (c) physical, economic,
environmental and political constraints applying to energy production and
consumption; (d) future prices of different energy sources; (e) future availability
of non-renewable resources and priorities for their use; (f) public reaction and
concern about nuclear power development etc.

TYPES OF NUCLEAR REACTORS

Nuclear reactors fall into two broad categories in terms of their basic
physical principles — "thermal" for those maintaining the chain reaction with
slow neutrons, and "fast" for those relying on fast neutrons. In the first category,
fission is induced by slow neutrons that have lost most of their original energy by
collision with nuclei of a "moderator". Such slow neutrons readily fission U–235,
whilst avoiding capture by the large excess of U–238 in the fuel. Under these
conditions a chain reaction can be induced in natural uranium with a heavy water
or graphite moderator. If ordinary water (light water) is used as moderator, as in
general practice, the fuel must be enriched in U–235 content from the 0.7% value
found in nature to about 2–3%. In "fast" reactors, the concentration of fissile
nuclei is high enough that the chain reaction can be sustained by the fast neutrons
emitted in fission and no moderator is needed. "Fast" reactors can operate on
about 20% enriched uranium but normally will use an equivalent amount of
Pu–239 produced initially in thermal neutron reactors, but ultimately bred within
the fast reactors themselves.

Gas-cooled Thermal Reactors include the Magnox Reactor, the Advanced
Gas-cooled Reactor (AGR) and the High Temperature Gas-cooled Reactor
(HTGR). In the first the fuel rods of natural uranium metal encased in "magnox",
a magnesium alloy, are inserted in a structure of graphite blocks which

constitutes the moderator. Pressurized carbon dioxide is the coolant. The AGR is a development of the Magnox reactor in which the fuel is slightly enriched (to about 2 per cent U–235) and is in the form of uranium dioxide (UO_2). The HTGR is a different type of reactor. The fuel is U–233 (dioxide or carbide) mixed with a thorium oxide or carbide fertile component, with graphite as moderator. The thorium captures neutrons to become U–233, which is fissile and makes an increasing contribution to the reactor's fuel supply. Helium is used as the coolant gas.

In *Water-cooled Reactors,* either light or heavy water is used as coolant. Most of the world's commercial power reactors use light water and are referred to as LWRs. The simplest of all types of reactors is the Boiling Water Reactor (BWR) in which water serves as coolant and moderator and directly provides the steam that drives the turbine. The fuel is slightly enriched (typically 2.4% U–235) uranium dioxide. In the Pressurized Water reactor (PWR) pressurized water serves as coolant and moderator. The hot water coolant is used to produce steam in a second circuit at lower pressure. The fuel is also slightly enriched uranium dioxide (about 3% U–235). The Pressurized Heavy Water Reactor (PHWR) uses pressurized heavy water as both moderator and coolant and produces steam in a second circuit in the same way as the PWR. The uranium dioxide fuel in the PHWR is, however, not enriched. The most common type of these reactors is the Canadian Deuterium Uranium Reactor (CANDU). The Steam-Generating Heavy Water Reactor (SGHWR) would use heavy water as moderator and light water as coolant. The latter boils, as in the BWR, to provide steam directly for the turbine. In this type of reactor, the fuel is uranium dioxide enriched to about 2% U–235.

Fast Reactors as built or proposed to date are also breeders. The Fast Breeder Reactor (FBR) uses a fuel consisting, typically of a mixture of about 20 per cent plutonium oxide with natural or depleted uranium oxide. In the leading version of the FBR, as proposed to date, neutrons released by the reactor fuel convert fertile material (for example, U–238, located mainly in a blanket around the reactor core) to fissile material (Pu–239), hence the name "breeder". The common type of the FBR operating as prototype is the Liquid Metal Fast Breeder Reactor (LMFBR) in which liquid sodium is used as the coolant.

The most common type of reactor, at least up to the year 2000, is the LWR (see Table 32).

The LWR Nuclear Fuel Cycle

Present day fuel for LWRs consist of uranium that has had its content of fissile U–235 enriched to 2.5–3%. Other fissile materials such as Pu–239 or U–233 may also be used. The energy produced in a LWR comes from the fission of U–235 by thermal neutrons. However, thermal neutrons also interact with U–238 to produce fissile plutonium which in turn contributes to the production of energy by the fission reaction. The average conversion ratio for a LWR is about

Table 32. Distribution of Reactor Types up to the Year 2000.

Year	Total Nuclear Power GW (e)	LWR %	AGR %	HWR %	GCR %	FBR %	HTGR %
1977	87	85.1	3.5	4.6	6.9	—	—
1980	146	86.3	4.1	4.8	4.1	0.7	—
1985	278	90.2	2.2	4.7	2.2	0.4	0.4
1990	504	91.2	1.2	5.6	1.1	0.4	0.4
1995	750	91.1	0.8	6.7	0.5	0.5	0.4
2000	1000	90.1	0.6	7.5	0.2	1.0	0.6

— After OECD/NEA (1977); data do not include USSR, Eastern Europe and China.
— LWR Light Water Reactor (the ratio of PWRs to BWRs is about 60:40).
  AGR Advanced Gas-Cooled Reactor.
  HWR Heavy Water Reactor.
  GCR Gas-Graphite Reactor.
  FBR Fast Breeder Reactor.
  HTGR High Temperature Gas-Cooled Reactor.

0.6, which means that 6 atoms of plutonium are produced in the fuel for every 10 atoms of U–235 that are fissioned. At the time the spent fuel is discharged, more than half the fissions occurring in the fuel are from self-produced plutonium rather than U–235.

Fuel is considered *spent* when it can no longer sustain the neutron chain reaction at economic power levels because of the depletion of fissile materials and the accumulation of neutron-absorbing by-products in the fuel and fuel hardware. At that point, however, the fuel still contains appreciable quantities of fissile isotopes, typically about 260 kg of fissile plutonium and 350 kg of unused U–235 in 44 tonnes of spent fuel, the approximate amount per GW(e)y. After the spent LWR fuel is removed from the core, several options are possible:

(a) the spent fuel rods can be disposed of without reprocessing. This is the *non-recycle option*, with two variations depending on whether the disposition is regarded as permanent (throwaway) or recoverable (stowaway);

(b) uranium can be recovered by reprocessing the spent fuel. Such uranium has a slightly higher U–235 content than natural uranium and can be further enriched for making new fuel. This option is called *Uranium-only* recycle. At this stage, there are two variations with respect to the plutonium contained in the spent fuel: permanent disposal (throwaway) or recoverable storage (stowaway).

(c) plutonium can be recovered from spent fuel and combined with uranium having a low U–235 content or with U–238 to form plutonium-uranium mixed-oxide or "mixed oxide" fuel. In this case the plutonium had been *"recycled"*.

These options of the nuclear fuel cycle of LWR are illustrated in Figs. 4, 5 and 6.

Fig. 4   LWR Fuel Cycle: No re-cycle (APS, 1978).

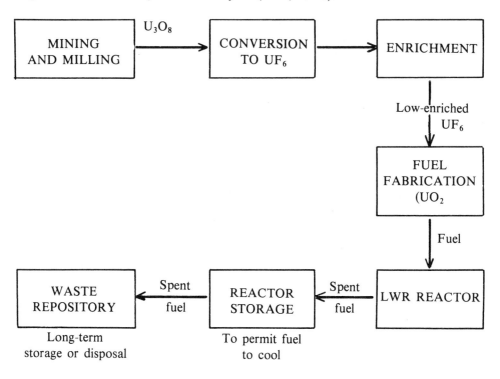

Fig. 5   LWR fuel cycle: Uranium recycle only (APS, 1978).

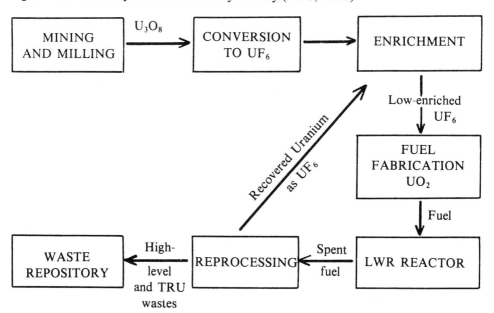

Fig. 6 LWR Fuel Cycle: Uranium and Plutonium Recycle (APS, 1978)

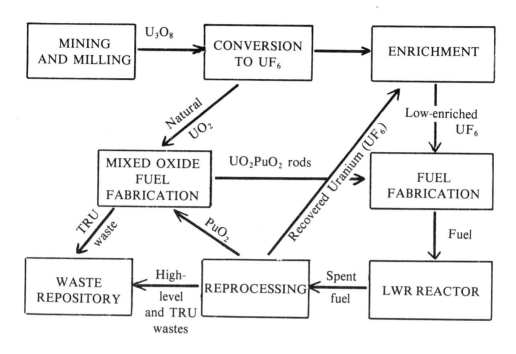

At present, the recycling of LWR fuel cycle has been indefinitely suspended in the U.S.A., and only a limited amount of recycling is underway elsewhere. Several studies have compared the different options (e.g. GESMO, 1976; APS, 1978) and it has been found that:*

(a) When both plutonium and uranium are recycled, enriched uranium supply requirements for the year 2000 are reduced by about 20% compared to the uranium-only-recycle option. U_3O_8 and UF_6 requirements are reduced by about 30% compared with the no-recycle option.

(b) The recycle of plutonium and uranium introduces a commercial trade in purified plutonium, with additional safeguard requirements.

(c) If all spent fuel is reprocessed and the plutonium recycled, the quantity of spent fuel plutonium discarded in various nuclear wastes is expected to be about 1–2% of what it would be without recycle. However, recycling the plutonium requires additional operations like reprocessing, mixed oxide fabrication, and management of various additional types of

* See also Chapter 13.

wastes. These will raise additional questions of economics, health and safety, and environmental matters, which differ from those encountered in the no-recycle option.

Other Fuel Cycles:

The substitution of thorium for U–238 in the conventional PWR and BWR technologies offers a real but marginal advantage for resource extension alone. Uranium-Thorium mixed oxide fuels have been fabricated and satisfactory performance has already been demonstrated in LWRs on an experimental basis. As mentioned earlier, the recycle option for LWRs leads to significant reduction in the 30 year reactor lifetime ore requirements (the net lifetime gain from U–Pu recycle versus no recycle is about 30%). The LWR–Th reactor offers a further saving in uranium ore of approximately 16%. From the use of thorium, benefits are possible with the redesign of the reactor core or of the moderator-coolant system to improve the conversion ratio.

The uranium ore requirements for CANDU reactors are less than for the light water fuel recycle. Substantial savings (about 50%) can be achieved in CANDU reactors with self-generated plutonium recycle. The fuelling of CANDU with U–235, thorium and recycled uranium leads to further savings in the consumption of uranium ore (39 to 45% less for such cycle than for uranium fuelling with self-generated plutonium recycle).

The High Temperature Gas Reactor (HTGR) is a helium cooled graphite moderated reactor using natural Th–232 as a fertile material and highly enriched uranium as fissile material. The reactor is fuelled with thorium, make-up U–235, and uranium (U–233, U–234, U–235, U–236) recovered from the discharged fuel and recycled. An alternative means of realizing the fuel value of plutonium recovered from discharged LWR fuel would be to recycle this plutonium in the HTGR in lieu of highly enriched U–235 make-up normally used. The uranium produced as a result of neutron absorption in thorium is recycled, along with the plutonium remaining in the discharged fuel.

The LMFBR programme is aimed towards the development of first generation breeders fueled with uranium and plutonium. In the Pu–U fast breeder the core consists of plutonium-uranium mixed oxide fuel surrounded by a blanket of depleted uranium which absorbs neutrons leaking from the reactor core to produce additional plutonium. Cooling is accomplished by means of liquid sodium in the case of the LMFBR or helium gas in proposed gas-cooled designs. The only ore requirements attributable to the breeder are those associated with the production of plutonium for start-up loadings of the first-generation breeders. This plutonium must be obtained from thermal reactors; these reactors will then require more ore because they are thereby deprived of the benefits of plutonium recycle. When operating without Pu recycle, a 1 GW(e)y of LWR operation produces about 260 kg of fissile plutonium. LWRs must operate for 30 GW(e)y without Pu recycle in order to produce about 7,500 kg of fissile Pu required to start-up a 1 GW(e) LMFBR (3000 kg Pu are required for the initial core and 4500 kg for replacement loadings before Pu in discharged fuel is recycled).

Since light water reactors will continue to be the main type of reactors in use at least until the year 2000, this study is mainly devoted to reviewing the environmental impacts — other than socio-economic — of the fuel cycle of LWRs. Concentration has been made on the *no-recycle* option which is the predominant one. However, some countries have re-processing facilities in operation; others are considering the re-processing option. Therefore, a separate chapter on this stage of the nuclear fuel cycle is also given. In this study, the environmental impacts of the different stages of the fuel cycle are classified into: (a) non-radiological; and (b) radiological. Whenever feasible, quantification of the environmental impacts is made on the basis of 1 GW(e)y.

The LWR fuel cycle requirements for the production of 1 GW(e)y are summarized in Table 33.

Table 33   LWR Fuel Cycle Requirements for 1 GW(e)y — no recycle*

Fuel cycle component	Unit	Requirement
Mining**	tonne U_3O_8	340
Milling	tonne U_3O_8	307
UF_6 Conversion	tonne U	266
Enrichment	tonne (SWU)***	132
UO_2 Fuel Fabrication	tonne U	43
Irradiated fuel storage	tonne/y (heavy metal)	41.2
Transportation	km	9.4×10^4
Waste management	ha	5

* After GESMO, Vol. 3, p. IV J(E)–17.
** The amount of uranium ore required varies according to ore grade. The average ore grade for GESMO scenario is about 0.1%; the ore mined is 340,000 tonne/GW(e)y.
*** A separative work unit (SWU) is a measure of the work expended to separate a quantity of uranium of a given assay into two components, one having a higher percentage of U–235 and one having a lower percentage.

CHAPTER 8

Mining and Milling of Uranium Ores

MINING OF URANIUM ORES

URANIUM is a fairly common constituent of the earth's crust, with an average abundance of 3.4 ppm (Fairbridge, 1972). Uranium is mined as an ore, usually containing greater than 0.1% U_3O_8 by weight. Uranium can also be recovered from "unconventional" sources; for example, as a by-product of phosphoric acid production, from phosphate rock, marine black shales, coals and lignites, special types of granites etc. Although such unconventional sources are very large in terms of a resource, their potential for contributing large increments of annual production is in most cases limited. The constraints vary from one case to another and include factors such as technological, economic as well as environmental.

The nuclear power industry's need for uranium in the short-term are considerable, and indicate the need for accelerated efforts to expand current resources. The requirement for uranium in 1977 was slightly less than 30,000 tonnes (U) and are expected to reach 178,000 tonnes (U) in the year 2000 (OECD, 1978). Table 34 gives the annual and cumulative requirements of natural uranium for the no-recycle scenario.

Table 34.   World Annual and Cumulative Requirements for Natural Uranium, in Tonnes (U)

	1985	1990	1995	2000
Annual	71,000	102,000	134,000	178,000
Cumulative	423,000	873,000	1,477,000	2,276,000

After OECD "present trend growth" (OECD, 1978)

These cumulative requirements are within known world uranium resources, presently estimated to total about 4 million tonnes in the cost category less than $ 130/kg U. However, considerable exploration effort will be required before such quantities of uranium can be classified as "Reserves", the resource category normally considered for the purpose of production. The cumulative uranium requirements by the year 2025 will reach 9 million tonnes U for LWRs without recycle (OECD, 1978) which greatly exceeds the presently estimated world resources of 4 million tonnes. At present, there is no concensus as to whether low-cost uranium to this amount actually physically exists in the uppermost part of the earth's crust from which it could be economically produced. In any case, accelerated efforts for the exploration and exploitation of new resources to meet the projected increasing demands for the nuclear industry seem inevitable.

Uranium ores are generally mined by underground or surface mining depending on the geological setting of the ore. Surface and underground mining accounted for nearly 60 and 40 per cent, respectively, of uranium ore production in the U.S.A. in 1975, but these proportions are expected to be reversed by the year 2000 (USERDA, 1976). In general 50 per cent of the uranium ore is mined by underground mining. Solution mining, in which the ore is leached in situ from the underground deposit and the product solution recovered by pumping to a treatment plant accounted for less than 2 per cent of uranium production in the U.S.A. in 1974 (USERDA, 1976).

The total uranium ore required by a given LWR over its operating life must include the ore to supply the start-up fuel inventory as well as the cumulative replacement loadings over the operating life. The amount of uranium ore required (0.1% U_3O_8) to produce 1 GW(e)y is about 340,000 tonnes.

ENVIRONMENTAL IMPACTS OF URANIUM MINING:

Non-Radiological Impacts

The mining of uranium ore, by either underground or open-pit techniques, results in significant disruption of natural conditions, the disturbance being greater in open-pit mining. Land use data indicate that about one-third of the total land involved is disturbed for the actual mining operation, while the remaining two-thirds remain undisturbed (USAEC, 1973). The land permanently committed to uranium ore mined amounts to 1.1 ha for the production of 1 GW(e)y.

A significant environmental impact associated with uranium mining (for example, in Northern Australia and Western U.S.A.) may result from dewatering of either underground or open pit mines (Rouse, 1978). This entails significant modification of the groundwater flow regime for those parts of the mine below the natural water table. Dewatering of underground operations is accomplished by a ring of dewatering wells, by use of sumps within the mine, or a combination of the two methods. In all cases, the result is a lowering of the water table and exposure of mineralized rocks to non-saturated ground-waters. This can result in increased oxidation and dissolution of toxic materials contained within the ore. In addition, experience has shown that significant amounts of toxic minerals are leached from ore particles suspended in the mine water as it flows from the face to the mine sump. In the case of open pit mines, precipitation falling on the exposed ore and spoil can leach toxic elements leading to increases in the amount of toxic pollutants in nearby ground waters.

Underground mining operations must maintain adequate ventilation to protect miners from health hazards associated with dust. Sparse data exist on the surface impact of vent shafts on nearby residences, but the siting and orientation of vent shafts will require increased consideration in the future as more underground mines are constructed and ventilation flows are discharged to the surface, often in proximity to residences. The primary chemical gaseous effluents from the mining of uranium derive from the burning of fossil fuels for the required power and the use of diesel oil for driving mining equipment.

Solid waste produced in surface mining operations consists mainly of the earth overburden above the ore-body and the barren rock in which the ore is dispersed. An average waste to ore ratio of about 30 has been reported for US mines (USAEC, 1973); ratios of about 4 and 14.5 have been estimated for two Australian uranium deposits (Ranger, 1974; Pan Continental, 1977). Initially, waste rock from excavation, if free from potential contaminants such as pyritic material, may be used for construction of earthworks, foundations and roads. The unused waste rock can be transported to a nearby dump.

Radiological Impacts

Exposure to radon daughters is considered the most important radiological occupational hazard in uranium mining. Following inhalation of radon and its daughter products from radon which has diffused into the mine air from the ore body, the alpha-emitting daughters irradiate the tissues and cells of the lung and respiratory tract increasing the incidence of cancer (Lundin *et al.*, 1969; Muller and Wheeler, 1976). Some smoking miners in the U.S.A. have experienced an incidence of lung cancer up to 10 times greater than for non-smoking miners. These observations were derived from past experience in mines not operating to current standards of radiological protection.

Historically, the radon daughters concentration in uranium mines has been expressed in terms of the working-level (WL) defined as any combination of radon daughters in one litre of air that will result in the ultimate emission of 1.3×10^5 MeV of alpha energy. Inhalation of air with a concentration of 1 WL for 170 working hours results in exposure of 1 *Working Level Month* (WLM) (UNSCEAR, 1977 p. 69). A lung tissue carcinogenic risk of $2.0 - 4.5 \times 10^{-4}$ per WLM/person (for 40 years) was regarded as probable by UNSCEAR (1977, p. 398). UNSCEAR (1977, p. 233) uses a miner productivity of 3 tonnes U_3O_8 per year and an annual exposure of $1.3 - 1.9$ WLM. If the annual mined U_3O_8 requirements for production of 1 GW(e)y is 340 tonnes, the total exposure per GW(e)y is about 147–215 WLM person. The estimate of incidence of fatal lung cancer is, therefore, 0.03 – 0.1 per GW(e)y.

Pochin (1976) assessed the risks of death from accidents and from ill health resulting from exposure to dust and silica in underground mines as 1.5×10^{-3} and 5×10^{-4} per person per year, respectively. These figures are based on past experience and it is to be expected that attention to industrial safety can reduce these risks. Taken together the figures are larger than the inferred radiation induced cancer mortality. The total mortality risk from all sources for underground miners is estimated to be 2–3 × 10^{-3} per person per year. Assuming that 113 miners are required to produce uranium ore for the production of 1 GW(e)y, the number of occupational deaths would be 0.23 to 0.34 per GW(e)y. The contributions from radiation exposure in these numbers are 0.03 and 0.1 deaths per GW(e)y, respectively.

In some uranium mines, exposure to gamma radiation can be significant. At Cluff Lake mine in Canada, it has been estimated that gamma radiation at the ore face can amount to about 5 mrem/hr. In such cases, shielding for workers may be necessary.

UNSCEAR (1977, p. 167) citing US data reported a radon release from mining and milling of 170 Ci Rn–222 per GW(e)y. The GESMO study (p. S–A–3 and p. IV JE–17) using more recent data, estimated a radon release during mining of about 5100 Ci per GW(e)y, calculated from measurements of radon concentrations in underground mines. This value corresponds to a release of about 2.4 Ci of Pb–210. Using these values, the collective dose commitment and inferred health effects to the public are given in Table 35.

Table 35. Collective Dose Commitment to Public and Inferred Cancer Mortality Resulting From Radon and Its Daughters Released During Mining of Uranium Ores (per GW(e)y)

Radon release during mining Ci/GW(e)y	Man-rem/GW(e)y					Cancer mortality
	Segmental Bronchioles	lung (whole)	Bone Marrow	Bone Lining Cells	Gonads whole body	
5100	2500	650	50	170	50	0.02

The number of early fatalities estimated from the UNSCEAR based figures is approximately one fifth of those based on the GESMO model. The Ford Foundation study (1977) has observed that the GESMO model appears to be conservative and, therefore, overestimates exposure.

After the mining operation has been completed, radon will continue to be released from an open-pit mine unless remedial action is taken (cover the pit and waste rock by soil or fill the pit with water etc.). Radon releases from an underground mine would be very much smaller since there is no longer forced ventilation. The annual release of radon from a worked out open-pit mine may be of the order of 100 Ci/GW(e)y (Wilde, 1978). Table 36 shows the incomplete collective dose commitment and inferred health effects to the public over a period of 100 years.

Table 36. Incomplete Collective Dose Commitment to Public and Inferred Cancer Mortality Resulting From Radon and Its Daughters Released After Completion of Mining of Uranium Ores (per GW(e)y)

Radon release after mining Ci/GW(e)y	Man-rem/GW(e)y					Cancer Mortality
	Segmental Bronchioles	Lung (whole)	Bone Marrow	Bone Lining Cells	Gonads whole body	
100	4900	1270	100	340	100	0.04

MILLING OF URANIUM ORES

In the milling operation the uranium ore is processed mechanically and chemically to extract the bulk of the uranium content and to produce an impure concentrate of uranium oxides, called the "yellowcake". The yellowcake contains between 60 and 90% U_3O_8. There are essentially two alternative methods for uranium milling, uranium is leached from the finely ground ore either by sulphuric acid or by sodium carbonate, the choice being determined mainly by the acidity or alkalinity of the host rock. The acid leach process is suitable for the majority of ores, while carbonate leaching is applicable to some deposits in Western Australia, Canada and the U.S.A.

Non-Radiological Impacts

Most of the land area (80%) required for milling is devoted to ponds for the permanent disposal of mill tailings. In effect, nearly the entire mass of ore processed by the mill ends up in the tailings ponds. Although the tailings pond area should be rehabilitated after the milling operation ceases, the land will most likely be removed from further unrestricted use. Table 37 gives the mass, area and volume of mill tailings generated in the production of 1 GW(e)y from LWR with no recycle.

By its very nature, milling is designed to change the minerological and chemical characteristics of mined ore and put the desired uranium product in a more soluble, concentrated form. Unfortunately, such milling operations also tend to increase the solubility of contaminants and toxic substances associated with the ore. The water pollution hazards associated with uranium milling operations depend on the milling process. For example, the alkali process tends to

Table 37. Mass, Area and Volume of Mill Tailings Generated in the Production of 1 GW(e)y from LWR with no recycle*

Ore mines (0.1% U_3O_8)	340,000 tonne
Mill tailing area	1.42 ha
Mill tailing mass	328,000 tonne
Mill tailing volume	328,000 m³

* After GESMO, Table IV–H–6.
Assuming 90% uranium recovery.
Average thickness of tailings about 12 m.
Density of tailings 1 g/cm³ (the density of compacted tailings is less than that of the original rock which may have been in the range of 2.0–3.0 g/cm³). A typical value for Australian acid-leach tailings is about 1.4 g/cm³. However, the value of 1 g/cm³ given by GESMO is used in this Table.

produce a set of conditions which are ideally suited to the dissolution of selenium, while the acid process tends to result in dissolution and mobilization of radium decay products, and various toxic heavy metals present in the ore. The dissolved contaminants are discharged with waste effluents to the tailings pond. Management of liquid wastes is very dependent on climate and ranges from treatment prior to discharge to almost total retention. Dissolved toxic substances may then have the potential for percolation into the ground water, or for direct seepage to near-by surface waters. About 4×10^5 tonne of liquid effluents are produced from a mill processing uranium ore for the production of 1 GW(e)y. Most of this water is dissipated from the tailings pond through evaporation.

The main air pollution impact from milling operations result from fossil fuel consumed locally in the operations and the subsequent generation of SO_x and NO_x. In addition, particulates are emitted by ore stockpiles and transfer stations. Generally, milling operations do not constitute a significant non-radiological air pollution source.

Radiological Impacts

The releases of radioactive material from uranium milling operations include airborne particulates and gases. Dusts containing uranium and uranium daughter products (thorium-230 and radium-226) are released from ore piles, the tailing retention system, and the ore crushing and grinding ventilation system. Natural uranium is released from the yellowcake drying and packaging operations as entrained solids (USEPA, 1973). Radon is released from the leach tank vents, ore piles, tailings retention system and the ore crushing and grinding ventilation system. The liquid effluents from a mill (acid-leach process) consists of waste solutions carrying radium-226, thorium-230, uranium and small concentrations of radon decay products. The isotopic composition and activity of mill tailings generated in the production of 1 GW(e)y of electricity from LWRs with no-recycle are given in Table 38. The estimated radioactive airborne effluents from milling operations are given in Table 39.

Table 38. Isotopic Composition and Activity of Mill Tailings*

Radionuclide	Activity (Ci/GW(e)y)	Radionuclide	Activity (Ci/GW(e)y)
U–238	9.2	U–235	0.4
Th–234	9.2	Th–231	0.4
Pa–234 m	9.2	Pa–231	4.3
U–234	9.2	Ac–227	4.3
Th–230	95.3	Th–227	4.3
Ra–226	95.3	Ra–223	4.3
Rn–222	95.3	Rn–219	4.3
Po–218	95.3	Po–215	4.3
Pb–214	95.3	Pb–211	4.3
Bi–214	95.3	Bi–211	4.3
Po–214	95.3	Po–211	4.3
Pb–210	95.3		
Bi–210	95.3		
Po–210	95.3	Total activity	1016

* For 0.1% U_3O_8 content in ore there will be 280 pCi of each daughter product activity per gm of ore assuming secular equilibrium (Swift *et al.*, 1974). For 340,000 tonnes of ore there will be a total of 95.3 Ci/GW(e)y activity for each daughter.

Table 39. Estimated Radioactive Airborne Effluents from Milling Operations

Radionuclide	Release from mill and active tailing area (Ci/y)*		Normalized release (Ci/GW(e)y)**	
	Low	High	Low	High
U–238	2.2×10^{-6}	9.1×10^{-2}	5.6×10^{-7}	2.3×10^{-2}
U–234	2.2×10^{-6}	9.1×10^{-2}	5.6×10^{-7}	2.3×10^{-2}
Th–234	1.4×10^{-7}	1.0×10^{-2}	3.6×10^{-8}	2.5×10^{-3}
Th–230	1.4×10^{-7}	3.0×10^{-2}	3.6×10^{-8}	7.7×10^{-3}
Ra–226	3.9×10^{-8}	3.2×10^{-2}	1.0×10^{-8}	8.2×10^{-3}
Pb–210	3.5×10^{-8}	3.0×10^{-2}	9.0×10^{-9}	7.7×10^{-3}
Bi–210	3.5×10^{-8}	3.0×10^{-2}	9.0×10^{-9}	7.7×10^{-3}
Po–210	3.5×10^{-8}	3.0×10^{-2}	9.0×10^{-9}	7.7×10^{-3}
Rn–222	97	5,800	25	1,487

* After ORNL/TM–4903(1975), p. 168–171. The high estimate indicates current practice, while the low estimate indicates advanced future practice.
** The output of the ORNL model mill is 1195 MT/y U_3O_8 using data in ORNL/TM–4903/1, p. 22 and assuming a recovery efficiency of 90%. Using a requirement of 307 tonne U_3O_8/GW(e)y, the ORNL model mill would supply about 3.9 GW(e)y.

The relevant radionuclides for occupational exposure in milling include the radon daughters and in addition uranium and Th–230. Although Stuart and Jackson (1975) have shown that in the lungs of animals Th–230 fractionates from the uranium and remains in the lung, limited attempts to measure *in vivo* in mill workers have failed to detect Th–230. The levels of radon are much lower in the mills than in the mines. Accordingly, the occupational collective dose from milling is insignificant compared to that from mining.

Under the assumption that adequate waste treatment prevents seepage of process liquid effluents into water sources, the local collective dose commitments from milling operations are expected to depend on the airborne releases, which contribute the external exposure from deposited material and to internal exposure via inhalation and ingestion pathways. The collective dose commitment from direct inhalation of particulates released from milling operations has been recalculated using the data given by UNSCEAR (UNSCEAR, 1977, p. 168–169) and the normalized releases given in Table 39; the results are given in Table 40.

Table 40. Estimated Collective Dose Commitments From Airborne Particulate Releases from Milling Operations*

Nuclide	Normalized activity released Ci/GW(e)y	Collective effective whole body dose commitment (man-rem/Ci)		Incomplete collective whole body dose commitment (man-rem/GW(e)y)
		Ingestion	Inhalation	
U–238	2.3×10^{-2}	6.2×10^{-1}	3.2	8.8×10^{-2}
Th–230	7.7×10^{-3}	4.6×10^{-1}	182.4	1.4
Ra–226	8.2×10^{-3}	9.4×10^{-1}	9.2	8.3×10^{-2}
Pb–210	7.7×10^{-3}	1.5	7.3	6.8×10^{-2}

* – Values for ingestion are incomplete collective dose commitments due to deposition on the ground incurred mainly through the ingestion pathway but also including a contribution from inhalation of resuspended material.
– Data were calculated from UNSCEAR (1977, p. 169). All doses (except external whole body in UNSCEAR were multiplied by a quality factor of 20 to convert from rad to rem, and individual doses have been combined into an "effective whole body dose" according to ICRP–26 weighting factors. The values obtained were normalized using the normalized activity released in Table 39.
– Using the total incomplete collective whole body dose commitment of 1.64; the calculated cancer mortality rate per GW(e)y is equal to 0.0002/GW(e)y.

According to UNSCEAR (1977), the radon release from a uranium mill and active tailings area is 170 Ci per GW(e)y. This value is much lower than the estimate in GESMO (930 Ci per GW(e)y) or that calculated in the present report (1,487 Ci per GW(e)y for present milling practices; Table 39). Using this value, the collective dose commitment and inferred health effects which might result from this release are given in Table 41.

Table 41. Collective Dose Commitment to Public and Inferred Cancer Mortality Resulting from Radon During Milling (per GW(e)y)

Radon Release during milling Ci/GW(e)y	man-rem/GW(e)y*					Inferred Cancer Mortality
	Segmental Bronchioles	Lung (whole)	Bone Marrow	Bone Lining Cells	Gonads whole body	
1,487	740	190	15	50	14	0.005

* includes effects from Pb-210.

Inactive Mill Tailings Piles:

Mill tailings piles present a potential for exposure to radiation by several pathways. The most important pathway is that of the radon–222. The radon–222 gas that is released into the spaces between the grains of tailings material diffuses toward the overlying tailings surface; some reaches the surface and some undergoes radioactive decay en route. The radon–222 that reaches the surface escapes into the air above, where it is mixed into the passing airstream by normal local air turbulences. The wind carries the radon–222 with it and continually decreases the concentration by further mixing and dilution; the concentration of Rn–222 decreases also by radioactive decay. Persons downwind of the tailings pile will be exposed to some concentration of radon–222 and its particulate radioactive decay products in the air they breathe. Some of the radioactive daughters of the radon–222 are retained in the tracheobronchial region of the lungs, irradiating the fluids and tissues, and thus increasing the risk of cancer formation there. Another pathway for exposure to radiation is that in which the wind lifts particles containing radionuclides from the surface of the tailings pile and carries them downwind, with simultaneous mixing, dilution, and deposition, until they reach persons inside or outside of buildings. Inhalation of the particles leads to exposure in several ways, but it is believed that the principal exposure is to the pulmonary region of the lungs in this case. The third principal pathway for exposure to radiation from the tailings piles is that in which the radionuclides in the pile emit gamma radiation which may penetrate the overlying material and air to interact with the body tissues of persons on or near the tailings piles.

The public health significance of the mill tailings has been recently estimated by USEPA (1976) on the basis of average conditions for 20 inactive mill sites in the U.S.A. Several important observations were made. First, both theoretical estimates and experimental evidence indicate that individuals in the general population may be receiving radiation doses to the lung as a result of the release of radon and its short-lived daughter products. Respirable airborne tailings particles bearing radionuclides provide another significant mode of potential radiation exposure at small downwind distances, as does the gamma radiation emitted from the tailings piles. Secondly, in order to accurately assess the potential public health significance of uranium tailings sites for purposes of determining remedial measures, a more comprehensive assessment of the problem using actual data applicable to existing sites rather than assumptions applicable to a model site is required.

The radiological assessment of the radon–222 emissions indicates that the estimated annual average dose from radon–222 decay products to individuals in dwellings in the vicinity of an average uranium tailings site (average uncovered dry tailings pile about 20 ha in area containing 500 pCi/g each of Th–230 and Ra–226) is approximately 8 rem to the lungs at about 50 m from the pile, 0.3 rem at 1 km and 0.1 rem at approximately 2.2 km. The radiological effects of wind blown particulates containing thorium–230 and radium–226 were found to be insignificant. For a typical uncovered tailings pile, the external gamma dose rate is estimated to be 10 rem/y at 0.9 m from the surface of the pile, decreasing to less than 1 mrem/y at 1.0 km.

It is estimated that the average individual exposed continuously over a lifetime to a dose equivalent of 8 rem/y to the bronchial epithelium region of the lung will incur about 70 per cent increase in the risk of bronchial cancer. Individuals exposed to an average dose equivalent of 0.3 rem/y and 0.1 rem/y would increase their risk of bronchial cancer by about 3 per cent and 1 per cent, respectively. The relative risk from particulate material blown from the pile is not as critical, because the radiation dose from inhaled radioactive particles is lower and, in addition, is delivered to the pulmonary region of the lung. Irradiation of the latter is believed to be less likely to produce a health effect compared to an identical dose equivalent delivered to the bronchial epithelium. The relative increase in risk caused by gamma radiation from the average tailings pile is also estimated to be less significant than the increase in risks caused by radon–222 from the pile.

Provided no remedial action is taken to reduce radon emanation, dry tailings piles would release about 100 Ci radon per year per GW(e)y. Incomplete, 100 year dose commitments, for this release are given in Table 42 together with the inferred cancer mortality.

It should be noted that the rate at which radon gas leaves the tailings pile decreases with increasing the earth cover depth over the pile and increasing moisture content of the cover (ORNL–TM–4903, 1975). Possible alternatives to reduce mill tailings emissions utilizing practical available technology have been summarized in ORNL–TM–4903 (1975) and in NUREG–0129 (1977). The utilization of a tailings pond with a compacted clay bottom liner as well as a 25–30 cm clay cover is considered environmentally-sound and a reliable

Table 42. Incomplete Collective Dose Commitment to Public and Inferred Cancer Mortality Resulting from Radon Releases After Milling (per GW(e)y)

Radon Release after milling (no radon abatement) Ci/GW(e)y	man/rem/GW(e)y*					Inferred Cancer mortality
	Segmental Bronchioles	Lung (whole)	Bone Marrow	Bone Lining Cells	Gonads whole body	
100	4980	1280	100	340	100	0.04

* Includes effects from Pb–210.

technique. A radon emission reduction of 98.6% (a 100% reduction connotes radon emanation identical to background) was envisioned using this method. Another option is the underground disposal of mill tailings in abandoned mines.

As with radon, the other tailings isotopes will be in equilibrium with the Th–230 parent, so that the activity in the pile will diminish by a factor of 2 over 80,000 years, the half-life of thorium. Continued physical stability and integrity of tailings piles is unlikely over these long time periods, and accordingly the emission rate to the atmosphere is not accurately predictable. The highest exposures to the general public from tailings have come from the inappropriate practice of using mill tailings as fill for foundation materials for buildings, including homes, schools, and stores. The most obvious example of this was found in Grand Junction, Colorado, where such practice was extensive (Seik, 1977). The resulting indoor radon concentrations ranged from background ($\sim$ 0.1 pCi/1) to 100's of pCi/1 (Seik, 1977) and an extensive remedial programme had to be undertaken. This seems clear evidence that the relative accessibility of tailings is the most hazardous aspect of past practice. Therefore, the subject of mill tailings management warrants special attention. Sound techniques for accomplishing the long-term stability and isolation of tailings have not yet been demonstrated.

CHAPTER 9

Uranium Hexafluoride Production

THE U_3O_8 concentrate extracted from the ore in the milling process must be purified and converted to the volatile compound uranium hexafluoride (UF_6), the feed stock for uranium enrichment plants. Two processes are used for purification and UF_6 production. The hydrofluor process consists of continuous successive reduction, hydrofluorination and fluorination of the ore concentrates followed by fractional distillation of the crude uranium hexafluoride to obtain a pure product. The second method employs a wet chemical solvent extraction step at the head end of the process to prepare a high purity uranium feed prior to the reduction, hydrofluorination and fluorination steps.

Non-Radiological Impacts

Of the land commitment to hexafluoride production (about 1.3 ha per GW(e)y; WASH–1248), approximately 10 per cent is disturbed for roads, fills and plant structures. About 1 per cent is committed for waste burial. The amount of water used with hexafluoride production is approximately 125,000 m^3 per GW(e)y, about 90% of which is returned to the source from which it came (the water is primarily used for process coolant; a fraction is used in the wet solvent extraction step as process water). The remaining water is discharged to the air through evaporation from the holding ponds.

The effluents from the wet and dry processes differ substantially. Most of the impurities entering with the yellowcake are rejected in the wet process raffinate solution from solvent extraction, whereas in the dry process, most of the yellowcake impurities are contained in solid wastes from the fluorination and distillation stages.

Effluents from a typical wet process consist of: neutralized aqueous raffinate from solvent extraction, caustic effluents and residual fumes from recovery of nitric acid, hydrogen fluoride and treatment of general off-gas streams, and a small quantity of solid calcium fluoride from the fluorination step (assuming fluorination in a fluidized bed). The raffinate stream amounts to about 5 m^3/tonne of uranium processed (WASH–1248). It may contain substantial dissolved solids, radium–226 and thorium–230 entering with the yellowcake feed, and about 0.2 per cent of the uranium processed. This stream is neutralized and impounded in a retention pond. Disposal of this effluent is a major problem associated with the wet process (USERDA, 1976; NUREG–75/007, 1975). Proposals in the U.S.A. call for evaporation of the supernatant water and burial of the sludge, or its transfer to a uranium mill retention system (NUREG–75–007, 1975). Some scrubber effluents are treated with lime to precipitate fluoride ion in settling ponds for packaging and burial as calcium fluoride.

Historically, analyses of airborne concentrations of fluoride as HF in air and concentrations in forage in the vicinity of a wet solvent extraction plant indicate fluoride levels below those expected to cause deleterious effects on human health or grazing animals (WASH–1248). In addition, analyses of water samples in the vicinity of the plant showed concentrations of fluoride and nitrate within the recommendations for drinking water sources.

Most of the chemical solid effluent from the dry process occurs as non-volatile ash containing iron, calcium, magnesium, copper and other fluorides. This residue can amount to about 0.1 tonne per tonne of UF$_6$ produced (WASH–1248; Sears *et al.*, 1977). After recovery of remnant uranium, the final residue is packaged and buried as low activity solid waste. Scrubber effluents arising from treatment of the hydrofluorination off-gas stream are treated similarly to those in the wet process.

Table 43 summarizes the non-radiological environmental considerations of

Table 43. Non-radiological Environmental Considerations of UF$_6$ Production (per 1 GW(e)y)*

I.	*Natural Resource Use*	
	Land (ha)	
	Temporarily committed	1.5
	Undisturbed	1.4
	Disturbed	0.12
	Committed for waste	0.01
	Water (1000 m^3)	
	Discharged to air	18.0
	Discharged to water bodies	132.0
	Total	150.0
	Fossil Fuel	
	Electricity Energy (1000 MW–h)	2.5
	Equivalent coal (1000 MT)	0.96
	Natural gas (million scf)	30.0
II.	*Effluents* (tonne)	
	Gases	
	SO$_x$	43.0
	NO$_x$	15.0
	Hydrocarbons	1.0
	CO	0.4
	Fluoride	0.16
	Liquids	
	Fluoride	26.0
	SO$_4$	6.7
	Nitrate	0.1
	Chloride	0.4
	Sodium	5.2
	Ammonia	2.4
	Iron	0.06
	Solids	60.0

* From WASH–1248, p. C–2, adjusted to 1 GW(e)y.

uranium hexafluoride production, assuming that one half of the output is by the hydrofluor process and the second half is by the wet solvent extraction process.

Radiological Impacts

Table 44 gives the radionuclides released from uranium hexa-fluoride production (airborne and water-borne).

Table 44. Radionuclides Released from UF_6 Production (per 1 GW(e)y)*

Radionuclide	Airborne emission Ci/GW(e)y	Water borne emission Ci/GW(e)y
U–238	7.7×10^{-4}	2.9×10^{-2}
U–234	7.7×10^{-4}	2.9×10^{-2}
U–235	3.3×10^{-5}	1.3×10^{-3}
Th–234	4.8×10^{-4}	3.0×10^{-2}
Th–230	2.1×10^{-6}	8.8×10^{-3}
Ra–226	2.2×10^{-6}	3.3×10^{-4}
Rn–222	7.9×10^{-6}	
Total	2.1×10^{-3}	9.9×10^{-2}

* From GESMO, Table IV F–9, p. IV F–37, based on NUREG–75/007, 1975 p. V–I and Allied Chemical Corporation, Docket No. 40–3392, 1975, p. 46, scaled to 1 GW(e)y.

Although UF_6 operations contribute to the radioactive content of the environment, the incremental exposure is insignificant; the occupational dose commitments and the collective dose commitments are given in Table 45.

Table 45. Occupational and Collective Dose Commitments from UF_6 Operations (per 1 GW(e)y)*

	Collective Dose Commitment to Workers man-rem/GW(e)y	Collective Dose Commitment to Public man-rem/GW(e)y
Total Body	0.98	9.5
G.I. Tract	0.6	1.2
Bone	11.4	22.6
Liver	1.2	0.2
Kidney	2.9	2.6
Thyroid	0.98	0.01
Lung	7.6	0.2
Skin	2.8	0.009
Inferred Cancer Mortality per 1 GW(e)y**	0.0004	0.001

* After GESMO, Tables IV–J(E)–9 to 16.
** Estimated from ICRP–26 risk coefficients.

CHAPTER 10

Uranium Enrichment

ISOTOPIC enrichment of uranium is necessary to provide fuel for LWRs and advanced gas cooled reactors. The concentration of U–235 in natural uranium is about 0.7% and the enriched uranium content for the current generation of LWRs is 2–4%. About 130,000 kg of separative work units (SWUs)* are required to prepare enough uranium for the production of 1 GW(e)y (LWR with no recycle). The enrichment facilities are large in size because a large number of separation stages are required to attain the necessary enrichment. In current gaseous diffusion plants, about 1,700 stages are needed to produce 4% enriched UF_6. The capacity of world uranium enrichment facilities was about 18 million SWU in 1975 and is estimated to have reached about 36 million SWU/y by the end of 1980 (ERDA–1543, 1976).

Large scale enrichment technologies are based on gaseous diffusion or centrifugation of uranium isotopes in the form of UF_6. Most of the current plants work with gaseous diffusion. In this process, gaseous UF_6 is compressed and passed over a porous membrane. Molecules of UF_6 containing the lighter isotope, U–235, diffuse through the membrane more rapidly than those with the heavier uranium–238 isotope; consequently, the UF_6 passing the membrane has a slightly greater proportion of molecules containing uranium–235. The degree of enrichment for one membrane is minute and many successive diffusion stages are necessary to change the proportion of U–235 from the naturally occurring level of 0.7% in the feed stream to the approximately 3% required in the product stream for LWR fuel with about 0.25% in a reject "tails" stream. Each stage requires recompression of the gaseous hexafluoride. This process requires large quantities of electrical energy. About 36.5 MW(e) of electricity would be needed by a gaseous diffusion plant to enrich the uranium for the generation of 1GW(e)y of electricity in a LWR (NUREG–0002, 1975). This electricity is generally provided by fossil-fuel operated plants, although at the Eurodif enrichment plant in France, it will be provided by a near-by nuclear power station.

In the gas centrifugation process molecules of UF_6 containing the heavier isotope U–238 migrate preferentially to the wall of a rapidly rotating cylinder. There is a consequent enrichment in the lighter uranium–235 isotope in the gas near the tube axis. The separation factor is greater than in the diffusion process, and the two streams removed from the tube axis and wall require only tens of stages arranged in a cascade to produce the required percentages of U–235 in the product and reject tails. High centrifugal stresses limit the size of the equipment, and many parallel cascades involving hundreds of thousands of centrifuges are required to achieve the separative capacity of a commercial enrichment plant. However, a gas centrifuge plant uses less than one tenth of the electrical energy required by a diffusion plant of similar capacity. Uranium enrichment can be also achieved with laser, which involves the selective dissociation of volatile uranium compounds or selective ionisation of uranium atoms to a degree controlled by

---

* A separative work unit (SWU) is a measure of the work expended to separate a quantity of uranium of a given assay into two components, one having a higher percentage of U–235 and one having a lower percentage. Separative work is generally expressed in kilogramme units.

differences in physical properties of uranium—235 and uranium—238 isotopes. This technology is, however, still at the laboratory stage.

Non-Radiological Impacts

Table 46 summarizes the resources required for gas diffusion and gas centrifugation enrichment plants.

Table 46. Summary of Resources Required for Gas Diffusion and Centrifugation Enrichment Plants*
(per 1 GW(e)y)

	Gas Diffusion	Gas Centrifugation
Natural Resource use		
Land (ha)		
Temporarily committed		
Disturbed	2.5	2.2
Undisturbed	2.2	2.2
Committed to landfill disposal		
of U–contaminated material	0.01	0.07
of uncontaminated material	0.01	0.01
Water (m³)		
Total required**	4.4×10^5	5.1×10^4
Discharge to air		
Cooling towers	3.4×10^5	1.5×10^4
Steam plant	1.5×10^3	1.5×10^3
Discharge to streams		
Plant operation	9.2×10^4	3.1×10^4
Sanitary waste	2.0×10^3	4.1×10^3
Energy requirements		
Coal (for heat and process steam), tonnes	792	792
Electric power, MW(e)	36.2	3.6
Gasoline and diesel fuel (litre)	10.0×10^3	10.0×10^3

* From ERDA–1543 (1976) p. 2.2–6 and 2.2–16. Normalized to 1 GW(e)y by dividing requirements for model enrichment plant by 66.3 (GESMO, p. IV J(E)–17).
** A water reserve capable of 38 m³/minute for 4 hours is also necessary for fire protection.

The primary source of environmental impact associated with the enrichment of uranium is the emission of particulates and oxides of nitrogen and sulphur from the generation of electrical energy through fossil fuel combustion. Table 47 gives the gaseous effluents from gaseous diffusion enrichment plant normalized to 1 GW(e)y.

Table 47. Gaseous Effluents from Gaseous Diffusion Enrichment Plant (per 1 GW(e)y)*

Effluent	Total process	Steam Plant**	Cooling Tower
HF (tonne/y)	3.6×10^{-2}		
NO_x (tonne/y)	1.7×10^{-2}	7.9	
SO_x (tonne/y)	6.9×10^{-5}	9.9	
Particulates (tonne/y)	6.0×10^{-3}	0.5	
CO (tonne/y)		0.16	
Hydrocarbons (tonne/y)		0.08	
Heat to stack (J)		5.6×10^{12}	
Drift (1/min.)			1.4

* From ERDA 1543, p.2.3–145 normalized to 1 GW(e)y by dividing releases from model enrichment plant by 66.3 (GESMO, P.IV J(E)–17).
** 792 tonne coal/y.

A number of chemical species are present in the liquid effluent stream from the gaseous diffusion enrichment plant. The concentration of chemicals undergoes considerable dilution before reaching the receiving water. With water treatment to reduce chromium concentrations, additional dilution within the receiving water body reduces all incremental concentrations resulting from the discharge to a small fraction of the recommended permissible water quality

Table 48. Liquid Effluents Discharged from Gaseous Diffusion Enrichment Plant (per 1 GW(e)y)*

Flow (1/day)	257,918
Main Constituents (g/day)	
Nitrate**	1,400
Aluminium	10
Fluorine	18
Uranium	0.13
Nitric Acid	1,300
Aluminium nitrate	72
Tributylphosphate	2.6
Phosphate	180
Chlorine	74
Chromium	12
Zinc	120

* From ERDA–1543, p. 2.3–146. Normalized to 1 GW(e)y by dividing releases from model enrichment plant by 66.3 (GESMO, P.IV J(E)–17).
** Total equivalent to nitric acid and aluminium nitrate.

standards. Table 48 gives the liquid effluents discharged from the gaseous diffusion enrichment plant primary holding pond.

Radiological Impacts

Table 49 gives the radionuclides discharged from gaseous diffusion enrichment plant.

Table 49. Radionuclides Discharged from Uranium Enrichment (per 1 GW(e)y)*

Radionuclide	Release Rate Through Liquid Effluent Ci/GW(e)y	Release Rate to Atmosphere Ci/GW(e)y
U–234	2×10^{-5}	7×10^{-4}
U–235	8×10^{-7}	2×10^{-5}
U–238	2×10^{-5}	9×10^{-5}

* After GESMO, p. IV F(A)–3.

The occupational dose commitments and the collective dose commitments to the public from the enrichment process are given in Table 50 together with the calculated inferred cancer mortality per GW(e)y.

Table 50. Collective Dose Commitments to Workers and the Public from Uranium Enrichment Process (per 1 GW(e)y)*

	Dose Commitment to workers (man-rem/GW(e)y	Collective Dose Commitment to public (man-rem/GW(e)y)
Total Body	0.7	0.02
G.I. Tract	0.4	1.4
Bone	5.8	0.2
Liver	0.7	0.004
Kidney	1.6	0.1
Thyroid	0.7	0.005
Lung	14.0	0.1
Skin	2.5	0.004
Inferred cancer Mortality (per GW(e)y**	5×10^{-4}	6×10^{-6}

* After GESMO, Table IV J(E)–9 to 16.
** Estimated from ICRP risk coefficients.

It should be noted that the occupational hazards are increased by the use of large quantities of electricity in gaseous diffusion enrichment. If two-thirds of the electricity is provided by a coal fired power station and if the occupational risk of producing this electricity is 0.5–5 deaths per GW(e)y (Comar and Sagan, 1976), it follows that electricity use in gaseous diffusion enrichment, results in another 0.01 to 0.1 deaths per GW(e)y. This risk is decreased in gas centrifugation plants which require about 1/10 of the electricity needed by diffusion plants.

CHAPTER 11

Fuel Fabrication

THE feed material for the fabrication of fuel for LWRs is UF_6 enriched to about 3 per cent in uranium–235. The UF_6 is converted to UO_2 and the UO_2 is formed into pellets and sintered to achieve the desired density. Finished pellets are loaded into Zircaloy or stainless steel tubes, fitted with end caps and welded to form a fuel pin. The completed fuel pins are assembled in fixed arrays to be handled as fuel elements. The currently dominant method of UF_6 to UO_2 conversion is a wet process involving the use of ammonium hydroxide to form as intermediate ammonium diuranate (ADU) compound prior to final conversion to UO_2.

Non-Radiological Impacts

Table 51 gives a summary of the non-radiological factors of fuel fabrication.

Table 51. Summary of Non-radiological Factors of Fuel Fabrication (per 1 GW(e)y)*

I.	*Natural Resources use*	
	Land (ha)	
	Temporarily committed	0.13
	Undisturbed	0.09
	Disturbed	0.02
	Commited for waste	0
	Water (m³)	
	Discharged to water	28,000
II.	*Energy Requirements*	
	Electrical Energy (MW-hr)	2.4×10^3
	Equivalent coal (tonne)	0.9×10^3
	Natural Gas (scf)	5.2×10^6
III.	*Effluents* (tonne)	
	*Gaseous***	
	SO_x	32
	NO_x	9
	Hydrocarbons	0.09
	CO	0.21
	F^-	0.006
	Liquids	
	NH_3	12.3
	NO_3	7.4
	Fluoride	5.7
	Solids	
	CaF_2	37

* After WASH–1248. Normalized to 1 GW(e)y.
** SO_x, NO_x, Hydrocarbons and CO are effluent gases from combustion of coal for power generation.

The land use for the fuel fabrication plant is essentially temporary; almost all of the land disturbed can be ultimately reclaimed for other uses. The water demand of about 28,000 m³ is mainly used to cool equipment and is returned to the biosphere after a holding period of days to weeks (WASH–1248).

The most significant effluents from the standpoint of potential environmental impact are chemical in nature. Nearly all of the airborne chemical effluents result from the combustion of fossil fuels to produce electricity to operate the fabrication plant. The only significant airborne chemical effluent from the process operations of the fabrication plant is fluorine (as hydrogen fluoride). The fluorine used for manufacture of UF_6 in the uranium conversion process becomes a waste product during the process of the production of UO_2 powder. The gaseous fluorine wastes generated are effectively removed from the air effluent streams by scrubbing and filtration systems.

The most significant chemical species in liquid effluents are nitrogen compounds that are generated from the use of ammonium hydroxide in the production of UO_2 powder and from the use of nitric acid in scrap recovery operations. Water from the air scrubber systems is combined with process liquid wastes and treated with lime to form a calcium fluoride precipitate which is filtered and stored on site. The small percentage of fluorine which is not removed by the lime treatment is released from the liquid waste holding ponds at a concentration of about 200 mg/litre, about two orders of magnitude above the concentration allowed by drinking water standards (WASH–1248).

Radiological Impacts

The radionuclides released from a fuel fabrication plant are given in Table 52.

Table 52.   Radionuclides   Released   from   Fuel   Fabrication   Plant
(per 1 GW(e)y)*

Radionuclide	Gaseous Effluents	Liquid Effluents
Th–234	3×10^{-5}	7×10^{-3}
U–234	2×10^{-4}	5×10^{-2}
U–235	6×10^{-6}	10^{-3}
U–238	3×10^{-5}	7×10^{-3}

* After GESMO, p. IV F (A)–3.

The occupational dose commitments and collective dose commitments to the public are given in Table 53 together with the inferred cancer mortality per GW(e)y.

Table 53.   Collective Dose Commitments to Workers and the Public from Fuel Fabrication (per 1 GW(e)y)*

	Dose Commitments to workers (man-rem/GW(e)y	Collective dose Commitment to public (man-rem/GW(e)y
Total body	11	0.6
G.I. Tract	12	0.6
Bone	15	9
Liver	11	0.001
Kidney	12	1.5
Thyroid	11	0.001
Lung	450	0.03
Skin	11	0.001
Inferred cancer mortality	0.01	2×10^{-4}

* After GESMO, Tables IV J(E)–9 to 16.

Such doses are very low and it is unlikely that the fuel fabrication activity would increase levels of exposure; the population doses are, therefore, insignificant (see also UNSCEAR, 1977).

CHAPTER 12

Nuclear Power Plant Siting, Design and Operation

LIKE any other power station, the interaction between nuclear power plants and their surroundings has recently received considerable attention. The impact that a power station may have on the environment depends to a large extent on its location with respect to the load centre, populated areas, open water, agricultural land, etc.

For nuclear power plants, siting considerations are stringent because of the release of radioactive substances into the environment during normal operation, potential release in an accident situation, and the post-operational phase when decommissioning may be undertaken. The selection process must satisfy the technical, socio-economic and environmental safety requirements which are the most important of the various factors considered for site studies. The methodology for site selection involves an iterative approach to test various alternatives as to location, type of power plant, cooling system, land requirements and exposure of the public to radiation from normal operation and potential exposure in the case of accidents, or radioactive effluents arising therefrom. After the preliminary screening, there may remain only a few alternative sites which must then be evaluated using weighting factors (Site Selection Criteria – see Fischer and Ahmed, 1974). The "best available" site will rarely be ideal in all aspects. In most cases the result of balancing the benefits against the cost and detriments that the society will be prepared to accept, will determine the suitable site. A reactor site should, therefore, be seismically acceptable, have an adequate supply of cooling water available and be easily accessible.

NON–RADIOLOGICAL IMPACTS

Land Requirements:

For nuclear power plants, land requirements may vary from plant to plant depending on the situation of the plant with regard to open water, population areas, the type and size of reactor, and meteorological conditions. Typical values for site requirements have been 120-240 ha for 2000 MW plants without cooling ponds comprising two units each; small plants do not necessarily require proportionally less land. The average requirement for nuclear power plants in the U.S.A. has recently been estimated at 41 ha per 1000 MW(e) (NUREG, –0002, 1976). Additional land is needed for the switchyard and transmission lines.

Not all land in an exclusion area is necessarily unproductive. Access limitations at some plants have turned land into wildlife refuges, in other cases limited farming and cattle grazing have been possible, though adequate radiation monitoring was needed. Such farming uses are likely to expand as the number of plants increase and land values go up. Similarly, at shore locations the availability of heated water will enable use of some of the sequestered shore for fish farming.

A long-range land-use problem may arise from the need to control population patterns near large nuclear power plants. Environmental impact assessments attempt to allow for long-term population trends; however, nuclear site selection usually assumes that low population density areas near a plant will retain their current character even though the availability of large amounts of power and the availability of waste heat may attract additional industries and their employees to the neighbourhood.

Thermal Discharges:

The heat produced by nuclear fission in the reactor core is extracted through suitable coolants and then runs a turbine to operate the generator to produce electricity. At the exhaust of the turbine the steam is condensed to water to maximize the energy conversion and then is returned to the boiler or reactor to repeat the cycle. A large amount of heat is rejected in the condensing process, and the rejected heat is substantially greater than the heat equivalent of the electric energy generated. The thermal efficiency of light-water reactors at present is approximately 33%* which means that almost two-thirds of the heat energy generated in a reactor core has to be rejected into the environment, in the vicinity of the power station.

The bulk of the waste heat is transferred from the steam to the cooling water in the condensers. Water is commonly used as the absorbent because of its general abundance, low cost, high specific heat, and ability to dissipate heat in the evaporation process. A LWR nuclear power plant of 1,000 MW(e) with a thermal efficiency of 33%, discharges in the condenser about 2,000 MW(t). If the cooling water flow rate is 50 m^3/s, the water temperature at the condenser outlet will have an increase of about 10°C. The cooling water is extracted from some suitable source (river or lake etc.), passed through the condenser where its temperature is increased by about 10°C, and returned to the source body of water. Eventually the warmed sink gives up this extra heat to the atmosphere. Such a system, which is referred to as "once-through" cooling, may cause unacceptable environmental changes (for the environmental aspects of thermal discharges see Chapter 5).

Transmission Lines:

The environmental impacts of high voltage transmission lines have been considered in Chapter 5 from six aspects: aesthetic considerations, land requirements, communications hazards, ozone and habitat effect.

---

*The thermal efficiency of heavy-water reactors is about 30%, of Magnox reactors about 33%, and of AGRs, HTGCRs and LMFBRs about 40%, the latter is equal to that of a modern fossil-fueled power plant.

Aesthetic and Recreational Factors:

Power plants vary widely in the visual impression they make on the viewer. Depending on one's point of view their appearance may be judged as blending harmoniously with the surroundings or as an insult to attractive countryside (see Chapter 5).

Radiological Impacts

During the operation of nuclear power reactors, radionuclides are formed by fission of the nuclear fuel and by neutron activation of structural materials, corrosion products, and impurities in reactor coolant water. Most of the fission products remain in the fuel elements, but a fraction can escape through the fuel cladding fissures into the coolant. Most of the radioactive-isotopes formed either by fission or activation process and released into the coolant or moderator are removed by gaseous and liquid water processing systems. Part of the radioactive material, nonetheless, may eventually be released into the environment.

From the enormous variety of fission and activation products generated by reactor operations, some are particularly relevant, either because of the ease with which they appear in effluent streams or because of their radiological significance. In this study, emphasis will be given to the environmental impacts of fission noble gases, particularly krypton–85, some activation gases, iodines, tritium, carbon–14, radionuclides in particulate forms released into the atmosphere and water bodies. Special consideration will be given to tritium and krypton–85 as both radionuclides are long-lived and are distributed worldwide. Krypton–85, particularly, deserves special attention because of the inherent difficulty of its control and its essentially non-reactive and mobile nature in the environment (USEPA, 1977). Because of its long-half-life, the radiation exposure commitments resulting from the environmental build-up of Carbon–14 are considerably larger than those from noble gases and tritium and, therefore, proper consideration of this nuclide is also necessary. Several radioiodines, particularly I–131, are radiologically significant in the local environs. Iodine–129 is also relevant as far as the global effects are concerned.

AIRBORNE EFFLUENTS

Fission Noble Gases:

Noble gases (krypton and xenon) may escape from fuel-cladding fissures and are difficult to retain. Noble gases escape out of the reactor primary coolant boundary either by leakage from both coolant and/or moderator system; from effluents of the side stream of the coolant and/or moderator which is continually diverted into a sub-system for control and purification; from the main condenser exhaust; from secondary coolant blow-down; or by continuous removal of non-condensable gases, in the steam flow of BWRs via the main condenser air-ejector system and eventually other leakages through the turbine gland-seal system, the condenser mechanical vacuum pump etc. (NUREG–0017, 1976).

A minor fraction of noble gases escaping the primary coolant boundary is directly released into the environment; most of this fraction passes through the reactor stack although an insignificant amount may reach the environment via building leakages. The major amount of radioactivity is, however, treated by a gaseous radioactive waste treatment system. Cryogenic systems, charcoal-adsorption systems and chemical methods can be used for separation and containment of noble gases. While charcoal beds provide a wide range of delay times to permit decay of the short-lived rare gases, cryogenic systems and absorption in liquid fluorocarbons minimize the volume of long-lived separated rare gases for long term storage. The fraction of noble gases which do not decay during the delay time imposed by the treatment system are released into the environment through the reactor stack.

Table 54 gives the noble gases discharged in airborne effluents from LWRs. The composition of the release from PWRs is primarily Xe–133 (half-life time 5.3 d). In BWRs, the important components released are Xe– 135, Xe– 138, Kr– 88, Kr– 87 and Kr– 85m.

Table 54. Noble Gases Discharged in Airborne Effluents from LWRs (after UNSCEAR, 1977, p. 176–177) Normalized to 1 GW(e)y

	Half-Life	PWR $Ci/GW(e)y$	BWR $Ci/GW(e)y$
Xe–133	5.27 d	1.5×10^4	2×10^5
Xe–135	9.2 hr	540	3.6×10^5
Xe–138	14.2 min	190	1.9×10^5
Kr–85	10.57 y	170	3×10^4
Kr–88	2.79 h	150	2×10^5
Xe–135 m	15.7 min	130	6×10^4
Kr–85 m	4.4 h	100	8×10^4
Xe–133 m	2.26 d	90	8×10^3
Kr–87	76 min	70	1.7×10^5
Kr–83 m	1.86 h	—	1×10^5
Xe–131 m	11.96 d	160	—
Ar–41	1.86 h	50	—
Xe–137	3.82 min	—	1×10^4
Others		—	4×10^3

Tritium:

Ternary fission in reactor fuels and neutron reaction with various isotopes dissolved in the coolant or moderator and with the coolant or moderator itself, gives rise to tritium. Most of the tritium activity released into the environment is converted to tritiated water (HTO) and participates in the normal water cycle. Most of the current tritium world inventory will be in the oceans as a result of transport by rain and by direct exchange of water vapour between air and sea water; once in the ocean, tritium will be dispersed through mixing processes. The average normalized release of tritium in airborne effluents is about 200 Ci/GW(e)y and 50 Ci/GW(e)y for PWRs and BWRs respectively (UNSCEAR, 1977, p. 179).

Carbon–14:

Carbon–14 is produced in LWRs by (n,α) reactions with oxygen–17 present in the oxide fuel and in the moderator, by (n, p) reactions with nitrogen–14 present in the fuel as impurities, and by ternary fission. Ternary fission production per unit electrical energy generated is practically independent of reactor design, while the normalized production of C–14 by the other reactions depends on the enrichment of the fuel, the relative masses of the fuel and moderator, the concentration of nitrogen impurities in the fuel and the temperature of the fuel and moderator. The average production of C–14 in LWRs has been estimated to be about 20 Ci per GW(e)y. The amount which can be assumed to be released at the reactor is about 30% of the total production, i.e. about 6 Ci per GW(e)y, (UNSCEAR, 1977, p. 181). However, higher releases have been reported: 13 Ci/GW(e)y for the Yankee Rowe PWR in U.S.A. and up to 200 and 460 Ci/GW(e)y for PWR and BWR for the USSR (UNSCEAR, 1977, p 181).

Iodine:

A number of iodine isotopes are produced in reactors by fission and by decay of other fission products. Iodine–131, with a half-life of 8 d, is the main concern from an environmental point of view. I–129 (half-life: 1.6×10^7y) has not been identified in power reactor environs, and its significance, in terms of possible release activities, is much less than that of the other iodine isotopes. The amount of I–131 released in airborne effluents has been estimated to be about 2–5 Ci/GW(e)y from BWRs and $5–50 \times 10^2$ Ci/GW(e)y from PWRs (UNSCEAR, 1977, p. 184).

Particulates:

Most of the fission and activation products can form aerosols, which may be released with the airborne effluents. Particulates may arise either directly or as decay products of fission noble gases. Aerosols originate particularly from primary coolant leaks, but they can also be generated by working on contaminated components or cleaning contaminated surfaces. High efficiency particulate air filters (HEPA) retain 99.99 per cent of the particles and, hence, particulate activities are very low. The average total activity of particulates has been estimated to be 500 m Ci/GW(e)y from PWRs and 1,100 m Ci/GW(e)y from BWRs. Over 80% of the average PWR released activity is Rb–88 Other small contributions are Cs–134, Cs–137, Co–58, Co–60, and Mn–54. For the BWRs, Ba–140, La–140 and Sr–89 each account for about 30 per cent of the total activity, with I–131, Co–58, Co–60, Cs–134 and Cs–137 each contributing 10 per cent or less.

Liquid Effluents:

Radioactive liquids containing several radionuclides may be released from reactor operations. Such liquid effluents are produced not only by leakages from the reactor boundary but also due to various discharges, e.g. from laboratories, laundry, showers etc. In some cases, a substantial amount of activity may arise from spent fuel storage pools. The liquid effluents are normally treated by a liquid radioactive waste treatment system; residual radioactivity arising after treatment is ultimately discharged mixed with the cooling water stream, into the aquatic environment.

Table 55. Radionuclides other than Tritium Discharged in Liquid Effluents (after UNSCEAR, 1977, pp. 189–190; normalized to 1 GW(e)y)

	Half-life	PWR Ci/GW(e)y	BWR Ci/GW(e)y
I–131	8 d	2.3	3.6
Cs–137	30.3 y	1.7	25.0
I–133	20.5 h	1.3	0.15
Cs–134	2.1 y	1.0	17.3
Co–60	5.2 y	0.42	6.1
Cr–51	27.7 d	0.18	0.58
Mn–54	313 d	0.12	1.9
Ce–144	290 d	0.08	0.34
Ru–103	41 d	0.02	
Ag–110 m	270 d	0.02	0.006
Sr–89	51 d	0.01	0.80
I–132	2.4 h	0.01	
Fe–59	45 d	0.01	0.046
Nb–95	35 d	0.006	
Zr–95	63 d	0.004	0.012
Mo–99	68 h	0.003	0.16
Sr–90	28 y	0.002	0.098
Zn–65	244 d		0.23
Ce–141	32.8 d		0.009

Tritium in liquid effluents has been estimated to be about 2,000 Ci/GW(e)y for PWRs and about 100 Ci/GW(e)y for BWRs (UNSCEAR, 1977, p. 179). Besides tritium, a number of other radionuclides are encountered in liquid effluents. These are summarized in Table 55. The average normalized releases excluding tritium are approximately 8 Ci/GW(e)y for PWRs and 60 Ci/GW(e)y for BWRs. The radionuclides Cs–137 and Cs–134 account for about 30% of the total activity discharged by PWRs and for about 70% of that from BWRs. I–131 and I–133 make up 10–40% and Co–58 and Co–60 about 15% of the activity in liquid effluents from LWRs.

Solid Wastes:

Solid radioactive wastes may be produced in significant quantities by reactor operation; the nature and level of radioactivity in such wastes depend on the sources of production. In the case of no-recycle option the spent fuel elements which contain large quantities of fission products, together with transuranium elements and unspent uranium, will be categorized as solid wastes. Filters and resin columns used to remove radioactivity from liquid and gaseous effluent streams will form solid wastes with rather high concentration of radionuclides. Activated and contaminated reactor parts which are removed during repair and maintenance operations fall into the category of high level solid radioactive wastes. A less significant source of solid radioactive waste is the contaminated material such as gloves, laboratory ware etc., which are used in operations involving radioactive substances. The yearly cumulative amount of solid radioactive wastes generated by reactor operations excluding spent fuel elements has been estimated to be about 2,800 m^3 containing about 100 Ci of activity (Kibbey and Godbee, 1975). The environmental impacts of such solid radioactive wastes, is not considered to be significant in the short term.

Occupational Exposure at Nuclear Power Plants:

Occupational radiation exposure at LWRs is due primarily to external exposure to gamma radiations from fission and activation products. Less information is available than desirable on the breakdown of the exposure by type of operation, but much of it is associated with refueling, maintenance, inspection, and waste treatment operations. The size of plants on-line at present is increasing, and most new plants are in the 1,100 to 1,400 MW(e) range. There is insufficient operating experience with these larger plants as yet to accurately predict the collective doses incurred with operating them during their total design lives. The best estimate of expected performance may be taken from the record of current operational experience. The exposures for 1973 to 1975 from the collective U.S. licensed plants are presented in Table 56. At this time there seems to be no clear trend of increase or decrease in exposure with size or age of the units. During 1973–1975 a total occupational dose of about 400 to 500 man-rem resulted from operating a LWR for a year. Assuming the ratio of generated power to rated capacity was 0.7 for each year, the estimated average occupational exposures would be 1,200 man-rem/GW(e)y for the period 1973–1975.

Table 56. Estimated Collective Dose to LWR Operating and Maintenance Personnel
(NUREG 75/032, 75/108; Brooke, 1976)

Year	No. BWRs	No. PWRs	No. LWRs	Avg. Rated Capacity MW(e)	Total Annual Electric Energy Produced* MW(e)	Annual Average Collective Dose man-rem/ reactor	Average man-rem/ GW(e)y
1973	14	12	26	546	9932	542	1400
1974	14	18	32	581	13014	448	1100
1975			44	640	19712	482	1100

* Estimated using an average capacity factor of 0.7.

Exposures to individuals at reactor plants are generally kept below 5 rem/y (whole body dose); exposures exceeding that level are rare. Since the usual practice in quoting average dose is to include many workers receiving much less than 5 rem/y, an estimate of the average occupational dose is not significant without information on the dose frequency distribution in the population of workers.

Population Exposure:

Radionuclides released in airborne or liquid effluents during reactor operation undergo a series of complex physical, chemical and biological processes before reaching man. Such processes depend on the location of the reactor, meteorological conditions, and the different exposure pathways.

The noble gases released from reactors usually make the largest contribution to the dose to the local population. Using the average release composition based on operating experience in a dispersion model incorporating average weather conditions and assuming discharge from a 30 m effective stack height (Bryant and Jones, 1973), it is possible to estimate the dose per unit release as a function of distance from the stack. With the further assumption of a uniform population density of 100/km^2 in the region from 1 to 100 km and 25/km^2 at greater distances, the corresponding collective doses can also be estimated. For the average BWR total normalized release of 1100 x 10^3 Ci/GW(e)y, the local collective dose would be 500 man-rem per GW(e)y (to 100 km), and the additional regional contribution would be 50 man-rem per GW(e)y (100–1500 km), including the contribution from Xe–133 beyond 1000 km (UNSCEAR, 1977, p. 191). For PWRs, the total normalized release of noble gases is about 15 x 10^3 Ci/GW(e)y. The local collective dose would be 1.4 man-rem per GW(e)y to 100 km, with an additional regional contribution of 0.11 man-rem per GW(e)y (100–1500 km), including the extrapolated contribution from Xe–133 beyond 1000 km.

Tritium present in airborne effluents from reactors can contribute to the local collective dose primarily through the inhalation pathway and secondarily through the ingestion pathway. The local collective dose contribution after deposition through the ingestion pathway may be quite variable from site to site owing to differences in local hydrology and water usage. The intake of airborne tritium occurs by inhalation and by passage through skin, in nearly equal proportions. The behaviour of tritium in the environment and its effects on biota are far from being well understood. Extended studies are needed to understand the incorporation of tritium in biological compounds and to investigate the long-term components in tritium elimination by man and other mammals (Farges and Jacobs, 1979). The estimated local collective doses (1–100 km) for tritium are 0.04 man-rem per GW(e)y for PWRs and 0.01 man-rem per GW(e)y for BWRs (UNSCEAR, 1977, p. 194). Tritium in liquid effluent from reactors can contribute most directly to the local collective dose if the discharge is to a river that provides a drinking water supply. The local collective dose has been estimated to be 7 man-rem per GW(e)y for PWRs and 0.7 man-rem per GW(e)y for BWRs.

Carbon-14 makes a very small contribution to the local collective dose. It is more significant as a component of the carbon cycle contributing to the global collective dose. For a normalized C–14 release from a light-water reactor of 6 Ci per GW(e)y, the local collective dose commitments are estimated to be 0.7, 0.6, 2.2 and 2.4 man-rem per GW(e)y for the lungs, gonads, bone lining cells and bone marrow, respectively (UNSCEAR, 1977, p. 194). Regional contributions to the collective dose are very small compared to the global collective dose commitment from C–14.

Assessment of local dose from I–131 released from reactors is complicated by the various forms that iodine may take: particulate, elemental, organic or as hypoiodous acid. Elemental iodine readily deposits on forage and enters the cow-milk-man pathway. Organic iodine, however, is retained much less efficiently by vegetation, and its deposition velocity is 200–1000 times smaller than that of the elemental form. Particulate-associated iodine will be deposited at rates intermediate between those for the elemental and organic forms. The behaviour of hypoiodous acid is uncertain; it may simply decompose to the elemental and organic forms. Physico-chemical transformations occurring during atmospheric transport may also affect the amount of iodine in its various organic forms, since these are not stable in sunlight. The elemental form would be expected to become rapidly associated with airborne aerosols, so that deposition at distances beyond the immediate vicinity of the release would be largely governed by the particulate behaviour. This is also shown by fallout measurements. On the basis of average operating experience, the normalized release of I–131 for BWRs is 5 Ci per GW(e)y (UNSCEAR, 1977, p. 196). The thyroid collective dose commitment is, therefore, estimated to be 110 man-rem per GW(e)y. The normalized collective dose commitments for PWRs are smaller by two orders of magnitude.

The activities of particulates released in airborne effluents may be quite variable but are relatively low. From releases based on average operating experience, estimates may be made of the local collective dose commitments from immersion and from inhalation intake while the radioactive material remains airborne and from the ingestion pathway and the exposure to contaminated ground following deposition. The estimated collective dose commitments are somewhat higher for BWRs than PWRs, ranging from 0.03 to 0.2 man-rem per GW(e)y to the lung, due primarily to Co–60, and 3 to 10 man-rem per GW(e)y to total body, due to Co–60 from external exposure and to Cs–134 and Cs–137 from external exposure and through the ingestion pathway. The dose commitments to bone marrow and bone lining cells are due almost entirely to the small amount of Sr–90 released (UNSCEAR, 1977, p. 196).

Radionuclides discharged in liquid effluents may result in doses to man through the pathways of drinking water and fish consumption from releases to fresh water, and of ocean fish and shellfish consumption from releases to salt water. A portion of the population may also be exposed on shorelines to external irradiation from sediments. In areas where irrigation of crops is practised, especially when sprinkling methods are utilized, a significant contribution to the collective dose could result. This pathway should then be considered. Ingestion of water fowl and immersion exposure from swimming are other pathways, but these have generally been found to be insignificant contributors to the collective dose. The estimated collective doses of radionuclides other than tritium in liquid effluents are given in Table 57. The collective doses to the total body include

Table 57. Collective Doses to Particular Organs from Radionuclides other than Tritium in Liquid Effluents from Reactors (man-rem GW(e)y)* (After UNSCEAR, 1977, p. 200)

Reactor	Total body	Lower large Intestine	Bone	Thyroid
		Release to salt water		
PWR	$2 \quad 10^{-2}$	$7 \quad 10^{-2}$	$3 \quad 10^{-6}$	$1 \quad 10^{-2}$
BWR	$2 \quad 10^{-1}$	1	$2 \quad 10^{-4}$	$3 \quad 10^{-2}$
		Release to fresh water		
PWR	6	20	$6 \quad 10^{-2}$	2
BWR	80	3	3	4

*RBE $= 1$ for conversion to rem.

contributions from ingestion of Cs–137, Cs–134 and Zn–65, and from external exposure of Cs–137, Cs–134, Co–60, Mn–54, Ce–144, Cr–51, Sb–125, and Ru–106 in sediments.

REACTOR ACCIDENTS

In reactor accidents, the inventory of fission products and actinide elements in the reactor core would be the origin of the hazards and cause of environmental damage, although the fractions of these products released in such accidents would depend, of course, on the type of reactor, associated engineered safeguards, the behaviour of the containment systems, and the nature of the particular accident. Depending upon the reactor, the particular isotope under consideration, and accident conditions, the fractions of various isotopes released can vary from a value of no environmental significance to nearly 100%, the latter applying to the worst combination of circumstances associated with a low probability of occurrence. Examples of this variability of fraction release and probability are contained in WASH–1400, 1975.

After an accidental release of radioactive material to the atmosphere, there is the possibility of a radiological hazard to persons downwind of the source. Some of this hazard originates directly from the passing cloud of airborne radioactive material. External gamma radiation emitted by the cloud would deliver a wholebody dose, and external beta radiation would deliver a less serious dose to the skin. Those inhaling air from the cloud would retain material in the lungs, and so receive a dose to the lung, especially from insoluble material. Soluble nuclides would pass from the lung to the bloodstream from which they would be taken up by organs of the body, the fraction absorbed depending upon biochemical and other factors, e.g. iodine isotopes would give a dose to the thyroid gland. Passage of the cloud of radioactive material at ground level downwind of the sources would also contaminate the ground, vegetation, and buildings over which it passed. The dose of external gamma radiation from fission products deposited on the ground (and to a lesser extent beta radiation) received by persons who remained in the area must be considered, and temporary evacuation of the local population from the area may be required.

Since the iodine and tellurium are volatile, they are the non-gaseous nuclides most likely to be released in an accident. They have short half-lives; evacuation of the local population might be required for 2–3 weeks, after which they could return home. Caesium–137 is also volatile and has a radioactive half-life of 30 years. Thus, deposited caesium could deliver persistent levels of whole-body radiation, either at low surface concentrations or at high surface concentrations; it is dangerous to life if a way cannot be found of removing the contamination. Ruthenium–106 (half-life 1 year) and ruthenium–103 (half-life 41 days) may also be released in some conditions. When external radiation had subsided, inhalation of radioactive material re-suspended in the air by the wind and by the activities of man and animals could still be an important hazard. Re-suspension would probably be most important after release from a core-disruptive accident to a fast reactor, if finely divided plutonium-239 and other long-lived alpha-emitting actinides were deposited on the ground. Regarding contaminated environments in general, whatever isotopes are present, there would be risk of skin contamination, which reasonable care and regular washing should easily control. However, contamination of green vegetables and growing crops would probably necessitate control action in more contaminated areas. As is now well known, probably the most important deposited activity would be iodine-131, if people continued to drink milk, produced by cows grazing contaminated pastures. Above a certain level of iodine-131 in milk, consumption would have to be banned until the iodine had decayed to safe levels. The milk could be converted into cheese and stored until the radioactivity had declined sufficiently.

Depending upon radiation dose and a particular exposed organ, there could be "early" and "delayed" somatic damage in individuals of a population exposed after a reactor accident. Also, there could be hereditary or genetic effects among their descendants. These effects were quantitatively assessed in WASH–1400, 1975. The early effects of radiation are dose dependent; there is a dose, equivalent to 100 rads for whole body radiation, below which early clinical effects are unlikely, and another about 800 rads or above at which early death appears certain. In whole body radiation it is the dose and resultant damage to bone marrow that is critical in causing early death, moreover if the bone marrow alone were irradiated the same range of lethal doses would apply. After lung irradiation caused by inhaling insoluble nuclides, in the absence of other irradiation early death would follow only if the lung dose exceeded a few thousand rads. The main delayed somatic effect of radiation is the induction of cancer. Data available on cancer incidence after irradiation, shows that there is an initial period of 10 years during which there is no increased incidence of cancer, followed by a period of some 30 years during which cancer incidence increases to an approximately constant level and then falls away. For leukaemia incidence after a radiation dose the corresponding times are approximately 5 and 15 years respectively. In most cases the incidence (probability) of cancer is assumed to be directly proportional to the total dose to the organ. For thyroid cancer, the variation of incidence with dose is more complicated. The risk coefficients for cancers induced by radiation are given in ICRP, 1977.

The development and consequences of hypothetical accidents to nuclear plants are often and most usefully analysed on a probabilistic basis. Attempts are made to calculate the consequences of *all* foreseeable accidents *and* the frequencies (probabilities of occurrence on average per year) to be associated with these consequences. The most notable and complete analysis of this type published so far for reactor systems is WASH–1400, 1975. Of the many accidents studied in such analysis, the "design basis accident"* besides other

---

* Defined as the most serious postulated accident.

accidents of greater consequence (if lower frequency) are described and analysed. The final results of frequency-consequence studies reduce to the consideration of two concepts of risk, namely "Individual Risk" and "Community Risk". Individual Risk is the probability of death per year either to some "most exposed" person who lives near the installation, or an average figure for any individual in a nearby resident group. Community Risk can be expressed as a graph showing the frequency of occurrence (F/Y) of incidents resulting in N or more fatalities. Such a graph is known as a f–N line, and the community risk may also be defined as the product f times N.

Some examples of risk values estimated for LWRs are given in WASH–1400, 1975. The background of risk arising from many other causes, non-nuclear, man-made and natural, were given for U.S.A. conditions, for purposes of comparison with the risks estimated for reactor accidents. The reactor accident risks calculated were small compared to other risks. The Reactor Safety Study (WASH–1400, 1975) has recently been reviewed by Lewis *et al.,* 1978 who outlined the shortcomings of the calculations used in the study and concluded that the ranges of uncertainty in the probability estimates are wider than those given in WASH–1400, 1975. A further problem remains about people's willingness to accept risks — even the background of normal risks of everyday life. It may be significant that where there is a risk of death approaching 10^{-3} per year, steps are usually taken to reduce this if possible. As the risk diminishes, concern and counter-measures are less in evidence, and for risks less than about 10^{-6} per year, it seems that the individual does not think it necessary to take steps to reduce the risk further. This suggests that society in practice regards individual risks of about 10^{-6} or less per year as acceptable. A figure of this order (10^{-6} to 10^{-5} per year) has been selected as appropriate by ICRP, 1977. The risk of being killed by lightning, ignored by most people unless a storm is actually in process, is in the range of 10^{-7} to 10^{-6} per year for different countries. The degree of uncertainty and the magnitude of the consequences relate to the acceptability of risk. Society seems unwilling to accept what are calculated to be small risks of very large consequence accidents, when the calculations are subject to wide uncertainty.

Much thought has been given to ensuring operational safety and reliability of reactor systems under all conditions. These methods include the provision of engineering safety systems. Public concern about reactor operations has concentrated on the reliability and effectiveness of these systems (particularly on the emergency core cooling system), and more generally on the possibility of an accident leading to the release of radioactive material to the environment.*

Several studies have been made in which the probabilities of reactor accidents of various degrees of severity were estimated, using information on the failure rates of engineering components of the reactor and other scientific information, and in which the environmental impacts of the radioactive releases

* See El-Hinnawi (1980) for a discussion on nuclear power as a public issue and for a summary of the Three Mile Island accident which occurred in March, 1979.

associated with these hypothetical accidents were calculated. WASH–1400, 1975 estimates the probability of a meltdown in a pressurized water reactor as one in 20,000 per reactor per year, and that most meltdowns would not breach the main containment above the reactor. According to the study, the individual risk of early fatality from nuclear power plant accidents is 2×10^{-10} per year averaged over approximately 15 million people located within 40 km of the first 100 LWRs and 2×10^{-11} per year if averaged over the entire United States population of 200 million. The worst accident, which WASH–1400, 1975 estimates might happen once per 10,000,000 years if there were 100 reactors, might cause 3300 early fatalities, about 10 times that number of early illnesses, additional genetic effects and long-term cancers, and perhaps $\$14 \times 10^{9}$ in property damage. The worst accident considered might cause as many as 57,000 latent cancer deaths in the exposed population over 30 years. A large accidental release could cause large numbers of early fatalities and large numbers of latent cancers, but the probability of such a large release is very low. The Rasmussen report estimated roughly a factor of uncertainty of 5 in its estimates (WASH–1400, 1975 Appendix 6, pp. 13–42 to 13–48). Based on a recent review panel report (NUREG/CR. 0400, 1978) the U.S. Nuclear Regulatory Commission decided that these error bounds are understated (NRC, January 18, 1979).

Because of the main public concern about reactor accidents it seems appropriate at this stage to give a brief account of hypothetical accidents in reactors other than the LWRs. In gas-cooled reactors, such as Magnox, the most serious accident considered is a failure of the pressure circuit causing a rapid loss of coolant (Matthews, 1975). However, high reliability of shut-down is ensured by providing several automatic shut-down devices of different independent kinds. Studies of the different engineered safety systems of Magnox reactors with prestressed concrete pressure vessels show that if there should be an accidental depressurization of a reactor, by far the most probable outcome will be no release of fission products (or actinides) to the environment.

In the Advanced Gas-Cooled Reactor (AGR) with prestressed concrete pressure vessels, a sudden loss of coolant pressure due to an accident would be unlikely to lead to fuel melt-out; the maximum transient fuel temperature would be several hundreds of degrees Celsius below the melting point of stainless steel. There is a possibility, of a few pins being punctured under internal fission gas pressure, as the clad temperatures reached transiently. A small percentage of the inventory of gaseous and volatile fission products in the pins will have diffused to the fuel-clad interspace during normal operation and will be available for release. The manner and magnitude of fission product release from fuel to coolant in such conditions has been much studied (Findlay, 1973); their deposition in the reactor-gas circuit must also be taken into account. The estimated size of release to the atmosphere after a loss-of-pressure accident is generally less than 10 Ci iodine-131. The only environmental impact might be a temporary ban on consumption of cow's milk up to about 2 km downwind. Partly as a result of these favourable environmental features, some AGRs are located close to urban areas in the

United Kingdom. However, there is, as yet, an absence of published studies of AGR (and Magnox reactor) accidents similar to those in WASH–1400, 1975.

In the Liquid Metal-Cooled Fast Breeder Reactors (LMFBRs), refueling accidents probably offer the most easily envisaged route for escape of significant quantities of fission products to the atmosphere of the secondary containment (if not to the free atmosphere), although present designs now make this exceedingly unlikely. The situation usually examined is one in which there is an error in the transfer of an irradiated fuel sub-assembly from the coolant vessel or circuit to the irradiated-fuel examination booths or to the irradiated fuel store. As a result, the sub-assembly could overheat and melt, releasing gaseous and volatile fission products. At the time of reactor shutdown a typical sub-assembly would contain 2×10^5 Ci of iodine-131 and 5000 Ci of caesium-137, but since movement of the sub-assembly out of the coolant would be preceded by a stay of about 100 days in the coolant clear of the reactor core, only some 50 Ci of iodine-131 would remain. If this and the 5000 Ci of long-lived caesium-137 should escape into the secondary containment, its pathway to the atmosphere would lie through a battery of sprays, high-efficiency particulate filters and charcoal filters. The radioactive material released to the atmosphere, assuming all these safety features operate correctly, would contain less than 1 Ci of iodine-131 and 1–10 Ci of caesium-137. This would lead to very small environmental effects; it would be expected that significant levels of dose would not be reached beyond the boundary fence.

Considerations such as these show that it should be possible to operate fast reactors to at least the standard of safety already achieved in thermal reactor operations. The only safety feature, more difficult to establish for fast reactors than it is for thermal reactors, is the effect of a highly improbable nuclear excursion that might take place for example, as a result of a geometrical compaction of the core giving a reactivity increase. The whole design concept of the reactor seeks to avoid "whole core accidents" but these still attract much attention in the safety study of the reactor. Recently, the UK Nuclear Installations Inspectorate has commissioned the National Radiological Protection Board to undertake a "theoretical study of the possible outcome of a range of events extending to extremes in which all protective measures have failed" (NRPB R53, 1977). The preamble, written by the Nuclear Installations Inspectorate speaks of a "sudden and very serious release" which can only take place "if the containments fail catastrophically". The NRPB report concentrates on early deaths, early morbidity, cancers and hereditary effects. Assuming that 5% of the core of a 3000 MW(e) fast reactor is vapourized and escapes to the atmosphere, together with volatile fission products from the remainder of the core, typical results are as follows: on a remote site, there could be 300 early deaths, 200 injuries, 3900 deaths from cancers and about 60 hereditary effects within 30 years (NRPB R53, 1977). One can demonstrate that the calculated early consequences of a large release from a fast reactor can probably be considerably reduced if proper account is taken of plume rise, an aspect to which only limited attention is given (NRPB R53, 1977).

CHAPTER 13

Reprocessing

THE spent fuel rods from LWRs contain plutonium, some higher actinides, and highly radioactive fission product elements in addition to unburnt fissile and fertile uranium. The amount of spent fuel per 1 GW(e)y has been estimated to be about 15 m^3 (containing about 44 tonne of heavy metals) for LWRs with no recycle (GESMO, 1976 Vol. 3). The uranium and plutonium in this spent fuel are valuable energy resources. However, to be utilized they must be separated from the fission products. The uranium may be re-enriched, and formed into new reactor fuel elements or it may be used as the basis of "mixed oxide" fuel in which an appropriate amount of separated plutonium is added. Also, the fission products and residual transuranic elements from processing must be separated as waste, conditioned for storage and disposed of ultimately.

As mentioned earlier in this study there are three options for LWRs: no recycle, uranium-only recycle and plutonium-uranium recycle. When both plutonium and uranium are recycled, enriched uranium supply operations for the year 2000 are reduced by about 20% compared to the uranium-only recycle option; U_3O_8 and natural UF_6 requirements are reduced by about 30% compared with the no-cycle option. The three options raise a number of techno-economic, health and safety and environmental considerations that require different solutions. The recycling of plutonium and uranium introduces a traffic in purified plutonium, which will require safeguards in addition to those in effect. If all spent enriched uranium fuel is reprocessed and the plutonium recycled, the quantity of spent fuel plutonium discarded in various nuclear wastes is about 1–2% of what it would be without recycle. The re-cycled plutonium is essential for the development of breeder reactors.

There are many technical difficulties associated with reprocessing operations on a scale sufficient to support the nuclear power industry, either at today's level or especially at the levels projected for future years. However, many believe that these difficulties can be solved by current technology. The separations must be performed in a chemical plant of special and secure design; fuel-rod refabrication plants must be designed and operated; the plutonium product must be protected from theft and sabotage; the fission products must be put in a form suitable for handling and isolation; and secure, safe, long-term waste isolation must be provided.

The spent fuel as received by the reprocessing plant will have undergone storage at the reactor site for at least 3-6 months and most of the shorter-lived fission and activation products will have decayed. Chopping and dissolving the fuel elements put the remaining radionuclides into gaseous or liquid forms which are mobile. Technologies for their safe handling have been demonstrated. Particular attention must be paid to the plutonium and the higher actinides present in the plant, in solution form, in the final product, and in trace quantities in waste streams.

The Purex solvent-extraction process is considered to be the most reliable method and has been satisfactorily used since the early 1950s. A closely related process was used at the Hanford Plant, the Savannah River Plant, and elsewhere for the separation of plutonium for military use. Although there are small variations in the types of equipment used in some process steps in modern plants, the chemistry of the process has not changed much since its origin. The feed material to a commercial Purex plant would be typically a mixture of oxides containing 96% U, 1% Pu and 3% fission products as obtained from LWRs in which the "burn-up" of fuel has reached about 30-40 MW(t) day/kg.

Effluents From Reprocessing Plants:

In light water reactors, essentially all of the radioactive fission gases are trapped physically or chemically in the zirconium cladding, in the fuel matrix itself, and in the fission gas plenum at the end of each fuel pin. These gases are released in the early stages of reprocessing, usually during chopping or dissolution. Kr-85 is released in chopping and dissolving stages and goes into the off-gas system. Under current practice, some of the tritium is released from fuel or cladding during dissolution and goes into solution as HTO with only a small fraction (1 to 10%) going into the off-gas. The remainder of the H-3 is retained by Zircaloy Hulls. I-129 exists in a wide variety of chemical forms within the fuel and accordingly is released in many forms, mainly as elemental iodine or organic iodides. It has been found mainly in the off-gas, but significant amounts can enter a liquid effluent stream if one exists. C-14 is produced as an activation product from the action of neutrons on 0-17, naturally occurring as a constituent of UO_2, and C-13 and N-14 impurities. Its chemical form in the fuel or clad is unknown, although it is believed that its release to off-gas would be as CO or CO_2.

Based on experience gained at the Nuclear Fuel Services (NFS) plant in the U.S.A., the representative values for the principal radioactive gaseous effluents released in the reprocessing of LWR fuel are given in Table 58 (Bonka et al., 1974; ERDA-43, 1976; NUREG-0116, 1976). Techniques for removing each of these elements from the off-gas stream are known and have been reviewed in ERDA-43 (1976).

Table 58. Gaseous Effluents Released in LWR Fuel Reprocessing

Isotope	Range of Activities Ci/GW(e)y
H–3	$1.5 - 2.7 \times 10^4$
C–14*	15–40
Kr–85	$2.4 – 3.8 \times 10^5$
I–129**	2.3×10^{-3} to 3×10^{-2}

* The wide variation of C–14 is in part due to varying assumptions concerning the level of N–14 impurity in fuel and cladding.
** Value depends on the decontamination factor achieveable at the processing plant; the lower value corresponding to a decontamination factor of 500 represents current state of the art.

For a recent "model" reprocessing plant (U-recycle only) used in the GESMO study, the estimated release of effluents is given in Table 59 (NUREG-0116, 1976).

Table 59.   Releases from Reprocessing Plant (GESMO model plant)

Gaseous Effluents (Ci/GW(e)y)	
Uranium	3.9×10^{-5}
C-14	24
Tritium	18.1×10^3
Kr-85	4×10^5
I-129	0.03
I-131	0.83
Fission products	0.18
Transuranics	0.023
Liquids	None expected
Solids	
Other than high level	0.52

The Thermal Oxide Fuel Reprocessing Plant (THORP) to be constructed on the Windscale site (U.K.) has a designed maximum throughput of 1,200 tonnes of irradiated uranium per year (equivalent to about 40 GW(e)y). The estimates of the effluents to be discharged into the environment are given in Table 60.

Occupational Exposure:

From the limited information available from the operating experience at the Nuclear Fuel Services (NFS) plant, the average occupational dose was estimated to be 600 man-rem/GW(e)y (APS, 1978). The NRC generic estimate in GESMO is 25 man-rem/GW(e)y for a modern plant (NUREG-0002, 1975). An estimate, based on design objectives of the exposures of those to be engaged in the operation and maintenance of THORP was given as an average of 0.8 rem per year, which is 16% of the ICRP dose limit of 5 rem per year applicable to occupational exposure of the whole body. Taking the ICRP-26 dose-risk relationship of 10^{-4} per rem, the mortality risk for radiation induced cancers from exposure to 0.8 rem per year is 8×10^{-5} per year.

Population Exposure:

The effluents of the fuel reprocessing plant account for most of the population dose in the fuel cycle. The most important nuclides are C-14, Kr-85,

Table 60. Estimated Discharges from THORP

Atmospheric Discharges	Estimated Discharge Level Ci/GW(e)y	
	A	B
Kr-85	3.5×10^5	3.5×10^5
Tritium	1.2×10^3	2.5×10^2
C-14	15	15
I-129	0.015	3.8×10^{-4}
Ruthenium	0.1	10^{-2}
Sr-90	0.025	2.5×10^{-3}
Alpha emitters	0.0025	2.5×10^{-4}
Liquid Discharges (to sea)		
Cs-134/137	200	40
Sr-90	40	8
Tritium	2.5×10^4	2.5×10^4
I-129	1.2	1.2
Ruthenium-106	1.9×10^2	38
Zr/Nb-95	35	7
Alpha emitters	45	9

A: The data refer to the estimated maximum release rates which might arise from the operation of the THORP plant. In the case of Kr-85 and C-14 atmospheric discharges, no retention has been assumed; this also applied to liquid discharges of tritium and I-129 although in practice some will occur in the plant. Arrangements are being made to provide for a facility for the removal of Kr-85 subject to satisfactory development and to the expenditure being justified on a cost-effectiveness basis, as recommended by ICRP-26. This might lead to a reduction in this discharge by approximately a factor of 10.

B: For the other isotopes where the discharges are being specifically limited, the figures in column B represent the current flow sheet estimates. The figures in column A allow for a substantial margin which was considered prudent at this stage of design.

I-129, and H-3. Typically, Kr-85 and C-14 produced during reactor operation are released from the fuel at the reprocessing plant during chopping and dissolution. Kr-85 mixes rapidly with the troposphere in its hemisphere of release and more gradually with the global atmosphere. H-3 and I-129 transfer to the surface waters and eventually to the oceans. For C-14, neither the production rate in the fuel cycle nor the fraction released is well known. Magno (1974) estimates that 30 Ci are produced per GW(e)y and most of this is released at the fuel reprocessing step.

An estimate of the cumulative global public dose to all future generations, resulting from current or planned release of radioactivity is difficult and of uncertain significance. The difficulties include incomplete knowledge of the population sizes and living habits far into the future. The uncertainty is in assigning significance to the numeric value of an integral obtained by integrating a nearly zero rate over an infinite interval. This problem has generally been dealt with by analyzing the dose commitment for a finite time into the future. Integration times chosen have been 50 years (NUREG-0002, 1975), 100 years (US EPA, 1977) and 500 years (UNSCEAR, 1977).

To avoid these difficulties the APS (1978) calculated the Dose Rate Increment for operation of a reference reprocessing plant for arbitrary but finite time T at constant effluent release rates corresponding to predictions for production and release. The reference reprocessing plant processes 1,500 tonnes per year of LWR spent fuel. It is assumed that the releases from this plant are:

	Estimated Discharge Level
Atmospheric Discharges	*Ci/GW(e)y*
Kr-85	1.9×10^5
C-14	13.2
H-3	1.6×10^4
I-129	0.02

Table 61 gives the whole-body dose rate increment after T years of continued operation at constant release; the results are given for T = 1, 10, 50, 100, 500 and ∞ years.

Table 61. Per Capita Dose Rate (rem/y) after T Years of Operation from a Continuous Release from a Reference 1,500 tonnes LWR Fuel-reprocessing Plant with no Engineering Retention of Kr-85, C-14 and H-3 (APS, 1978; normalized to 1 GW(e)y)

T(y)	*Kr-85*	*C-14*	*H-3**	*Sum***	*H-3****	*Sum*	*I-129 (Thyroid)*
1	1×10^{-9}	12×10^{-9}	2.6×10^{-9}	15.6×10^{-9}	2×10^{-10}	1.6×10^{-8}	6×10^{-11}
10	8×10^{-9}	1×10^{-8}	12×10^{-9}	30×10^{-9}	12×10^{-10}	3.1×10^{-8}	4×10^{-10}
50	16×10^{-9}	3×10^{-8}	16×10^{-9}	62×10^{-9}	24×10^{-10}	6.4×10^{-8}	14×10^{-10}
100	18×10^{-9}	4×10^{-8}	16×10^{-9}	74×10^{-9}	26×10^{-10}	7.7×10^{-8}	16×10^{-10}
500	18×10^{-9}	6×10^{-8}	16×10^{-9}	94×10^{-9}	26×10^{-10}	9.7×10^{-8}	2×10^{-9}
∞	18×10^{-9}	46×10^{-8}	16×10^{-9}	49×10^{-8}	26×10^{-10}	4.9×10^{-7}	24×10^{-7}

*   Indicates atmospheric release of H-3.
** Sum of dose rates from the individual nuclides if H-3 is released to atmosphere.
*** Indicates ocean release of H-3.

After 50 years of operation the global whole-body dose rate would be 6×10^{-8}, about half from C-14 and the remainder distributed between H-3 and Kr-85. Operation at the same level for 500 years would increase the dose rate by only about 50 per cent.

Estimates of the collective dose commitments from routine discharges from THORP are given in Table 62.

Table 62 Estimated Collective Dose Commitments from THORP

Radionuclide	Collective dose commitment* (man-rem/GW(e)y)	Tissues or organ irradiated	Effective dose commitment (man-rem/GW(e)y)
Atmospheric Discharges			
C-14	135	whole body	135
Sr-90	7.5	bone	0.3
Pu**	1.9	lung	0.5
Kr-85	43.8	whole body	43.5
Kr-85	9250	skin	92.5
		Sub total	272.1
Aqueous Discharges (to sea)			
Cs-134/137	53.5	whole body	53.5
Tritium	11.3	whole body	11.3
I-129	70	thyroid	2.8
Sr-90	40	bone	1.6
Pu (soluble)	0.03	bone	0
		Sub total	69.2
		TOTAL	341.3

* Collective dose commitments are normalized to GW(e)y (assuming 10 years of operation) and integrated over a period of 100 years.
** Assumed here to be insoluble.

If the estimated collective dose to the THORP work-force (20 man-rem/GW(e)y) is added to the above total of collective dose commitment for the world population, a total of 360 man-rem/GW(e)y is obtained. Using the risk factor, from ICRP-26, of 10^{-4} per rem the mortality risk from radiation induced cancer, relating to a collective dose commitment of 360 man-rem/GW(e)y, is about 0.04 per year.

Of the total collective dose commitment of 360 man-rem/GW(e)y quoted above only a relatively small proportion is genetically significant. The dose to the gonads is the sum of all the whole body doses contributed by those radionuclides which irradiate the whole body each of these being adjusted to account for the relationship between whole body dose and gonad dose. The gonad dose is then adjusted to take account of age structure and mean age of conception. The genetically significant fraction of the gonad dose is estimated to about 0.4 in the case of the general public and about 0.3 for those occupationally exposed.

The genetically significant collective dose commitment integrated over a period of 100 years is estimated at about 80 man-rem/GW(e)y to the world population and about 6 man-rem/GW(e)y to the THORP work-force. Authoritative estimates (NAS, 1972; ICRP, 1977; UNSCEAR, 1977) of genetic risk factors lie between 2×10^{-4} and 3×10^{-4} as the risk of serious genetic abnormality per genetically significant man-rem. This results in an estimate of about 0.02 serious abnormalities in the total of all subsequent generations, in the whole world population, per GW(e)y.

The collective gonad dose commitments to the population of the UK are given in Table 63 (integrated over 100 years).

Table 63.   Estimated Collective Gonad Dose Commitments to the Population of UK from THORP (man-rem/GW(e)y)

Radionuclide	Collective whole body dose commitment	Gonad dose whole body dose	Collective Gonad dose commitment
Kr-85	2.2	0.03	2.7
Tritium	5.3	0.03	5.3
C-14	8.8	0.02	5.3
Cs-134/137	30	0.03	30
		TOTAL	43.4

Assuming that the population of the UK is 55 million the collective gonad dose commitment of 43.4 man-rem/GW(e)y of THORP operation results in an average gonad dose of 0.0008 millirem/person/GW(e)y. This is very low in comparison with the average gonad dose from natural background radiation which is estimated to be about 87 millirem per year in the UK. (NRPB, 1974).

Risks to the THORP workforce, to members of the general public near Windscale and to wider populations have been numerically estimated. The risk estimates are summarized in Tables 64 and 65.

Environmental Impacts of Accidents:

While the total long-lived radionuclide inventory of a reprocessing plant complex (including storage of spent fuel and highly active fission products) may well exceed that of a reactor, the following characteristics of a reprocessing plant act in its favour when compared with a reactor:

(i) Temperatures are in general much lower.
(ii) Pressures are in general much lower.
(iii) The non-nuclear potential energy of the plant is much lower.

Table 64. Predicted Risks to Individuals in THORP Workforce and in the Critical Groups of the General Population

Group	Nature of effect	Risk (per person per year)*	Equivalent Risks (per year of exposure to the risk)**	
Workforce	Death from radiation induced cancer	10^{-4}		1/5 of risk of accidental death.
			or	1/25 of risk from regular smoking of 10 cigarettes per day.
			or	1/20 of spontaneous cancer
Individual member of groups	Death from radiation induced cancer	2.5×10^{-6}		1/200 of risk of accidental death.
				1/1200 of risk from regular smoking of 10 cigarettes/ day.
			or	1/1000 of spontaneous cancer

* Assuming operation of THORP at design capacity, 1,200 t/y. 37,000 MWD/t, cooled fuel (equivalent to 40 GW(e)y). Includes risk from Magnox reprocessing.
** After Royal Commission on Environmental Pollution (1976).

Table 65. Predicted Risk to World Population from THORP Operation

Group at risk	Nature of effect	Risk* (for each year of operation)	Comment
World Population (including THORP workforce)	Death from radiation induced cancer	2 (in entire world population)	Death from spontaneous cancers are estimated to be 8 million
World Population (including THORP workforce)	Serious genetic defect	1 (in entire world population)	Applies to total of all subsequent generations

* Risk based on integration of dose over a total period of 100 years arising from one year of operation at design capacity (1200 t/y which is equivalent to 40 GW(e)y).

The inventory of radioactive material in a nuclear fuel reprocessing plant is considerable. It is necessary, therefore, to design and operate the plant in such a way as to prevent accidents which could result in a release of more than a small fraction of the inventory. Discussion, here, will be confined to possible accidents in reprocessing plant which might result in a release of radioactive material beyond the site boundary.

In evidence given at the Windscale Inquiry (1978) three cases were presented and examined in some detail. These were:
(i) Release of activity from a highly active liquid fission product storage tank due to loss of coolant.
(ii) Release of activity from a spent fuel storage pond due to loss of coolant.
(iii) Arbitrary release of plutonium to atmosphere.
Each of these will be considered in the following.

Loss of Cooling Water to Highly Active Storage Tanks (Windscale):

The storage tanks were chosen for study because they have a bigger inventory of fission products than other components of the reprocessing plant complex, and also because complete release of fission products could in theory occur, through failure of the tank. Heat generated by the radioactive decay of the fission products makes it necessary to provide means of cooling of the highly active liquor and to ensure that the cooling capabilities are always in operation. Cooling is carried out by means of cooling coils and jackets through which cooling water is circulated. The number of cooling systems provided is greater than actually required so that failure of one or more cooling systems would not result in an unacceptable loss of cooling capacity. An additional precaution is that coolant water may be supplied in four independent ways. The chance of all these cooling methods failing is negligibly small (it has been assessed as about one chance in 1,000,000 years). Nevertheless if it is assumed that the cooling capability fails, and cannot for some reason be restored, and that it proves impossible to transfer the contents of the affected tank to spare tanks held in readiness, then after some hours the liquid would boil. This would result in the transfer of some activity from the boiling liquid to the vapour above the liquid with consequent discharge of radioactivity from a stack via the vessel ventilation system.

The fraction of the total activity in the liquid which could enter the vapour phase has been determined as being not more than one part in ten thousand (10^{-4}) (by laboratory experiments on inactive simulates and by measures made on actual plants where radioactive liquids are evaporated). Thus of the 10^8 curies in the liquid in the tank not more than 10^4 curies would be available for release in this way. It was pessimistically assumed that despite the presence of an efficient off-gas clean-up system, the whole of the 10^4 curies was released via the stack.

The consequences of such a release were assessed by the Safety and Reliability Directorate (SRD) of the UKAEA making use of a computer programme (Kaiser, 1976). The imputs to the programme were the quantity and type of material released, the duration of the release, the stack height, weather conditions, deposition velocity and population distribution. These calculations indicated a possible need for evacuation, for up to a few days, of the small number of people who live within a mile of the Windscale plant, and a temporary ban of up to a few weeks on the consumption of foodstuffs (milk, free range eggs, etc.) produced within about 15 km radius. The programme was also used to obtain an estimate of the number of people who might, in the most unfavourable circumstances contract cancer as a consequence of the postulated release. In the worst case this might be about 10.

Further evidence was given on the calculated times to reach defined conditions assuming complete loss of cooling with no remedial action. These are indicated in Table 66.

Table 66 Calculated Times to Reach Defined Consequences due to Loss of Cooling Water to Storage Tanks at Windscale.

		PWR Fuel 1 year cooled	PWR Fuel 5 years cooled
Fuel and Waste Liquid Parameters:			
Irradiation	MWD/T	33,000	33,000
Rating	MW/T	37	37
Cooling time	years	1	5
Concentration	1/t*	888	200
Heat content	MW	1.5	1.1
Fission product content	t(U)	146	650
Initial bulk liquor temp.	^{0}C	60	60
Time in days at which:			
Boiling point is reached		0.4	0.6
Boil-away is complete		2.6	3.5
Conversion to oxides is complete (temp. taken at 1000^{0}C)		3.6	8.0
Melting point of stainless steel is reached (temp. taken as 1450^{0}C)		3.9	8.5

* The concentration of the stored liquor is expressed in litres of liquor per tonne of uranium in the spent fuel from which the fission products were extracted.

Further evidence was given regarding the possible sequence of events which might follow those set out above. Three possibilities were postulated:

(i) The concrete cell in which the storage tank is housed continues to provide a barrier against release of activity.
(ii) The cell walls and roof collapse but the base remains intact.
(iii) The cell walls and roof remain intact but the base is penetrated by molten mass.

Calculation of temperature profiles indicated that, of these hypothetical events the third was the most probable. It was suggested that fission products would become intimately mixed with the molten steel and concrete; the degree of binding being quite high. It is not possible to quantify either the fraction or the rate of subsequent leaching by ground water resulting in conveyance to the nearby river or to sea. However, it was suggested that such release would certainly be less than 50 mega-curies of radio-caesium considered by the Royal Commission on Environmental Pollution and which is estimated to lead eventually to 50 deaths. As the hot mass penetrates the ground it would drive off ground water as steam which, it might be argued, could carry away radioactivity which might reach the surface and be released to atmosphere. It was difficult to see how a greater amount of activity could be released via this route than that postulated above for the case of boiling of the liquor (10^4 curies). The problem was likely to remain one of ground contamination. With regard to atmospheric discharges via the stack it was concluded that the activity released was unlikely to be very different from the case of straightforward boiling of the liquor.

The low probability (10^{-6} per year) of the initiating event, (taking into account the system diversity and redundancy) must be emphasized. The probability that no remedial action would be taken is also low resulting in an extremely low overall probability. It is this element of the risk equation, when considered against the consequences indicated above, that makes the risk acceptably low.

Consequences of Aircraft Crash on Highly Active Storage Tanks Cells at Windscale:

The probability of an aircraft crash at any specific location in the U.K. on a random basis is less than 10^{-7} per year if it is assumed that the area affected is about $10^4 m^2$. This is based on data from both military and civil aircraft crashes. The probability will be less at Windscale which is away from airfields and recognized flight paths and which is also a restricted area. Furthermore, the Major Hazards Advisory Committee in the U.K. has concluded that "there is generally little to be gained by taking into account and guarding against events which have a very remote probability — although they could happen tomorrow. For example in the U.K. the probability of an aircraft crash affecting a site is less than once per million years per site". (Health and Safety Commission, 1976).

Despite the low probability of such an event the possible consequences of an aircraft crashing onto the cells housing the highly active storage tanks were considered in evidence given at the Windscale Inquiry. Consideration was given both to the "punch" penetration and also to the transfer of kinetic energy to the structure. In the first case material would break away from the point of impact, and also in some cases a "scab" would be formed on the rear face which would fall away from the wall. In the second case the absorbed energy would result in a vibration response, the damage caused being dependent on the availability of the structure to absorb such energy. A Nuclear Engineering Design formula was used to estimate both the penetration produced and the dynamic response of the structure (Nuclear Engineering Design, 1976). If it is assumed that an aircraft engine impacts on either the wall or roof of the cell, calculations show that it would just penetrate a wall of 1.1 m thickness. The walls of the cells are 1.7 m thick which provides an adequate margin of safety. Calculations also show that the box structure of the cell would ensure that the wall or roof would not collapse, and only local damage would occur. If it is assumed that the engine impacts on the roof in such a manner as to cause the breaking of all the external features of the cooling system on the top of the cell, the possible consequences and possible remedial actions would be as considered earlier.

Loss of Cooling to a Spent Fuel Storage Pond at Windscale:

The possible consequences of such an event, again of low probability, were also presented in evidence at the Windscale Inquiry. Assuming complete loss of cooling and no remedial action, the pond water temperature would increase to an equilibrium level and evaporation of the water would eventually expose the fuel element storage bottles (of stainless steel). Later, the water in the bottles would boil and in turn expose the fuel elements. This would lead to puncturing of fuel element pins (due to increasing pressure of the helium within the pins) resulting in the release of gaseous fission products. Such a puncture would not, however, cause sufficient damage to the pins to permit fuel pellets to fall into the bottles. The times calculated to reach these events are given in Table 67.

Calculations were also made to establish whether the fuel element reaches an equilibrium temperature before the fuel cladding melts, if the loss of cooling persists beyond the stages considered above. The equilibrium temperature of a Zircaloy clad PWR fuel element was calculated as 1072^0C which is to be compared with the melting point of Zircaloy of 1800^0C. In the case of stainless steel clad PWR fuel element, the equilibrium temperature would be less than 1072^0C which is to be compared with the melting point of stainless steel of 1400^0C. With regard to the impact of an aircraft engine, similar considerations to those outlined above for highly active storage tanks cells apply to fuel storage ponds. The ponds' walls are adequate to withstand the punch effect and it was calculated that the transfer of kinetic-energy would be such that a collapse of a wall would not occur.

Table 67.   Calculated Time for the Consequences of Loss of Cooling to Spent Fuel Storage Pond at Windscale

Cumulative time (in days) to:	Case 1 Heat Load 11 MW	Case 2 Heat Load 7.4 MW
− reach equilibrium temp.	4.3	7.0
− expose top of bottles	15.9	24.0
− puncture first fuel (approximately 700°C)	16.7	25.0

Case 1 assumed a storage pond containing 1000 t of PWR fuel with an average cooling period of 1 year. This case wäs considered to be unrealistic because it bears no resemblance to programme deliveries. Case 2 assumed that a 1000 t capacity pond would be filled at the rate of 620 t/y (i.e. filled in approximately 1.6 years) which corresponds to the maximum expected rate of delivery. It was further assumed that the pond would contain one-third CAGR fuel (average cooling 13 months) from U.K. CAGR stations and two-thirds PWR fuel (average cooling 16 months).

Consequences of an Accidental Release of Plutonium from the Windscale site:

The results of calculations were given in evidence at the Windscale Inquiry. Assumptions were as follows:
 (i) An *arbitrarily* assumed release of 15 kg of Pu-239 as oxide at ground level from the centre of the Windscale site. This assumption does not imply that such a release is possible.
 (ii) 1% of the plutonium is in respirable form.
The calculation given in Table 68 indicate the consequences assuming the duration of the release is between 0.5 and 2 hours.

Table 68. Calculated Consequences of Release of Plutonium at Windscale

Probability of one or more cancers for the general public.		0.04 − 0.3
Area/distance over which the initial airborne concentration due to resuspended activity exceeds the maximum permissible level for the general public	Area* (km²) Downwind distance (km)	0.3 − 5.8 1.1 − 18.6
Downwind distance for which the level of various organs is exceeded (km)	lung < whole body < Thyroid < Bone marrow < Bone cells	0.5 to 6.2 0.5 to 1.2 0.5 0.5 to 5.2 0.5 to 13

* Excluding area within the site (site radius 0.5 km).

CHAPTER 14

Ratioactive Waste Management

RADIOACTIVE wastes are generated in practically all areas of the nuclear industry and accumulate as either liquids, solids, or gases with varying radiation levels. The bulk of the wastes occur at the front end of the nuclear fuel cycle which includes mining and milling, while the more radioactive wastes occur at the back end of the cycle which includes reactor operation and fuel reprocessing (in case of re-cycling option). Table 69 gives the estimated amount of wastes associated with the production of 1GW(e)y from LWRs with no recycle; the values for uranium recycle and U–Pu recycle options are also given for comparison.

Assuming a 65% capacity factor, the OECD (1978) "present trend" generating capacity projection predicts the production of about 7500 GW(e)y of nuclear electricity from 1976 through the year 2000. Table 70 gives the estimated accumulated waste from the LWR nuclear fuel cycle with no recycle.

Since the wastes from the front end of the fuel cycle have been dealt with in a previous Chapter (see under mining and milling), this chapter deals exclusively with wastes from the back end of the cycle (reactor operation and re-processing). Radioactive wastes (in gaseous, liquid or solid form are generally considered as low, intermediate, high level and wastes contaminated with transuranic elements. Such wastes may be dealt with as follows: (a) Immediate disposal as they may arise, which applies to low level gaseous and solid wastes. (b) Intermediate level wastes may require conditioning and storage prior to disposal. (c) High level and transuranic wastes will require conditioning and extended storage prior to disposal. Safe and acceptable methods of disposal have been demonstrated and used for the first two categories of waste.

Low and intermediate level wastes are frequently disposed of by shallow land burial. The drums containing the waste are buried in excavated trenches and covered with soil. The most important pre-requisite in selecting a site for shallow burial is the hydro-geological set up; the shallowest water table should be at least 5–8 metres below the waste packages. The soil should be of low permeability and the site should have good drainage characteristics. The main environmental impact associated with shallow land burial of radioactive wastes is the use of about 0.05 ha of land per GW(e)y (NUREG–0116, 1976), although the land can be used for some selected purposes. Some countries dispose of low level wastes by dumping into the sea.

High-Level and Transuranic Waste Management:

In the case of LWRs with no recycle, the spent fuel represents the main source of high level radioactive wastes. If the spent fuel is reprocessed (U recycle or U–Pu recycle), the main high level and transuranic wastes are encountered in the reprocessing.

The conceptual system for high-level waste management includes the following steps: (a) short-term storage as liquid in tanks; (b) solidification; (c) short-term storage as a solid; (d) shipment to a repository; and (e) disposal. Temporary storage of liquid high-level wastes in tanks has been practiced for

Table 69. LWRs Fuel Cycle Wastes Produced from 1 GW(e)y.

Fuel cycle stage	Type of waste	No-recycle Volume (m^3)	No-recycle Activity (Ci)	U-Recycle Volume	U-Recycle Activity	U-Pu Recycle Volume	U-Pu Recycle Activity
Mining	Mine waste	Not estimated		Not estimated		Not estimated	
Milling	Mill tailings	1.7×10^5	1.0×10^3	1.7×10^5	1.0×10^3	1.7×10^5	1.0×10^3
UF$_6$ Production Dry process	Low level CaF$_2$	69	6.3	69	6.3	69	6.3
	Chemical waste	44	6.4	44	6.4	44	6.4
Wet process	same as dry process + sludge						
Enrichment Diffusion tech.	Low level waste	1.5	Not estimated	1.5	Not estimated	1.5	Not estimated
	same	85	Not estimated	85	Not estimated	85	Not estimated
Centrifuge tech.	Low level CaF$_2$	36	Negligible	36	Negligible	36	Negligible
Fuel Fabrication	Low level Misc.	550	2375	550	2375	550	2375
Reactor Operation PWR	same	1225	5125	1225	5125	1225	5125
BWR	Spent fuel	15					
	Fission products		1.6×10^7				
	α TRU		2.2×10^5				
	β TRU		2.8×10^6				
Fuel Reprocessing	Non-radioactive solids			7.7	Negligible	7.7	Negligible
	Calcined High Level wastes			2.4		2.4	
	Fission products				17×10^6		17×10^6
	α TRU				8.7×10^4		4.3×10^5
	β TRU				1.7×10^4		2.7×10^4
	Miscellaneous transuranic solids			44.2		44.2	
	Activation Products				1.9×10^5		1.9×10^5
	Fission Products				1.7×0^4		1.6×10^4
	α TRU				2.4×10^5		1.2×10^3
	β TRU				3×10^6		1.6×10^4

The cases of U recycle and U-Pu recycle assume no mixed oxide fuel is manufactured.
Where no U or Pu is recycled, the spent fuel is regarded as "waste".
Activities associated with reprocessing wastes assume 5 years decay.
In case of U recycle, all plutonium is considered waste.
TRU – transuranium elements.
The table is calculated from GESMO, Table IV H–4 and IV J(E) – 17 normalized to 1 GW(e)y.

Table 70.   Estimated Accumulated Fuel Cycle Waste from LWRs (1976–2000) with no Recycle; Total Power Generated 7500 GW(e)

	Volume (m³)	Activity (Ci)
Mining	Not estimated	
Milling	1.3×10^9	7.5×10^6
UF_6 Production		
Dry Process	5.2×10^5	4.7×10^4
Wet process	3.3×10^5	4.8×10^4
Enrichment		
Diffusion	1.1×10^4	Not estimated
Centrifuge	6.4×10^5	Not estimated
Fuel Fabrication	2.7×10^5	Negligible
Reactor Operation		
Spent fuel	$1.1 \times 10^{5*}$	
Fission products		1.2×10^{11}
α –TRU		1.7×10^9
β –TRU		2.1×10^{10}
Low level	$\sim 7 \times 10^6$	$\sim 3 \times 10^7$

*   It should be noted that several countries are considering reprocessing and therefore, this figure will vary.

over 30 years. Tank designs have been improved and the most modern designs have proved to be virtually free of leaks and operational problems. To prepare the high-level waste for shipment and disposal, and generally to reduce the risk of its dispersal, it should be solidified. A number of technologies exist for solidification, but reduction of the waste to a glass or ceramic form has been favoured because of its advantages in reducing the risk of dispersal. The production of glass or ceramic from liquid high-level wastes is usually a two-step process: calcination and then melting together with glass-forming material to form glass (or ceramic). The product is contained in a sealed canister ready for shipment storage, or disposal.

If the solidified high-level waste is not immediately shipped to a suitable repository, storage capability at the plant must be provided. Following solidification or interim storage, the solid wastes are shipped for storage or disposal at a repository. In the case of surface storage the design involves thick-walled, high-integrity overpacks inside concrete shields which provide shielding and channelling for natural-draft air cooling. Land would be committed only temporarily, and effluents from normal operation would be small (NUREG–0116, 1976).

The management of solid wastes contaminated with transuranic elements may involve the following steps: (a) incineration of combustibles; (b) treatment (e.g. solidification) of non-combustibles; (c) packaging; (d) interim storage on site, (e) shipment to the repository: and (f) disposal.

In the nuclear fuel cycle, high-level and transuranic wastes are the materials which have sufficiently persistent biological hazards and require special long-term considerations. The different options which have been proposed and considered for isolation of these wastes can be classified into three general categories: (a) elimination of portions of the waste from existence on earth, (b) storage and subsequent disposal of the wastes in various geological media, and (c) disposal on or into deep ocean bed.

Since the actinides are considered to be the main source of hazard in wastes after long periods of storage (1000 years), suggestions were made that the actinides be separated, or partitioned, from the fission-products and processing-chemical wastes at the time of fuel reprocessing. The total quantities of the actinides involved would be sufficiently low that one might imagine such "total" disposal schemes as ejection from the earth by rocket or elimination by transmutation* (ERDA 76–43, 1976; NUREG–0116, 1976). The efficacy of any such technique relies not only upon finding a satisfactory scheme for rocket ejection or transmutation, but also upon reducing greatly the actinides lost to low-level waste streams (TRU wastes) in comparison with current practice, and upon having an exceedingly efficient partitioning technology. Such technology is not anticipated to be available before the year 2000.

Sea-bed isolation, ice sheet isolation, and deep continental geologic isolation are future alternatives for disposal of radioactive waste on earth. Several concepts have been advanced for each of those alternatives, which are made specific by choice of site, wasteform, and emplacement medium and method.

Sea-bed disposal would involve the emplacement of solid wastes, suitably encapsulated, either on the deep ocean bed or in the sediments or rocks under the deep ocean. At present both options are being investigated. Evaluation will include investigation of the effectiveness of containment and studies of possible migration rates of radionuclides. These studies and evaluations are at an early stage and further results will be necessary before these operations can be adequately considered.

It has been suggested that radioactive wastes might be placed on, within, or under existing glaciers such as those of Antarctica. The disposal of radioactive waste in Antarctica has appeal primarily in that it offers geographical isolation in the largest area on earth that is essentially devoid of any flora or fauna and it is the most remote region from normal human contact. Disposal in ice had the additional apparent advantage that, ice is impermeable to water, and fractures developed in it are self-healing through recrystallization and plastic flow. Lack of

---

* Transmutation is the transformation of long-lived radioactive nuclides into short-lived or stable ones.

knowledge regarding climatic factors in the future and ice sheet stability constitute the major uncertainties related to this proposed option.

A variety of concepts for deep continental geological isolation have been advanced and include: (1) conventional mined cavities; (2) solution-mined cavities; (3) a matrix of drilled holes; (4) super-deep holes; (5) hydrofracture emplacement; (6) deep-well injection; (7) various rock-melting concepts. These alternatives differ in the nature of the wasteform (solid or liquid), the manner of emplacement, and the geologic medium in which the waste is to be emplaced. With the exception of conventional mined cavities, the remaining alternatives are disposal concepts in which retrievability of the emplaced waste, although possible, would be beyond existing technology, significantly more expensive, or appreciably more difficult than emplacement of the waste. If retrievability of the emplaced waste is to be considered alternatives 2–7 cannot be employed.

Conventional mined cavities constitute the only deep continental geological isolation concept based on a demonstrated technology for both the emplacement and the retrieval of radioactive wastes. In this option solid radioactive wasteforms are lowered through a vertical shaft to a mined cavity, located several hundred metres below ground level, and placed in a series of holes made in the floor of the cavity to receive the waste containers. The depth of the mined cavity would be dependent, in part, on the mechanical properties of the rock. Spacing of the holes would depend upon the thermal properties of the rock and on waste loading of the containers. Of primary concern are the potential modes of breaching the geologic confinement and possible modes of transport to the biosphere of the radionuclides once released from the containment.

Concepts utilizing solution-mined cavities, a matrix of drilled holes, and superdeep holes all involve emplacement of canisters containing the radioactive wastes. The first two alternatives involve emplacement at depths not likely to exceed 2 km; principal concerns are the nature of the groundwater regime and the physical integrity of the geologic unit in which the waste is emplaced. Little is known at present about the stability of solution-mined cavities that have dried out, the methods of drying them out, or about the optimal size or depth of such cavities. Furthermore, this concept is necessarily restricted to emplacement in a geologic medium that is soluble in water – viz., rock salt. Although the matrix of drilled holes is not restricted to a specific geologic medium, it does require identification of a geologic unit with limited or no fractures or other features capable of transmitting ground water at significant rates. Moreover, the probability of compromising the integrity of containment is greatly increased by the multitude of holes that would penetrate the disposal formation and that would have to be sealed. In both concepts of disposal, the number of canisters that could be placed in a given cavity or drilled hole and/or the waste loading of canisters would have to be restricted in order to prevent the melting of the emplacement medium by heat generated from the high-level waste. Some promise of the possibility for superdeep emplacement is provided by the fact that large-diameter holes have been completed to several thousand metres; extension of existing technology and the development of associated techniques to permit emplacement of waste canisters through such holes to greater depths would make this a highly attractive future disposal alternative. Such a mode of emplacement would eliminate the need for conventional mining to the desired disposal medium and, thus, also eliminate certain costs and environmental problems associated with conventional mining.

Hydrofracture emplacement and deep-well injection concepts both involve the pumping of liquid waste into deep geologic units. The technology for both alternatives is well established and routinely practiced by industry. The principal safety concerns prior to emplacement are containment during transportation and the integrity of all pipes and well casing. The major safety consideration after emplacement is to ensure the integrity of long-term isolation in the geologic medium. If retrievability is not required and an appropriate matrix (cement, resin, etc.) is developed to provide initial confinement, this alternative could be acceptable for disposal in favourable hydrogeologic regimes.

Rock melting concepts involve various methods of emplacement, in a variety of geologic media, of both liquid and solid wasteforms that have sufficiently high thermal power to bring the host geologic medium to its melting range locally. Waste having a thermal power of 10 KW/m^3, for example, would result in melting of typical basalt if emplaced in a cavity 1 m in diameter; some high-level wastes can generate up to 200 KW/m^3. Several specific rock melting disposal schemes have been formerly proposed (ERDA 76–43, 1976). Provided that an appropriate host medium is selected, such melting and subsequent crystallization of the melt upon cooling could provide an integral rock mass, with extremely low permeability, in which the radioactive waste form is stable in its environment. Certain rock types, such as shale and limestone, may not be suitable for such a disposal alternative because it is possible that vapours, particularly water and carbon dioxide, will be expelled due to decomposition of the rock. No disposal technique involving melting has yet been thoroughly investigated.

In principle, the integrity of a radioactive waste repository may be compromised by natural events such as tectonic or igneous activity, erosion, and meteorite impact. However, a suitably located deep (more than 300 m) geological repository would be reasonably protected against such effects. Most geologic processes operate at extremely slow rates and could require tens of thousands of years to effect the breaching of confinement. For buried radioactive waste the only important medium of radionuclide transport – i.e., capable of potential transfer of radionuclides from a repository to the biosphere – is ground water. A favourable groundwater regime can by itself provide effective waste isolation in that radionuclides would be transported so slowly that they could not reach the biosphere during the desired period of isolation. Because of the extreme importance of hydrogeology to high level waste management, the parameters that characterize the hydrogeologic system and an appropriate means of quantitatively analyzing the groundwater regime are pre-requisites in geological site selection.

Table 71.　Summary of Predicted Environmental Impacts of Radioactive Waste Management (per GW(e)y).

Natural Resources	
Land (ha)	
Temporarily committed	1.3
Undisturbed	1.2
Disturbed	0.1
Committed for waste disposal	1.1
Water (m^3)	
Discharged to air	1439
Discharged to water bodies	193
Discharged to ground	3636
Fossil Fuel	
Electrical energy (MW–hr)	0.62×10^3
Equivalent coal (tonne)	0.22×10^3
Natural gas (m^3)	$1 \quad \times 10^5$
Effluents (tonnes)	
Airborne	
SO$_x$	0.03
NO$_x$	0.03
Hydrocarbons	0.02
CO	0.007
Particulates	0.02
Cl	0.013
Solids	0.42
Radioactive Effluents (Ci)	
Airborne	
Rn-222	0.0071
Ra-226	5.3×10^{-7}
Th-230	5.3×10^{-7}
Uranium	7.9×10^{-6}
Tritium	$14 \quad \times 10^3$
Kr-85	$290 \quad \times 10^3$
I-129	1.3
Fission products	0.003
Transuranics	0.0014
C-14	19
Liquids	
Uranium and daughters	5.4×10^{-6}
Fisson and activation products	5.9×10^{-6}
Solids	
Other than high-level (shallow burial)	4700
TRU and high-level (deep burial)	1.1×10^7

After NUREG-0116 (1976).
Most radionuclide releases are attributable to the disposal of spent fuel (no-recycle option) and the conservative assumption of complete release of gaseous nuclides in the geologic repository.

Plutonium as a Waste:

In the case of no recycle and uranium-only recycle, the plutonium is not recovered. Instead it becomes a waste which must undergo much the same handling and disposal procedures as high-level or TRU wastes with additional considerations for criticality control and long-term heat effects. Because plutonium has been viewed as a valuable energy source, there are few analyses of its disposal in the literature (NUREG–0116, 1976). Two scenarios are generally discussed: (1) Leaving the plutonium with the high-level waste (in solution and on through solidification); and (2) separation of some of the plutonium, but not purifying it beyond the first separation. In the first case, the environmental impacts would be much the same as those for the handling of high-level waste except that the release of transuranic nuclides in the event of accident would be greater. If plutonium is separated from both uranium and the fission products, it can be handled in much the same way as presently proposed by the industry for handling recycle plutonium.

Table 71 summarizes the environmental impacts of radioactive waste management (NUREG–0116, 1976).

Tables 72 and 73 give the dose commitments to workers and the public from radioactive waste management, together with the calculated inferred cancer mortality.

Table 72. Dose Commitment to Workers from Radioactive Waste Management (man-rem/GW(e)y).

	No recycle	U-Recycle	U-Pu recycle
Total Body	0.35	0.34	0.32
G. I. Tract	0.35	0.34	0.34
Bone	0.35	0.34	0.34
Liver	0.35	0.34	0.34
Kidney	0.35	0.34	0.34
Thyroid	0.35	0.34	0.34
Lung	0.35	0.34	0.34
Skin	0.35	0.34	0.34
Cancer mortality (calculated)	← 4×10^{-5} →		

After GESMO, Tables IV J(E)-9 to 16.

Table 73. Dose Commitment to public from radioactive waste management (man-rem/GW(e)y).

	No recycle	U-Recycle	U-Pu recycle
Total Body	0.002	0.002	0.002
G.I. Tract	0.003	0.003	0.003
Bone	0.003	0.004	0.02
Liver	0.005	0.002	0.004
Kidney	0.004	0.002	0.003
Thyroid	0.007	0.001	0.001
Lung	0.002	0.001	0.002
Skin	0.001	0.001	0.001
Cancer mortality (calculated)	←	6×10^{-7}	→

After GESMO, Tables IV J(E)-9 to 16.

CHAPTER 15

Decommissioning of Nuclear Facilities

DECOMMISSIONING a nuclear facility can be defined as the measures taken at the end of the facility's operating lifetime to assure the continued protection of the public from the residual radioactivity and other potential hazards in the retired facility. Two basic approaches to decommissioning of each facility are generally considered (Smith and Schneider, 1978):

(a)   Immediate Dismantlement — Radioactive materials are removed and the facility is decontaminated and disassembled immediately following final shutdown. Upon completion, the property is released for construction of a replacement nuclear facility or for other purposes in which case authorization and special permission would be required.

(b)   Safe storage with or without Deferred Dismantlement — Radioactive materials and contaminated areas are secured and structures and equipment are maintained as necessary to assure the protection of the public from the residual radioactivity. During the period of safe storage, the facility remains limited to nuclear uses. Dismantlement is deferred to allow the radioactivity within the facility to decay to lower levels. Upon completion of dismantlement, the property is released for special uses as mentioned above.

Methods for decommissioning of nuclear facilities range from minimal removal and fixation of residual radioactivity with maintenance and surveillance (protective storage or Stage I according to current IAEA terminology) to extensive cleanup, decontamination and entombment (Stage II, according to IAEA). Each of these methods of safe storage requires surveillance and care during the holding period which may vary in length from a few years to decades. Each method ends with the deferred dismantling of the nuclear facility (Stage III, according to IAEA).

Entombment (Stage II) of a nuclear facility requires the encasement of the radioactive material in concrete or other structural material sufficiently strong and structurally long-lived to assure retention of the ratioactivity until it has decayed to levels which permit unconditional release of the site. The entombed structure must be designed to outlast any of the confined radiological or chemical hazards. Unless the structure is to be re-entered later and decommissioned further, the potential chemical and radiological hazards should vanish in no more than about 200 years in order to fulfil the bases for entombment. But this cannot be applied in practice due to the presence of long-lived radionuclides. In the United Kingdom, Stage II is regarded as a storage period, either for a defined period or of unspecified duration after which Stage III (total dismantling) is to be undertaken.

It was assumed possible to entomb the reactor pressure vessel and its internal structures within the biological shield since the principal source of radiological dose was Co–60 which decays with relatively short half-life (5.27 years). Thus, within about 100 years, the residual radioactivity will have decayed to levels indistinguishable from normal background, well within the safe structural lifetime of the entombment structure. The amounts of long-lived radionuclides formed in the relatively brief operating life of these early plants were sufficiently small as to present no significant hazard. However, in large power reactors that have operated for 30 to 40 years, the induced Nb–94 and

Ni–59 activities in the reactor vessel and its internal components are well above unconditional release levels and, since Ni–59 has an 80,000 year half-life and Nb–94 has a 20,000 year half-life, the radioactivity will not decay to unconditional release levels within the foreseeable lifetime of any man-made surface structure. It should be noted, however, that Ni–59 is a ß emitter of low energy and low radiotoxicity and might not pose a serious environmental problem on the long run.

In the case of a fuel reprocessing plant the problems of decommissioning are different from those of nuclear reactors. Such plants are designed to facilitate the decontamination of the plant vessels as an operational requirement, so that the plant can be decontaminated to very low levels. Additionally there is not the problem of neutron activation of the plant vessels.

Entombed facilities are a nuisance, subject to vandalism and intentional (but not necessarily malicious) intrusion by man. Any structure conceived by man can also be compromised by him. Extra hardening of entombed facilities such as filling them with concrete will prevent or retard deliberate unauthorized penetration for a period of time, but the facilities will still require some surveillance. It is also difficult to reverse this type of decommissioning if a decision is made later to remove the facility. Provisions for subsequent retrieval of radioactive materials under entombment could be done as an option but this would sacrifice penetrability protection. Furthermore, implementation of the entombment mode would significantly increase the total number of radioactivity-containing sites that require surveillance. This would contradict the philosophy of long-term protection of the public and non-proliferation of radioactivity-containing sites.

Several studies on decommissioning of nuclear facilities (Smith and Schneider, 1978; Nemec, 1978; IAEA, 1978; Detilleux and Lennemann, 1977; Gasch *et al.*, 1978; Essmann *et al.*, 1978; Auler *et al.*, 1978 and Martin *et al.*, 1978) have outlined the technical and regulatory aspects of the problem. Table 74 gives a summary of decommissioning data for a LWR (PWR, about 1 GW(e)).

Table 74. Decommissioning Data for LWR (PWR) 1 GW(e) (after Smith and Schneider, 1978)

	Immediate Dismantlement	Passive Safe Storage with Deferred Dismantlement After:		
		10 Years	30 Years	100 Years
Total Cost (Million 1978 $)	42	50	52	51
Time requirements (Years)	4	4.5	4.5	4.5
Occupational dose (man-rem)*	1326	792	470	445
Public dose (man-rem)*	22	8	3.5.	3
Serious accidents fatalities*	0.097	0.105	0.107	0.113
Manpower (Cumulative man-years)*	320	334	362	460

*Includes decommissioning operations, interim care, and transportation where applicable.

The volumes of radioactive material for disposal estimated to result from the dismantlement of this LWR are:

Activated material (m³)	1191
Contaminated material (m³)	16078
Radioactive wastes (m³)	618
Spent fuel (m³)	37
Truck shipments	1363
Rail shipments	28

All non-transuranic material was assumed to be buried in a shallow land burial facility, and all high-level material was assumed to be placed in deep geologic storage. Other estimates of radioactive material resulting from LWRs are lower than the figures given above.

The estimates of radiation doses to workers and the public resulting from decommissioning of LWR are given in Table 75. Nearly all of the radiation dose to the public results from transport of radioactive materials from the facility to the burial site.

Table 75. Estimated Dose Commitment to Workers and the Public from Decommissioning of LWR (after Smith and Schneider, 1978), in man-rem/GW(e)y.

	Immediate Dismantling	Passive Safe Storage with deferred dismantling after:		
		10 years	30 years	100 years
I. *Workers*				
Decommissioning	40.8	27.8	14.7	13.9
Transportation	3.4	1.3	0.5	0.5
Safe Storage	—	0.3	0.5	0.5
II. *Public*				
Decommissioning	0.3×10^{-5}	$< 0.3 \times 10^{-5}$	$< 0.3 \times 10^{-5}$	$< 0.3 \times 10^{-5}$
Transportation	0.7	0.3	0.1	0.1
Safe Storage	—	← negligible →		

Decommissioning of fuel reprocessing plants was studied by Smith and Schneider (1978) using the Barnwell Nuclear Fuel Plant as a model. Although the facility is not yet operating, it is believed to be representative of commercial fuel reprocessing plants. The main chemical process building, spent fuel receiving and storage station, liquid radioactive waste storage tank farm, and conceptual high-level liquid waste solidification facility were postulated to the decommissioned. The volumes of radioactive material to be disposed of from the plant are:

TRU-material (m³)	4600
Non-TRU material (m³)	3100
Truck shipments	160
Rail shipments	180

Table 76 summarizes the decommissioning data for this fuel reprocessing plant and Table 77 gives the estimated radiation doses from the decommissioning operations.

Table 76. Decommissioning Data for Fuel Reprocessing Plant (after Smith and Schneider, 1978)

	Immediate Dismantlement	Custodial Safe Storage With Deferred Dismantlement After			Passive Safe Storage With Deferred Dismantlement After		
		10 Years	30 Years	100 Years	10 Years	30 Years	100 Years
Total Cost (Million 1978 $)	67	78	95	157	74	77	90
Time Requirements (Years)	5.2	2.4	2.4	2.4	2.7	2.7	2.7
Occupational Dose (man-rem)*	532	453	333	179	445	312	137
Public Dose (man-rem)*	19	15	10	4	14	10	4
Accidents Fatalities*	0.021	0.025	0.034	0.060	0.023	0.024	0.030
Manpower (Cumulative man-years)*	423	510	693	1338	481	514	634

* Includes decommissioning operations, interim care, and transportation where applicable.

Table 77. Estimated Dose Commitments to Workers and Public from Decommissioning a Fuel Reprocessing Plant (after Smith and Schneider, 1978), in man-rem

	Immediate Dismantlement	Custodial Safe Storage With Deferred Dismantlement After			Passive Safe Storage With Deferred Dismantlement After		
		10 Years	30 Years	100 Years	10 Years	30 Years	100 Years
I. *Workers*							
Decommissioning	512	423	290	113	426	296	124
Transportation	20.2	16.7	11.6	4.7	16.7	11.6	4.7
Safe Storage	—	12.8	31.4	61.4	1.8	4.4	8.6
II. *Public*							
Decommissioning	10.2	8.2	5.1	2.0	8.2	5.1	2.0
Transportation	8.5	7.1	5.0	2.1	7.1	5.0	2.1
Safe Storage	—	←————— negligible —————→					

The issues involved in decommissioning nuclear facilities are complex— more so in some respects than in the construction and commissioning of plants. The various aspects of the problem — technical, economical, radiological, environmental and organizational are conflicting in many ways, and will not be resolved until waste disposal routes are defined. Setting aside the uncertainties of waste disposal, it is reasonably clear that the dismantling of nuclear plants is technically feasible.* Reactors are likely to present the greatest problems because of their large numbers, induced radioactivity and the large volumes of radioactive waste which will be produced. However, the technical problems will diminish as experience is gained on the first few plants. The major issue to be resolved is the timescale on which complete dismantling of a plant is to be undertaken. Although there may, in some cases, be pressures for early dismantling, this would result in economic and other penalties. In particular, pressure would in turn be placed on the establishment of suitable waste disposal routes, and a need for early decision on this problem could foreclose some options. Moreover, delaying dismantling for a period of 50–100 years could minimize the need for remotely controlled operations and permit a much greater degree of manual effort without significantly enhanced radiological detriment. The technical problems posed in the dismantling of fuel reprocessing or fabrication plants differ in a number of respects from those of reactors, and the balance between the costs and benefits will generally favour early dismantling.

There is at present some controversy as to whether consideration should be given at the design stage to introducing features that would ease decommissioning problems following a major accident. Clearly some thought must be given to the effects of relatively minor accidents, but major accidents must be very rare events and it would appear to be difficult to justify provisions to remedy or ameliorate the effects of such an accident during the decommissioning process. Rather it is felt that an ad hoc approach is probably the only feasible one as regards decommissioning the plant, though some thought should be given to the possible need to deal with large quantities of very highly contaminated waste.

---

* The ELK River nuclear reactor in the U.S.A. has been dismantled. The most difficult parts of the reactor decommissioning operation, namely the removal to safe storage of the reactor calandria, has already been successfully undertaken (Marchildon. 1971).

CHAPTER 16

Transport of Radioactive Material

THE complexities of the nuclear fuel cycle entail the movement of appreciable amounts of nuclear materials at various stages. While the actual quantities involved are small in comparison with the enormous transportation requirements of coal-fired power stations, which in fact account for a major environmental impact of such stations, the nature of the nuclear material tends to engender enough public concern to require a detailed environmental analysis. Table 78 gives the transport requirements for 1 GW(e)y for LWRs with no-recycle. The volume of transport of radioactive material has grown and will continue to grow in step with the growth of the nuclear power industry. Radioactive material arising in the nuclear fuel cycle are generally transported by truck, to a less extent by rail or sea.

Table 79 gives the estimated dose commitment to workers and the public from transport of nuclear fuel cycle material. A recent study on the

Table 78. LWR Nuclear Fuel Cycle Transport, No-Recycle (per 1 GW(e)y)

Material	From	To	Quantity	Activity (Ci)	No. of Shipments*
U Ore	Mine	Mill	3.4×10^5 MT	1.4×10^3	6300
Yellow Cake	Mill	Conversion Plant	307 MT U_3O_8	360	20
UF_6	Conversion Plant	Enrichment Plant	266 MT U	360	22
Enriched UF_6	Enrichment Plant	Fuel Prepar. Plant	43 MT U	62	6
UO_2	Fuel Prepar. Plant	Fuel Fabr. Plant	43 MT U	62	12
Unirradiated Fuel Assemblies	Fuel Fabr. Plant	Reactor	43 MT U	62	7
Irradiated Fuel	Reactor	Storage	16 MT	7.6×10^6	20
Assemblies			24 MT**	1.1×10^7	5
Low-level waste	Conversion Plant	Burial Site	280 m³		25–38
Low-level waste	Fabrication Plant	Burial Site	180 m³		19–25
Low-level waste	Reactor	Burial Site	100–1000 m³	2×10^3	60

* All shipments are by truck.

** Shipments by rail.

Sources (GESMO, IV J(E)–17, WASH–1248).

Table 79. Dose Commitment to Workers and the Public from Nuclear Fuel Cycle Transport, LWR with No-Recycle (man-rem/GW(e)y)

	Workers	Public
Total body	0.36	0.11
G.I. Tract	0.36	0.11
Bone	0.36	0.11
Liver	0.36	0.11
Kidney	0.36	0.11
Thyroid	0.36	0.11
Lung	0.36	0.11
Skin	0.36	0.11

After GESMO (Table IV J(E)–9 to 16.

environmental assessment of transportation of radioactive material (fuel-cycle and non-fuel-cycle material) in a densely populated urban area (New York City) showed that the radiological impacts of transportation of radioactive material were small; a total of about 800 man-rem was found to be attributable to the annual, accident free transport of this material. The fuel cycle shipments represented 8.2% of the shipments per year, and accounted for about 0.03% of the accident-free radiological impact (about 0.23 man-rem/y). The break-down of this radiation dose is given in the following (Ducharme *et al.*, 1978):

Truck Crew	0.17 man-rem/y
Pedestrians	0.006 man-rem/y
People in vehicles	0.029 man-rem/y
People in buildings	0.018 man-rem/y

The estimated maximum annual dose received by a truck driver transporting nuclear fuel cycle material is estimated to be 2.2 rem (based on a maximum dose of 73 mrem per trip and 30 trips per year) (NUREG–0170, 1977). The annual dose to the population is more difficult to define since it depends on the route of the trucks, speed and distance between the people and truck etc. The maximum annual dose received by an individual standing for a 3 minutes at 1 m distance from a truck carrying redioactive material is estimated to be 1.3 mrem/y (NUREG–0170, 1977). Assuming an annual traffic of 250 trucks at a speed of 50 km/h each, an individual at a distance of 30 m from such trucks will receive a dose of 0.009 mrem/y which is quite negligible.

The probability of an accident occurring in transportation is small, about 10^{-6} per vehicle mile and decreases with increased severity of the accident to an extremely small probability of about 10^{-13} per vehicle mile for extremely severe accidents (WASH–1248). A serious accident resulting in an accidental criticality event with concomitant radiation exposure such as might be related to the transport of enriched uranium materials is judged not to be credible. The packaging is designed to prevent criticality under normal and severe accident conditions. An accident which could lead to accidental criticality with UF_6 would require breach of an inner container, which is unlikely to happen other than in an extremely severe accident. In addition, the contents would have to be moderated, for example, water introduced into the package. All other packages of enriched uranium are subcritical even in the event they are accidentally moderated by water.

The other type of accident which could result in high radiation exposure in the vicinity of the accident relates to an event which might breach the shielding container for high-level solid waste. Based on a 2,000 mile shipment to a storage facility, and an accident rate of 0.8 accidents per million rail-car-miles, it is estimated that 0.048 accidents of any type, and 0.5×10^{-6} severe accidents might occur relevant to the shipment of high-level waste per 1 GW(e) LWR lifetime (WASH–1248). Spent fuel and high-level waste containers are designed and tested for their integrity to withstand accidents likely to be encountered in transport.

Considering the low probability of a uranium fuel cycle component being involved in a serious accident, the nature and form of the low specific activity materials, and the requirements for package design for UF_6, fissile, and high-level activity material, the risk to the environment from transportation of materials that form the uranium fuel cycle is small. In the U.S.A., over a period of 25 years, there have been only about 300 reportable accidents in transportation involving packages of radioactive material. Only about 30 involved any release of contents or increased radiation levels and none resulted in serious injuries from radiation (WASH–1248).

References

Allied Chemical Corporation (1975): Supplement Report in Support of Application for Renewal of Source Materials License SUB–526, Docket No. 40–3392.

APS (1978): American Physical Society; Study Group on Nuclear Fuel Cycles and Waste Management. *Review of Modern Physics* 50, No. 1 Part II (1978).

Auler, I. *et al.* (1978): Decay Behaviour and Structure of the Radioactive Inventory Considering the Decommissioning of a Nuclear Power Plant with a LWR. Intern. Symp. Decommissioning Nuclear Facilities, IAEA, Vienna Paper SM–234/1.

Beir Committee (1972): *The Effects on Populations of Exposure to Low Levels of Ionizing Radiation.* Nat. Acad. Sci., Washington, D.C.

Belter, W. G. (1975): Management of Waste Heat at Nuclear Power Stations; in *"Environmental Effects of Cooling Systems at Nuclear Power Plants"*, IAEA, Vienna.

Biswas, A. K. (1974): *Energy and the Environment.* Planning and Finance Service, Environment Canada, Ottawa, Report No. 1.

Biwas, A. K. and B. Cook (1974): *Beneficial Uses of Thermal Discharges;* Planning and Finance Service, Environment Canada, Ottawa, Report No. 2.

Bonka, H. K. *et al.* (1974): *Umwelbelastung durch Radiokohlenstoff ans Kerntechnischen Anlagen.* Reaktortagung Berlin, 2–5 April (1974)

Braatz, U. and H. Dibbert (1977): Kernbrennstoffversorgung, Jahr. der Atomwirtschaft, 1977.

Bridges, J. E. (1975): *Biologic Effects of High Voltage Electric Fields.* Electric Power Research Institute, Palo Alto, California Report 381–1.

Bridges, J. E. (1977): Environmental Considerations Concerning the Biological effects of Power Frequency (50 or 60 Hz) Electric Fields. *Proc IEEE* Paper F77–256–1.

Brooke, B. (1976): Personal Communication; see OECD, 1978.

Bross, I. D. and N. Natarajan (1972): Leukemia from Low-Level Radiation; identification of susceptible Children. *New England J. Medicine,* 287, 107 (1972).

Bryant, P. M. and J. A. Jones (1973): in: *Environmental Behaviour of Radionuclides Released in the Nuclear Industry.* IAEA Publication STI/PUB/345, Vienna.

Comar, C. L. and L. A. Sagan (1976): Health Effects of Energy Production and Conversion, *Annual Review of Energy 1,* 581.

Coutant, C. C. (1975): Temperature Selection by Fish, a Factor in Power Plant Impact Assessment. in: *Environmental Effects of Cooling Systems at Nuclear Power Plants,* IAEA, Vienna p. 575.

Coutant, C. C. (1977): Cold Shock to Acquatic Organisms; Guidance for Power Plant Siting Design and Operation. Nuclear Safety, 18, 329.

Detilleux, E. and W. L. Lennemann (1977): Criteria, Standards and Policies Regarding Decommissioning of Nuclear Facilities. Nuclear Power and its Fuel Cycle Vol. 4, IAEA, Vienna.

Ducharme, A. R. *et al.* (1978): Generic Environmental Assessment on Transportation of Radioactive materials near or through a Large Densely Populated Area. Sandia Laboratories, New Mexico.

Duret, M. F. *et al.* (1978): The contribution of Nuclear Power to World Energy Supply 1975–2000. World Energy Resources World Energy Conference, IPC Sci. Techn. Press, London 1978.

El-Hinnawi E. (Edit.) (1980): Nuclear Energy and the Environment. Pergamon Press, Oxford.

ERDA–43 (1976): Alternatives for Managing Wastes from Reactors and Post Fission Operations in the LWR Fuel Cycle. U.S. ERDA, Washington, D.C.

ERDA–1543 (1976): Expansion of U.S. Uranium Enrichment Capacity; Final Environmental Statement. U.S. ERDA; Washington, D.C.

Essmann, I. *et al.* (1978): Provision of Decommissioning of the German Utilities for LWR–power plants. Intern. Symp. Decommissioning Nuclear Facilities, Vienna, IAEA Paper SM–234/2.

Evans, H. J. *et al.* (1979): Radiation-induced Chromosome Aberrations in Nuclear Dockyard Workers. Nature, 277 531 (1979).

Fairbridge, R. W. (1972): *Encyclopedia of Geochemistry and Environmental Sciences.* Van Nostrand Reinhotel, New York.

Farges, L. and D. Jacobs (1979): Symposium on Tritium, *At. Energy Rev.* 17.

Findlay, J. R. (1973): *The Birth, Abundance and Movement of Fission Products in Reactor Circuits,* University of Birmingham.

Fischer, J. A. and R. Ahmed (1974): A Systematic Approach to Evaluate Sites for Nuclear Power Plants. *Trans. Amer. Nucl. Soc.* 19, Suppl. 1, 7.

Ford Foundation (1977): *Engineering Assessment of Inactive Uranium Mill Tailings New and Old Rifle Sites,* Rifle, Co. U.S. Dept. Energy, Washington, D.C.

Gasch, A. *et al.* (1978): Results of an Analysis of the Quantities of Radioactive Waste which Develop during the Decommissioning of Nuclear Power Plants. Intern. Symp. Decommissioning Nuclear Facilities, IAEA paper SM–234/3, Vienna.

GESMO (1976): Final Generic Environmental Statement on the Use of Recycle Plutonium in Mixed Oxide Fuel in Light Water Cooled Reactors. Vol. 3. U.S. Nucl. Regulatory Comm., NUREG–0002.

Gibbons, J. W. (1976): Thermal Alteration and the Enhancement of Species Population; in *"Thermal Ecology II"*, Edited by G. W. Esch and R. W. McFarlane, U.S. ERDA, Washington, D.C.

Giraud, A. (1976): World Energy Resources. Conference on World Nuclear Power. Atomic Ind. Forum November, 1976.

Health and Safety Commission (1976): Advisory Committee on Major Hazards; UKAEA

Hanrahan, E. J. *et al.* (1976): Demand for Uranium. Atomic Ind. Forum, Phoenix, USA.

Hochachka, P. W. and G. N. Somero (1973): *Strategies of Biochemical Adoption.* Saunder Publ. Co. Philadelphia.

Hutchison, V. H. (1976): Factors Influencing Thermal Tolerances of Individual Organisms in *"Thermal Ecology II"* Edited by G. W. Esch and R. W. McFarlane. U.S. ERDA, Washington, D.C.

IAEA (1975): *Environmental Effects of Cooling Systems at Nuclear Power Plants,* Vienna.

IAEA (1977): Urban District Heating Using Nuclear Heat. Proceed. of a Symposium, Vienna, Doc. STI/PUB/461, Vienna.

IAEA (1978): Intern. Symposium Decommissioning Nuclear Facilities. 205, Vienna.

IAEA (1979): Private Communication.

ICRP (1977): *International Commission on Radiological Protection Publ.* No. 26, Pergamon Press, Oxford, 1977.

Janes, D. E. (1977): Background Information on High Voltage Fields. *Envir. Health Persp.* 20, 141.

Josefssen, L. and J. Thunell (1967): Nuclear District Heating, a Study for the Town of Lund. in: *Containment and Siting of Nuclear Power Plants,* IAEA, Vienna.

Kaiser, G. D. (1976): A Description of the Mathematical and Physical Models Incorporated in TIRION 2. UK Atomic Energy Authority Report SRD R63.

Kaufman, G. E. and S. M. Michaelson (1974): Critical Review of the Biological Effects of Electric and Magnetic Fields; in *"Biologic and Clinical Effects of Low Frequency Magnetic and Electric Fields.* Thomas Publ. Co. Springfield, Ill.

Kibbey, A. and H. W. Godbee (1975): Solid Radioactive Waste Prechers at Nuclear Power Plants; Nuclear Safety 16, 581.

Kneale G. *et al.* (1978): IAEA Symposium Late Biological Effects of Ionising Radiation. March, 1978, IAEA, Vienna.

Kneale, G. W. *et al.* (1979): Radiation Exposure of Hanford Workers Dying from Cancer and Other Causes. *Health Physics,* 36, 87 (1979).

Kochupillai, N. *et al.* (1976): Down's syndrome and Related Abnormalities in an Area of High Background Radiation in Coastal Kerala Nature, 262, 60–61 (1976).

Krymm, R. and G. Woite (1976): *IAEA Bulletin,* 18, p.6.

Lee, S. S. and S. M. Sengupta (1977): *Waste Heat Management and Utilization.* Proceedings of a Conference, Miami Beach, 1976. Machan. Engin. Dept. Univ. Miami.

Lewis, H. W. *et al.* (1978): Risk Assessment Reviews Group. Report to U.S. Nuclear Reg. Comm., NUREG/CR–400, Washington, D.C.

Lundin, F. E. *et al.* (1969): Mortality or Uranium Miners in Relation to Radiation Exposure. *Health Physics,* 16, 571.

Magno, P. J. *et al.* (1974): Proceed. 13th Air Cleaning Conference, U.S. Atom. Energy Comm.

Mancuso, T. F. *et al.* (1977): Radiation Exposure of Hanford Workers Dying from Cancer and Other Sources. *Health Physics,* 33, 369 (1977).

Marchildon, P. (1971): The NRX Calandria Replacement; Chalf River Nuclear Lab., Canada.

Marks, S. *et al.* (1978): Cancer Mortality in Hanford Workers. IAEA Symposium on the Latent Biological Effects of Ionizing Radiation IAEA SM–224 (1978).

Martin, A. *et al.* (1978): Criteria for the Management of Redundant Nuclear Facilities, IAEA, Vienna, Paper SM–234/46.

Matthews, R. R. (1975): Nuclear Power and the Environment. *Intern. Symposium on Man and His Environment,* Birmingham Univ.

Messer, K. P. (1977): Uranium Demand as Judged by Electric Utilities. International Symp. Uranium Supply and Demand; Uranium Institute, London, June, 1977.

Mole, R. H. (1978): *The Lancet,* 1156 (1978).

Muller, J. and W. C. Wheeler (1976): Causes of Death in Ontario Uranium Miners. in: *Radiation Protection in Mining and Milling of Uranium and Thorium,* ILO, Occupational Safety and Health Series, Paper 32, ILO, Geneva.

Najarian, T. and T. Colton (1978): Mortality from Leukemia and Cancer in Shipyard Nuclear Workers. *The Lancet,* 1018 (1978).

NAS (1972): National Academy Science — see under BEIR Report.

Nemec, J. F. (1978): An Engineering Approach to Decommissioning. Intern. Symp. Decommissioning Nuclear Facilities, IAEA, Vienna, Paper IAEA–SM–234/15.

NRC (1979): Policy Statement on Reactor Safety Study; U.S. Nucl. Reg. Commission, Washington, D.C. 19 January.

NRPB (1974): Radiation Exposure of the Public — The Current Levels in the U.K.; U.K. National Radiological Protection Board.

NRPB R53, (1977): U.K. National Radiological Protection Board, R53.

Nuclear Engineering and Design (1976): 37, p. 183.

NUREG–0002 (1975): see GESMO.

NUREG–75/007 (1975): Final Environmental Statement, Sequoyah Uranium Hexafluoride Plant. Nucl. Regul. Comm. Docket 40–8027.

NUREG–75–032 (1975): Occupational Exposure at Light Water Cooled Power Reactors (1969–1974); U.S. Nuclear Reg. Comm., Washington, D.C.

NUREG–75/108 (1975): Seventh Annual Radiation Exposure Report; U.S. Nuclear Reg. Comm., Washington, D.C.

NUREG–0017 (1976): Radioactive Material Released from Nuclear Power Plants during 1974. U.S. Nucl. Reg. Commission, Washington, D.C.

NUREG–0116 (1976): Environmental Survey of the Reprocessing and Waste Management Portions of the LWR Fuel Cycle. Suppl. 1 to WASH–1248, U.S. Nucl. Reg. Comm., Washington, D.C.

NUREG–0129 (1977): Bear Creek Project, Rocky Mountain Energy Co.; U.S. Nucl. Reg. Comm. Washington, D.C.

NUREG–0170 (1977): Final Environment Statement on the Transportation of Radioactive Material by Air and Other Modes; U.S. Nucl. Reg. Commission, Washington, D.C.

OECD (1978): Nuclear Fuel Cycle Requirements, Paris

OECD–NEA (1976): Uranium Resources, Production and Demand, OECD, Paris

OECD–NEA (1977): Uranium Resources, Production and Demand, OECD, Paris

ORNL/TM-4903 (1975): Sears, M. B. *et al.,* Correlation of Radioactive Waste Treatment Cost and the Environmental Impact of Waste Effluents in the Nuclear Fuel Cycle for Use in Establishing Guidelines of Milling Uranium Ores. Oak Ridge National Lab.

Pancontinental Mining Ltd. (1977): The Jabiluka Project, Draft Environmental Impact Statement, Australia

Pochin, E. E. (1976): Estimated Population Exposure from Nuclear Power Production and other Radiation Sources, OECD, Paris

Ranger Uranium Mines Ltd. (1974): Environmental Impact Statement, Australia.

Reissland, J. A. and G. W. Dolphin (1978): National Radiological Protection Board, Harwell, May 18, (1978)

Rothman, K. J. (1977): J. Amer. Med. Assoc. 238, 1023 (1977)

Rouse, J. V. (1978): Environmental Considerations of Uranium Mining and Milling, Mining Engineering, 1433

Royal Commission Environmental Pollution (1976): *Sixth Report: Nuclear Power and the Environment.* HMSO, London

Sears, M. B. *et al.* (1977): Correlation of Radioactive Waste Treatment Costs and the Environmental Impact of Waste Effluents in the Nuclear Fuel-Cycle Conversion of Yellowcake to Uranium Hexafluoride. Part I, ORNL/NUREG/TM–7, Oak Ridge Nat. Lab.

Seik, R. D. (1977): Review of Environmental Aspects of Uranium Mill Operations. Workshop on Methods for Measuring Radiation in and Around Uranium Mines, IAEA, Vienna

Smith, P. G. *et al.* (1973): *Brit. Med. Journal,* 2, 482

Smith, R. I. and K. J. Schneider (1978): Analyses of the Decommissioning of a Pressurized Water Reactor and a Fuel Reprocessing Plant. *Intern. Symp. Decommissioning of Nuclear Facilities,* UAEA, Vienna, Paper IAEA–SM–234/16.

Stuart, B. O. and P. O. Jackson (1975): The Inhalation of Uranium Ores. Conference on Occupational Health Experience with Uranium, U.S. ERDA, Washington, D. C.

Sundaram: in UNSCEAR, 1977, p.432

Swift, J. J. *et al.* (1974): Assessment of Potential Radiological Impact of Airborne Releases and Direct Gamma Radiation from Inactive Uranium Mill Tailings piles. U.S. Environment. Protection Agency, Washington, D.C.

United Nations Conference on Human Environment (1972): Report A/CONF. 48/14/Rev. 1. United Nations, New York (1973)

UNSCEAR (1977): *Sources and Effects of Ionizing Radiation.* United Nations Scientific Committee on the Effects of Atomic Radiation. United Nations, New York.

USAEC (1973): *Final Environmental Statement Related to Operation of the Highland Uranium Mill.* U.S. Atom. Energy Commission Docket 40-8102

USEPA (1973): *Environmental Analysis of the Uranium Fuel Cycle.* Part I, U.S. Environment. Protection Agency, 520–9–73–3B, Washington, D.C.

USEPA (1976): Potential Radiological Impact of Airborne Releases and Direct Gamma Radiation to Individuals Living Near Inactive Uranium Mill Tailings Piles. U.S. Environment. Protection Agency, 520/1–76–001; Washington, D.C.

USEPA (1977): *Environmental Radiation Protection Standards for Nuclear Power Operations,* U.S. Environment. Protection Agency, Washington, D.C.

US ERDA (1976): *Final Environmental Statement, Light Water and Breeder Reactor Programme,* US ERDA – 1541, Vol 4. Washington D.C.

WAES (1977): *Workshop on Alternative Energy Strategies; Energy Global Prospects 1985–2000.* McGraw-Hill, New York

WASH–1248 (1974): *Environmental Survey of the Uranium Fuel Cycle,* U.S. Atomic Energy Commission, Washington, D.C.

WASH–1400 (1975): *Reactor Safety Study (the Rasmussen Report);* U.S. Atomic Energy Comm., Washington, D.C.

Williams, K. R.: *Projected Energy Requirements up to the year 2000.* International Symp. Uranium Supply and Demand; Uranium Institute, London, June, 1977

Windscale Inquiry (1978): Report by J. Parker; HMSO, London.

Wilde, 1978.

PART III

RENEWABLE SOURCES OF ENERGY

CHAPTER 17

Geothermal Energy

GEOTHERMAL energy is based on the natural heat of the earth. The upper crust has a mean temperature gradient of 20-30°C/km depth, and White (1965) estimated the heat stored, beyond surface temperatures, in the outer 10 km of the earth's crust to be about 12.6×10^{26} J. This resource base is equivalent to the heat content of 4.6×10^{16} tonne of coal (assuming heat content of coal to be 27.6×10^9 J/tonne), or more than about 70,000 times the heat content of the technically and economically recoverable coal reserves in the world*. However, geothermal heat in the outer 10 km of the earth's crust is diffuse to be an exploitable energy resource on a worldwide basis. Resources suitable for commercial exploitation may be defined as localized geological deposits of heat concentrated at attainable depths, in confined volumes, and at temperatures sufficient for electric or thermal energy utilization.

TYPES OF GEOTHERMAL RESOURCES

From the geological point of view, geothermal resources can be classified into: hydrothermal convection systems, hot igneous systems and conduction-dominated systems. Fig. 7 is a map illustrating the geographical distribution of geothermal fields.

Subsurface reservoirs of steam or hot water, which may display such surface characteristics as boiling springs, sulphurous mud flats and fumaroles, are categorized as hydrothermal convection systems. The creation of such systems begins with a source of heat, hot or molten rock, that lies relatively close to the earth's surface. Overlying this high temperature rock zone is a permeable rock formation containing water, largely of meteoric origin which rises upward as it is heated by the rock below (Fig. 8). Above the permeable rock is a cap of impermeable rock, which traps the super-heated water, but cracks or fissures in it allows the fluid to rise to the earth's surface either as steam (a vapour-dominated or dry steam system) or hot water (a liquid-dominated system). Hydro-thermal convection systems are usually associated with the earth crustal tectonic plate boundaries and related volcanic activity.

Hydrothermal convection systems can be subdivided into hot water-dominated or vapour-dominated systems, depending on whether the near-surface permeable zones produce principally water or steam when tapped by production wells. Liquid-dominated systems (or hot-water systems) which are far more common than vapour-dominated ones, vary greatly in their chemical and thermal characteristics by site. Based on the temperature of the liquid, these systems can be classified into:

(a) High-temperature; greater than 150°C
(b) Medium-temperature; 90°C – 150°C
(c) Low-temperature; less than 90°C.

---

* The coal reserves recoverable under current economic and technological conditions are estimated to be 6.3×10^{11} tonne (Peters et al., 1978).

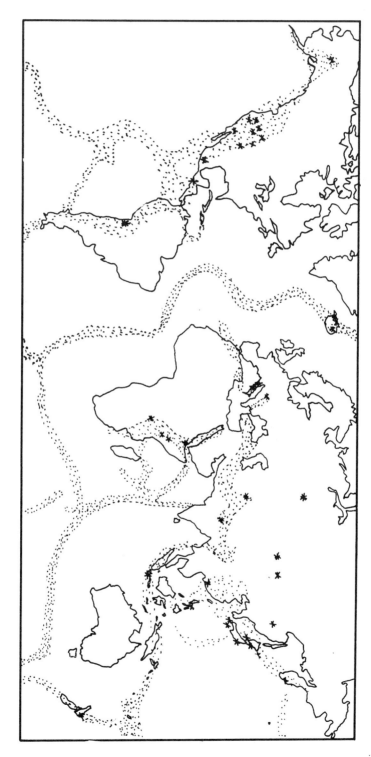

Fig. 1 Geographical distribution of geothermal fields. Shaded area: Geothermal regions; * Geothermal areas (Compiled after Mehta, 1976, Geonomics, 1975)

Fig. 8 Schematic diagram of a geothermal reservoir

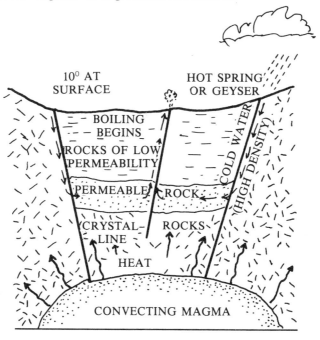

High-temperature systems may be further divided by the characteristics that affect their performance (salinity, chemical composition, stratigraphic set-up, permeability of reservoir ... etc.). These characteristics are important to determine the commercial viability of a hot-water system, either for electricity generation or for other purposes. Generally, the flashed steam process is used for electricity generation from hot-water systems. In this process, as the hot water (which is under high pressure) is withdrawn from the reservoir by wells and nears the surface, the pressure decreases, causing about 20% of the fluid to boil and "flash" into steam. Separators separate the steam from the water; the former is then directed to the turbines. The water leaving the separators is available for further processing depending upon its mineral content. The water may be reinjected into the local rock formation, it may be desalinated for further use or prior to reinjection (minerals may be economically recoverable). Examples of hot-water geothermal fields are Wairakei and Broadlands in New Zealand, Cerro Prieto in Mexico, Salton Sea in California, Otake in Japan. Another method for the production of electricity from high- or medium-temperature geothermal waters is the binary cycle process. In this process, the water withdrawn from the reservoir is used to heat a second fluid (Freon or isobutane) having a low boiling point. The vapour thus generated by boiling the second fluid is used to drive the turbine. Once used, the vapour is condensed and recirculated through the heat exchanger in a closed system, where it may be heated and used again. Equipment using Freon as a second fluid is now commercially available in the range of temperatures between 75-150°C and for capacities of 10 to 100 kW(e). This can be used for production of electricity at suitable sites, especially in remote rural areas (Phéline, 1979).

Vapour-dominated systems (also known as dry steam systems) produce superheated steam (about 240°C) at high pressures (up to 35 kg/cm²) with minor amounts of other gases but little or no water. Since the steam is usually of high quality — that is, it contains few particles or other substances that must be extracted before use — it can be fed directly into a conventional steam turbine to generate electricity. Upon issuing from the turbine, the steam is directed through a condenser where it is condensed; a part of that liquid may be used as a coolant for the condenser and the remainder may be reinjected into the rock formation. Examples of dry steam geothermal fields are: Larderello, Italy; The Geysers, California and Matsukawa, Japan.

The second type of geothermal resources (hot igneous systems) include both magma and hot impermeable dry rock (the solidified margins around magma and the overlying roof rock). The recovery of geothermal energy directly from magma is not yet feasible. The technology required to utilize hot dry rock is beginning to be developed. Preliminary engineering approaches to tapping the energy potential are focusing on the design of a circulatory fluid flow loop through the rock (Fig. 9). First a well is drilled into the hot formation; then cold water is injected under high pressure to fracture the formation, and a second well is drilled to intersect the fractured zone. Finally, cool surface water is injected into the first well, passed over the hot dry rock, and withdrawn through the second well in the form of steam or hot water. The heated fluids generated could then be processed using the flashed steam or binary cycle process.

A third type of geothermal system arises where a deep sedimentary basin occurs in a zone of high heat flow. In situations such as the Paris Basin or Hungarian Basin, water may be tapped by wells at temperatures up to about 100°C. A special category of this type occurs in regions where the normal heat flow of the earth is trapped by insulating impermeable clay beds in a rapidly subsiding geosyncline or downward bend of the earth's crust. Geopressured deposits are hotter than normally pressured deposits because upward loss of the included water has been stopped for millions of years. The high temperatures and pressures have resulted in a natural cracking of the petroleum hydrocarbons in the sediment and in geopressured zones, reservoir fluids commonly contain 1-2 m³ of natural gas per m³ of fluid. These dissolved hydrocarbon gases would be a valuable by-product of fluid production. Temperatures of produced water from geopressured geothermal systems would range from 150°C-180°C; well head pressures would range from 280-560 kg/cm². Production rates could be several million cubic metres of fluids per day per well and perhaps about 30,000 m³ of natural gas per day per well (Hickel, 1972). Geopressured reservoirs have been found in many countries while searching for oil and gas; for example in the U.S.A., Mexico, S. America, Far East, Middle East, Africa, Europe and in the U.S.S.R. The exploitation of such reservoirs for energy has not yet been demonstrated.

Fig. 9 Schematic diagram of a hot dry rock heat recovery process.

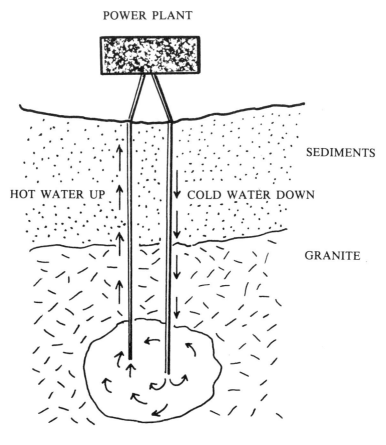

POWER PLANT

SEDIMENTS

HOT WATER UP   COLD WATER DOWN

GRANITE

THERMAL REGION

USE OF GEOTHERMAL ENERGY

The utilization of geothermal energy for the production of electricity and the supply of domestic and industrial heat dates from the early years of the twentieth century. For 50 years the generation of electricity from geothermal energy was confined to Italy (the first experimental electric power generator was operated at Larderello in 1904 and power has been produced since 1913) and interest in this technology was slow to spread elsewhere. In 1943 the use of geothermal hot water for space heating was pioneered in Iceland. More recently intensive exploration work was undertaken in New Zealand, Philippines, Japan, the U.S.A.

and some other countries, which led to the commissioning of a number of geothermal power stations in these countries. Table 80 summarizes the world geothermal electrical generating capacity by the middle of 1979. It should be noted that geothermal fields are at various stages of exploration and development in Chile (El Tatio), Nicaragua (Momotombo), Turkey (Kizildere, Afyon), Kenya (Olkaria), Ethiopia, Guadalupe, Western U.S.A. and India. In addition, geothermal exploration is being carried out or considered in more than 50 countries.

Table 80. World Geothermal Electrical Generating Capacity (MWe)* (1979)

Country	Field	Installed	Planned
U.S.A.	The Geysers, Cal.	660 (S)	460
	Imperial Valley, Cal.	—	150
	Roosevelt, Utah	1	50
Italy	Larderello	381 (S)	—
	Travale	15 (S)	—
	Monte Amiata	25	—
New Zealand	Wairakei	192	—
	Kawerau	10	—
	Broadlands	—	100
Japan	Matsukawa	22 (S)	—
	Otake	11	—
	Onuma	10	—
	Onikobe	25 (S)	—
	Hatchobaru	50	—
	Kakkonda	50	—
	Nigorikawa	—	50
Mexico	Pathé	3.5	—
	Cerro Prieto	150	—
El Salvador	Achuachapán	60	60
Iceland	Námafjall	3	—
	Krafla	—	55
Philippines	Tiwi	110	110
	Macban	110	110
U.S.S.R.	Pauzhetsk	5	7
	Paratunka	0.7	—
Turkey	Kizildere	0.5	10
Chile	El Tatio	—	15
Indonesia	Kawah Kamojang	—	90**
	Total	1,894.7	1,267

* Compiled after Muffler (1975); Ellis (1975); UN Committee on Natural Resources, E/C. 764 (1977); Ellis (1978).
 (S) dry steam fields, the rest are hot water dominated fields.
** Arismunandar (private communication).

The non-electrical applications of geothermal energy are widespread. Early uses included medicinal mineral baths which are well-documented. The use of geothermal energy for space heating has become widespread in many countries; the most significant district heating system is the well-known system of Reykjavik in Iceland. Other countries using geothermal energy for space heating include U.S.A., U.S.S.R., New Zealand, Japan, Hungary, France (Howard, 1975; Banwell, 1975; El-Hinnawi, 1977; Pheline, 1977; Energies Nouvelles, 1978). In the latter, some 25,000 – 30,000 apartments near Paris are heated with geothermal energy and there are plans to extend this system to cover some 500,000 apartments by 1985 – 1990. Hot geothermal waters are also used extensively in agriculture, especially in greenhouses, in connection with animal husbandry, in aquaculture and in controlled growth of single-cell proteins. Agricultural applications of geothermal energy currently constitute the major part of the non-electrical applications. Over 90 per cent of this capacity is associated with large acreages of greenhouses in the U.S.S.R. (about 25 million m^2 of greenhouses are operated with geothermal waters producing about one million tonnes of vegetables per year). There is also a wide range of use of geothermal energy in industry: from the drying of fish, earth and timber, to pulp and paper processing. The two largest present industrial applications are a diatomaceous earth plant in Iceland and a pulp paper and wood processing plant in New Zealand. It should be emphasized that there is a wide range of uses of geothermal fluids for non-electrical purposes, and that several schemes are being developed for combined production of power and process heat. Table 81 summarizes the major non-electrial uses of geothermal energy.

Different estimations have been made for the capacity of geothermal energy most likely to be harnessed by the year 2000. Table 82 gives an estimate of geothermal capacity that is likely to be harnessed by 1985 and the year 2000; the geological potential of favourable areas is given for comparison. Assuming that the total primary energy demand by the year 2000 is $600 \times 10^{18}J$ (WEC, 1978), geothermal energy that can be harnessed by that time can provide 0.8% of this demand.

ENVIRONMENTAL IMPACTS OF GEOTHERMAL ENERGY*

The widespread belief that geothermal resources represent a relatively "clean" non-polluting energy source recently has played an important role in heightening public interest in geothermal development. However, some recent studies (e.g. Bowen, 1973; Axtmann, 1975 A, B; Kruger, 1976; El-Hinnawi, 1977; EPA, 1977) have reviewed the environmental aspects of geothermal energy development and have highlighted a number of potential impacts. Although

---

* In this chapter the environmental impacts are quantified on the basis of 100 MW(e)y rather than on 1 GW(e)y.

Table 81. Some Major Non-Electrical Developments of Geothermal Energy*

Country	Area	Fluid** Temperature (°C)	Approx. Harnessed Energy (MWt)	Use
France	Paris Basin	60–80 (w)	30	Heating
Hungary	Szentes-Szegred	80–90 (w)	350	Heating, Agriculture
Iceland	Reykjavik	80–130 (w)	350	Heating
	Hveragerdi	180–(w and s)	20	Heating and hot houses
	Namafjall	185 (s)	10	Diatomite drying
Japan	Okawa	70 (w)	2	Heating
	Various areas	70–100 (w)	5	Horticulture
New Zealand	Kawerau	195 and 165 (s)	125	Kraft paper mill
	Rotorua	100–175 (s and w)	50	Heating, air-conditioning
U.S.A.	Boise Idaho	77 (w)	10	Heating
	Klamath Falls Oregon	40–110 (w)	6	Heating, greenhouse
U.S.S.R.	Makhachkala	60–70 (w)	25	Heating
	Zugdidi	80–100 (w)	60	Heating
	Cherkessk	80–100 (w)	25	Heating
	Various areas	60–100 (w)	500	Greenhouse

* Calculated after Howard (1975).
** Fluid utilized – (w) = water; (s) = steam.

Table 82. Geothermal Energy Scenario until the Year 2000

		1979	1985	2000
I.	Geothermal Electricity estimated from collected data [MW(e)]*	1 895	5 000	50,000
II.	Geothermal non-electrical energy estimated from collected data [MW(t)]**	2 000	10,000	100,000

* Geothermal Electricity calculated according to geological potential is about 5×10^5 MW(e); see Auer et al., 1978.
** Geothermal non-electrical energy calculated according to geological potential is about 8×10^7 MW(t); see Auer et al., 1978.

knowledge of these environmental impacts is still incomplete, geothermal resources do appear to offer several significant environmental advantages over fossil and nuclear energy.

Since geothermal energy must be utilized or converted in the immediate vicinity of the resource to prevent excessive heat loss, the entire fuel cycle, from resource extraction to transmission, is located at one site. Unlike fossil fuel or nuclear power production, geothermal energy is not a technology that requires a massive infrastructure of facilities and equipment and large amounts of input energy (the fuel cycle for fossil fuel and nuclear power productions includes mining, refining, transportation, fuel processing and waste disposal which are not located at one site but in many cases involve widely separated locations). Another environmental benefit arises from the fact that those geothermal power stations that use steam as a working fluid to drive a turbine do not need an external source of water for cooling purposes, because the condensed steam is recycled for that purpose. Thus, they do not place additional demands on scarce water supplies.

However, not all the environmental effects of geothermal energy are positive. There are a number of adverse local effects on land, air and water. These effects are site specific varying according to the geochemical properties of the hydrothermal reservoir and the exploitation history of the resource. Accordingly, generalizations about geothermal emissions are difficult to propound, particularly since the effluent characteristics may be affected strongly by individual plant designs and power cycles (Axtmann, 1975 B). In the present study use is made of the extensive information available for The Geysers, U.S.A. and the Wairakei, New Zealand geothermal fields as examples of dry steam and hot water dominated fields, respectively. It should be emphasized that the environmental impacts to be encountered in geothermal fields in other sites have to be assessed for each site.

Impact on Land

The development of geothermal resources has a number of land-use impacts. Prospecting takes place in a wide range of situations, including open farm land, forest, and near closely settled areas, as well as in rough volcanic terrain. The geological, geophysical and geochemical survey work that precedes a geothermal development requires widespread access. Survey lines and vehicle access tracks may have to be opened up in virgin areas. However, the work needs no massive excavations or general clearance. Test drilling in promising areas will require a few access roads for a drilling rig and clearances of about 1000 m^2 for each drilling site.

The intensive development of a geothermal area carries with it the problems of any large civil engineering operation (e.g., road construction, accommodation for workers, and the accumulation of equipment on the site). Patterns of work and way of life in small communities may be radically changed. Noise from vehicles, diesel-powered drilling rigs, and other mechanical operations can be expected.

As many geothermal areas are associated with soft rocks such as pumice or hydrothermally altered rock, development can cause land erosion and slips if care is not exercised. Significant land erosion occurred at Wairakei and at The Geysers (Reed and Campbell, 1975; EPA, 1977), but with experience better techniques have been evolved.

The average overall area involved in a high temperature geothermal development is of the magnitude of 1 km² for 25-50 MWe (Banwell, 1975; Reed and Campbell, 1975). However, the amount of surface land required for the development of a geothermal field differs from one location to another depending mainly on the topography and the well spacing density. From the area presently used at The Geysers (dry steam field), it is estimated that a 1000 MWe facility consisting of ten 100 MWe units with a well spacing of 1 well per 0.23 km² would occupy 21-40 km² of land. Of this amount, an average of 20 per cent of surface area would be disturbed physically through clearance of vegetation, grading, and paving (EPA, 1977). For a more densely developed Wairakei type hot-water system, a 1000 MWe facility would require 16 km² for 381 wells (EPA, 1977).

Another important land-use issue is the extent to which such development is compatible with surrounding land uses. Possible adverse effects to adjacent land could result from the changes in the use of the land at the site; e.g. human activity in relation to noise and pollutant emissions. Furthermore, such impacts are likely to be long-term in relation to the life of a geothermal field. Many geothermal developments for electric power production have occurred in passive multipurpose land uses such as wildlife reserves, cattle grazing, and watersheds. At The Geysers, wilderness surrounding the development area has remained largely unaffected (EPA, 1977). The impacts of geothermal development on land fertility appear to be minimal. During most of the 60 years of field development at Larderello, Italy, for example, the surrounding land has had varied agricultural uses. Today, pipelines traverse vineyards, orchards and farmlands with no known detrimental effect.

Land subsidence resulting from the withdrawal of geothermal fluids from the reservoir is a potential serious impact of geothermal development in some areas. Land subsidence does not appear to be a problem in the development of dry steam (vapour dominated) geothermal fields, e.g. at The Geysers or in Larderello, Italy. One of the fundamental conditions considered necessary to formation of a vapour-dominated system is a "competent" host rock; that is, rock not subject to compaction and, therefore, not subject to subsidence (Bowen, 1973; Swanberg, 1975). In contrast, the problem of land subsidence is more acute for liquid-dominated geothermal fields. In Wairakei, New Zealand, where waste geothermal water is discharged to a river rather than reinjected, total maximum vertical movement has exceeded 3.7 metres since 1956, and has affected an area of 65 km² but the effect grades to zero at the perimeter of the affected area. Horizontal movement also has been recorded (Axtmann, 1975; Swanberg, 1975; EPA, 1977). Significantly, the area of maximum subsidence may occur outside the production field, which means that subsidence could affect the property of adjacent land-owners more than the immediate development area.

In Cerro Prieto, Mexico, and in a hot water field located near the Imperial Valley in California, subsidence was recorded some 10 km outside the well field even before extensive production began (Dutcher *et al.*, 1972). At this site, geothermal waters have been discharged to an evaporation and sedimentation pond rather than reinjected.

The problem of land subsidence can possibly be alleviated or avoided altogether in one or more of several ways. Reinjection of geothermal fluids to deep wells following power production is highly promising but has some practical limitations. Some geothermal liquids contain a large amount of dissolved solids, such as silica or calcium, which may clog in reinjection pipes. Such problems could be solved by chemical additives to keep dissolved solids in solution; but their use may create other environmental problems. Reinjection may also lower the temperature and hence the energy potential of subsurface geothermal waters (Einarsson *et al.*, 1975). This effect has been observed in Japanese hot water fields. On the other hand, reinjection is potentially an important factor in increasing the productive life of geothermal field (Einarsson *et al.*, 1975), as it is considered a "recycling" of the geothermal waste water through the reservoir. With surface disposal, the residual heat in the water, after separation of the steam, is wasted and its disposal may cause thermal pollution of surface waters. By reinjection this heat energy, which represents roughly 50% of the heat produced by the borehole, is also recycled to the reservoir.

The unstable conditions in the earth's crust that create geothermal resources are conditions that also produce seismic activity. In fact, geothermal resources currently being developed are located in seismically active zones and micro-earthquakes (less than 4 on the Richter scale) have been observed near many major geothermal areas around the world. It has been mentioned that reinjection of fluids into tectonically active areas induces seismic activity (Swanberg, 1975). Although this has been the case with deep well disposal of other liquid wastes, reinjection experiments in geothermal fields in Italy have produced no seismic or micro-seismic effects (El-Hinnawi, 1977). Monitoring at The Geysers, showed no evidence to indicate that geothermal development has increased the seismicity of the area (EPA, 1977).

Noise

The development of a geothermal field creates considerable noise, particularly at the drilling and well-testing stages. Drilling operations, with noise from diesel engines and other heavy equipment, can create localized noise levels of 80-90 dBA for 24 hours a day. However, this problem is not unique to geothermal development and is encountered in other construction operations. The noise from the first discharging of wells is intense and can create annoyance at distances of several kilometres. Suter (1978) surveyed the effects of geothermal development on wildlife behaviour, while Andersen (1975) noted the effect of noise on farm animals. Unsilenced geothermal wells may produce noise levels up to 120 decibels in their near vicinity, and to prevent damage to hearing, workers must wear ear protectors.

Once cleared from drilling mud and rock cuttings, wells can be discharged through efficient silencers that reduce the noise level to below 100 dBA near the well, and at a distance, the sound is reduced to a low rumbling which as noted by Armstead (1975) has a soporific rather than a distressing effect. In addition to lowering the overall noise, the silencers remove the high frequencies that are typical of a geothermal well blasting steam and water into the atmosphere at a supersonic speed. By reducing the time of first blowing of wells to an absolute minimum and with early installation of efficient mufflers, geothermal noise can be reduced and is not likely to be a major environmental problem.

In rare situations ground vibration may be produced within about a hundred metre radius from a well that has got out of control during drilling or through a break in the casing. As pointed out by Jhaveri (1975) continuous low-frequency vibration can cause irritability and lowered work performance. "Rogue" wells are a temporary phenomenon and are either quenched by a secondary drilling operation or may with time tend to quench themselves by rock slumping. The local ground vibration created by a "rogue" well at Wairakei became a popular tourist curiosity while it persisted for a few years.

Airborne Effluents

The major sources of air pollutants emitted during geothermal power production are: direct releases of geothermal steam during all stages of development and releases of non-condensable gases during plant operation. The types of pollutants likely to result from geothermal power development are primarily determined by the chemical composition of the geothermal fluid at a site. Both the total quantity of gases in the fluid and the relative concentration of their constituents depend on the geochemistry of the underground reservoir.

The compositions of geothermal steam in various areas are summarized in Table 83. In steam-dominated fields (The Geysers and Larderello), the discharged steam composition corresponds to that at depth. However, in high temperature water-dominated fields the proportion of gas in the steam depends on the extent to which steam has flashed from the original high-temperature water. The gases (except ammonia) are predominantly concentrated in the steam phase and the gas/steam ratio decreases with increasing steam proportion in the discharge. The gas contents vary widely, and occasionally beyond the range in Table 83. For example, at Ngawha, New Zealand and at Monte Amiata, Italy, wells discharged steam containing up to 50% gas (mainly CO_2), but the fields were not major producers (Ellis and Mahon, 1977). In this high gas situation the H_2S/CO_2 ratio is usually low. Carbon dioxide is usually the predominant gas, but the proportions of other gases vary widely. In extreme cases in active volcanic areas, e.g., Namafjall, Iceland, and Tatun, Northern Taiwan, gases may contain 10-30 % hydrogen sulphide. There are other areas

Table 83. Concentration of Gases in Geothermal Steam (mg/kg)

Area	Separation Pressure (bars abs.)	CO_2	H_2S	NH_3	CH_4	H_2	N_2	H_3BO_3	Ref.
Larderello (average)	·	47500	540	150	215	50	240	220	(1)
The Geysers (average)	·	3260	222	194	194	56	52	91	(2)
Cerro Prieto (average)	10.5	14100	1500	110	·	·	·	<0.1	(3)
Wairakei (average)	10.9	2300	72	4.8	2.5	0.5	17	<0.1	(4)
Broadlands Well 8	5.8	30000	350	34	420	3	230	<0.1	(4)
Otake Well 7	3.0	2400	16	·	←——————Remainder 67——————→			·	(5)
Hveragerdi Well 4	1	1210	36	·	1.6	2.5	·	·	(6)
Salton Sea (average)	1	3000	2.6	35	10	·	·	·	(7)

References
(1) ENEL (1970)
(2) Reed and Campbell (1975)
(3) Mercado (1976 A)
(4) Ellis (1978)
(5) Noguchi (1966)
(6) Arnorsson (1974 A)
(7) Gudiksen, et al. (1978)

with extremely low H_2S concentrations in steam, as at Salton Sea, California and Ahuachapan, El Salvador. In the fields so far developed, in both water- and steam-dominated areas, changes with time in the gas content of the steam flow tend to be negative, so that the initial assessment gives the most pessimistic picture of the pollution potential (Ellis and Mahon, 1977).

Table 84 gives the amounts of the main airborne effluents for some geothermal plants. A 100 MWe plant in a steam-dominated area requires about 10^7 tonnes of geothermal steam per year; in water-dominated fields this steam would be accompanied by between 2 and 10 times the amount of hot water.

Table 84.   Main Airborne Effluents (tonne per 100 MWe(y)*)

	H_2S	NH_3	CO_2
The Geysers	2000	1700	30000
Larderello	5000	1300	400000
Wairakei	300	50	8000
Broadlands	6000	500	400000
Cerro Prieto	4000	800	150000

* After Ellis (1978)

Hydrogen sulphide is the main airborne effluent of real concern at present. Standards for air quality vary, but in populated areas the limits of H_2S in air are often close to the limit of detection for the human nose; by volume 0.002 ppm (Axtmann, 1975 A). Although H_2S is a poisonous gas at high concentrations, the main problem in geothermal development is simply its objectionable smell. However, traces in the air may also create problems with electrical equipment, cause general corrosion of metals, and blacken lead-based paints. Sub-ppm levels do not appear to present a health problem, and communities have lived for generations with local air containing 0.01-0.1 ppm (e.g., Rotorua, New Zealand).

As H_2S is a heavy gas it tends to concentrate around geothermal plants in enclosed areas, in well-head pits and in underground cable-ways. H_2S becomes a health hazard at concentrations of over 10 ppm. At Cerro Prieto there was trouble with hydrogen sulphide wafting to the ground near the plant from the 40 m gas stacks, and fans and ducting were installed to remove the gas from the vicinity (Mercado, 1975).

Hydrogen sulphide passing through a power station with direct condensers distributes itself between the gas ejectors, the cooling tower plume and the reject condensate in a manner which varies with the design and operation of the plant. In the early Geysers power stations for example, about equal quantities of H_2S were emitted through each route (Bowen, 1973). The atmospheric disposal of waste gases was considered satisfactory in the early days of the geothermal development at The Geysers, but when H_2S emissions rose from 14 kg/hour in 1960 to 750 kg/hour in 1975, widespread complaints were received from nearby Lake County residents (Leibowitz, 1977). In the Wairakei station about 30% of the inflow H_2S exists from a 30 m gas exhaust chimney and the dispersal of the gas leads to concentrations of up to 1 ppm H_2S in the air within a 0.5 km radius. There has been little comment from the nearest permanent residents (0.8 km). The remaining H_2S enters the river water/condensate, and together with the reject hot water causes the odour of H_2S to persist in the Waikato river for a distance downstream.

Various processes have been proposed for limiting H_2S emission from geothermal stations, and several have been tried in The Geysers plants (Laszlo, 1976). These include several modifications of the Claus reaction (Allen and McCluer, 1975; Axtmann, 1975 B). The Jefferson Lake modification appears to be the most likely of this type, and Velker and Axtmann (1978) consider that recovery of sulphur may give break-even economics in the U.S.A. where plants emit more than about 20 tonnes per day of H_2S. The "iron catalyst" modification is in operation on existing Geysers units with efficiencies of about 40-70%. It involves the gases pumped from direct condensers being burnt to form SO_2, and the scrubbing of the gases by cooling tower water to which an iron catalyst has been added. It is not altogether satisfactory because of corrosion and sludge formation in the plant. At The Geysers the Stretford process is now favoured, and it is being installed in the newest unit. Shell and tube stainless steel condensers provide a highly concentrated gas stream to the process which involves washing the gases with a solution of sodium carbonate, sodium ammonium polyvanadate and anthroquinone disulphonic acid. The hydrogen sulphide becomes oxidized to sulphur by the vanadate and the reduced vanadium is restored to vanadate by air. High-quality sulphur is recovered. The economics of operation are not yet established, but Laszlo (1976) considered that the process may raise the cost of the power plant by 45%. The level of current control technology suggests that hydrogen sulphide abatement levels of over 90% are possible.

Recent work by Thompson and Kats (1978) showed that constant H_2S concentrations of 0.3 ppm or over affected the foliage and growth of California buckeye, lettuce, alfalfa, sugar beet, grapes, ponderosa pine, and Douglas fir. Rapid growing crops such as grapes, alfalfa, and lettuce were more affected than the slower-growing buckeye and pine. Douglas fir was more sensitive than ponderosa pine. Low concentrations of H_2S at 0.1 and 0.03 ppm had little effect, and actually stimulated the growth of the field crops. These effects would probably be the maximum to occur, as field concentrations would be highly variable according to wind conditions. A practical example is provided by the Rotorua, New Zealand, public gardens where a wide range of healthy trees and flowers has existed for over 100 years within a few hundred metres of thermal activity causing local H_2S concentrations of about 0.1 ppm. Hydrogen sulphide oxidises slowly to sulphur dioxide over a period of about a day. By this time the sulphur gases would be well dispersed into the atmosphere. Radiata pine shows good tolerance to foliar concentrations of sulphate at many hundreds of ppm, but the long-term effects of consistent SO_2 air concentrations at the 0.01 ppm and of the possibility of formation of acid rains near geothermal plants need further investigation.

Other gaseous effluents include carbon dioxide and ammonia. Carbon dioxide enters mainly the exhaust gas stream and does not constitute an environmental problem near the geothermal plant, although it adds to the global atmosphere CO_2 budget (Reed and Campbell, 1975). However, at Wairakei, where river water is the condenser water supply, 30-35% of the carbon dioxide passes into the reject condensate and hence to the Waikato river where it may contribute to increased weed growth (Axtmann, 1977). Ammonia emitted in the exhaust gas stream is rapidly diffused by atmospheric processes and the values are rapidly lowered to acceptable values. However, if ammonia reacts with other chemicals (such as hydrogen sulphide to form ammonium sulphate), harmful environmental impacts on human health and certain plant and animal species may result (EPA, 1977).

Radon-222 is found in trace amounts in the non-condensable gas portion of geothermal steam. While the values vary widely, both between and within fields, they are in the range 3-200 nCi/kg of steam (see Table 85).

Table 85. Concentration of Rn-222 in Geothermal Steam

		nCi/kg of steam	Ci/100 MW(e)y*
The Geysers	(1)	20	180
Larderello	(2)	20-200	180-1800
Wairakei	(3)	10-100 (variable)	~900
Broadlands	(3)	1-5	9-45

(1) After Reed and Campbell, 1975, Stoker and Kruger, 1975.
(2) After D'Amore (1975).
(3) After Ellis (1978).

n = nano, i.e. 10^{-9}.
*   Assuming that 10^7 tonnes of steam are required to produce 100 MW(e)y and that 90% of the radon enters the exhaust vapour.

The levels of radon measured in the ambient air outside the geothermal plant at The Geysers are in the range of 0.03-0.15 pCi/1** (Reed and Campbell, 1975; Stoker and Kruger, 1975), well below the average for air over continental land areas which is about 0.3 pCi/1 (Stoker and Kruger, 1975).

Several geothermal areas are situated in districts with known mercury mineralization (e.g., The Geysers, California and Ngawha, New Zealand) and

** pCi = pico Curie = 10^{-12} Ci.

higher than normal regional concentrations of mercury occur in the air (Siegel and Siegel, 1975). In geothermal systems, mercury concentrates with great preference into the steam and gas phases rather than into the hot water. The mercury is present mainly as elemental mercury in spite of the presence of sulphur (Robertson *et al.*, 1978). Table 86 gives the concentrations of mercury in geothermal steam of various areas. In a power station most of the mercury from well discharges passes through the turbines into the condensers. Between 10 and 50% of the mercury is dissolved in the condensate; the remainder passes into the exhaust gas. At Cerro Prieto, for example, 6.8 g/hour is discharged into the air of a total 7.4 g/hour throughput. With good dispersal into the atmosphere, e.g. by injecting the waste gases into the cooling tower plume, it can be calculated that mercury concentrations in the immediate vicinity of the station would be about 0.01-0.1 $\mu g/m^3$ (Ellis, 1976). These levels are unlikely to present a health hazard (although Schroeder (1971) considered that prolonged exposure to atmospheric levels of mercury in excess of 0.1 $\mu g/m^3$ may be harmful, the adopted maximum level for 8-hour industrial exposure is 50 $\mu g/m^3$ (Sax, 1975)).

Table 86. Concentration of Mercury in Geothermal Steam

		$\mu g/kg$ in steam	$kg\ Hg/100\ MW(e)y$
The Geysers	(1)	5	45
Matsukawa	(2)	50-60	450-540
Wairakei	(2)	1-4	9-36
Broadlands	(2)	4-8	36-72
Salton Sea	(3)	10-20	95-180
Cerro Prieto	(3)	7-9	63-81
Namafjall	(4)	0.2	1.8
Otake	(2)	0.3-2	2.7-18

(1) Reed and Campbell (1975).
(2) Ellis (1978).
(3) Robertson *et al.* (1978.
(4) Olafsson (1977).

Liquid Effluents

In steam-dominated systems (e.g. at The Geysers), the steam after passing through the turbines is condensed by contact with cooling water. Cooling water consists normally of steam condensate which has passed through an open cooling tower cycle. Since the amounts of steam condensate are much higher than the needs for cooling, a substantial fraction has to be disposed of as liquid effluent. A 100 MWe power station at The Geysers produces about 3800 m³/day of waste water, the composition of which is given in Table 87.

Table 87. Composition of Condensate return from a 100 MWe Geothermal
Plant (The Geysers)*

	tonne/y
Nitrate	0.15
Magnesium	1.5
Sulphate	197.0
Chloride	5.3
Silica	5.5
Calcium	8.0
Sulphur (free)	13.3
Boron	25.9
Ammonia	223.5
Organics	311.9
Alkalinity (HCO_3)	648.0

* After EPA (1977).

In liquid-dominated systems the mixed steam/water discharge from a well passes at a controlled pressure into cyclone separators. Gases ex-solved from the original deep high-temperature water concentrate mainly in the steam phase (CO_2 + H_2S + minor gases), while the water becomes slightly concentrated through steam loss, and rises in pH through the loss of acidic gases. The separated water phase is discharged through silencers (open towers acting as atmospheric pressure separators), emerging as an effluent at about 100^0 (current practice at Wairakei). Alternatively, the separated water, still under pressure, may be disposed of into reinjection wells (as practiced at Ahuachapan and several Japanese stations) or under the surface of a disposal pond (as at Cerro Prieto). The water mass flows are considerable (approximately twice to 10 times the steam flow). The overall efficiency in converting the thermal energy of well discharges to electricity is about 7-10%, lower than for the steam-dominated areas (about 15%). This neglects the possible secondary utilization of separated hot water for heating purposes, or the return to the aquifer of the heat contained in the reinjected water.

A hot water geothermal electric station therefore may have four major effluent streams; hot saline water, surplus condensate retaining some dissolved gases, a waste gas flow, and impure vapour rising from cooling towers. The chemical constituents present in the deep water are distributed finally between each flow in a manner related to their relative solubilities and fractionation kinetics.

The quality of the geothermal water — its physical and chemical characteristics — varies widely. While some geothermal hot waters contain relatively few pollutants, most contain a relatively large amount of dissolved solids and heavy metals because the high temperatures of the brines increase the dissolution rate of solids and heavy metals in the rock. Table 88 gives the

Table 88. Compositions of Some High-temperature Geothermal Waters (concentrations in ppm in waters collected at atmospheric pressure

Source	Approx. Temp. (°C)	pH	Li	Na	K	Cs	Mg	Ca	F	Cl	Br	I	SO₄	HCO₃	B	SiO₂	H₂S	NH₃	P	Ref.
Well 24, Wairakei, N.Z.	250	8.3	13.2	1250	210	2.5	0.04	12	8.4	2210	5.5	0.3	28	24	29	670	1	0.2	0.01	(1)
Well 8, Broadlands, N.Z.	260	7.9	11	975	232	1.6	0.1	3	8.4	1858	6.3	0.6	3.5	157	53	796	2	2.3	—	(1)
Well 9, Otake, Japan	215	8.15	5.2	936	131	0.8	0.19	12	4.7	1474	3.4	0.26	136	47	20	310	—	0.06	0.3	(2)
Well 15A, Mexicali, Mexico	250	—	15	6000	1125	—	—	321	2.2	11500	—	—	0	43	9	678	—	22	—	(3)
Well 7, El Tatio, Chile	255	7.3	48	5000	840	18	0.1	203	2.5	9100	—	—	29	44	210	810	—	3.1	—	(1)
Well 1A, Kizildere, Turkey	200	9.0	4.5	1280	135	0.3	0.1	3.0	24	117	1.2	—	770	1350	26	263	—	2.5	—	(1)
Well 1, Ngawha, N.Z.	230	7.4	12	950	77	0.4	—	28	0.85	1625	—	—	17	86	1200	460	—	46	—	(1)
Well 8, Reykjanjanes, Iceland	290	7.1	6.6	12730	1990	—	9.8	2249	0.14	25054	87	0.27	2.4	—	11	943	—	2.0	—	(4)
Well 205, Matsao, Taiwan	245	2.4	26	5490	900	9.6	131	1470	7	13400	—	—	350	—	106	639	—	36	—	(1)
Well IID-1, Salton Sea, California*	340	4.7	215	50400	17500	14	54	28000	15	155000	120	18	5	7100†	390	400	16	386	—	(5)

References: (1) Ellis (1978); (2) Koga (1970); (3) Mercado (1976 B); (4) Olafsson and Riley (1978); (5) Muffler and White (1969).

* Concentration and pH in deep aquifer.
† CO₂ concentrations.

Table 89. Trace Elements (μg/kg) Concentrations in Geothermal Well Waters

	Fe	Mn	Cu	Pb	Zn	Cd	Tl	As	Sb	Ref.
Wairakei, N.Z.	12	0.7	2	4	2	0.5	—	4700	100	(1)
Broadlands, N.Z.	360	13	1	4	1	0.02	7	3300	200	(1)
Mexicali, Mex.	200	640	5	5	6	—	—	2000	400	(2)
Matsao, Taiwan*	148000	28000	35	500	8800	—	—	—	—	(1)
Salton Sea, Cal.	2.3×10^6	1.4×10^6	8000	1×10^5	5×10^5	2000	1500	12000	400	(3)
Reykjanes, Iceland	—	2600	14	—	92	1.1	—	146	0.6	(4)

References: (1) Ellis (1978)
(2) Mercado (1967)
(3) White (1968)
(4) Olafsson and Riley (1978).

* A very low pH water (pH 2.9).

Table 90. Hot Waters Representative of Types Used for Community or Agricultural Heating (Concentrations in ppm)

Source	Temp. (°C)	pH	Li	Na	K	Ca	Mg	Fe	F	Cl	Br	I	SO_4	HCO_3	B	SiO_2	H_2S	NH_3	P	NO_3	Ref
Rotorua, Well 231	110	8.9	1.8	345	17	1.5	0.5	—	6.9	343	1.0	0.5	22	194	5.1	274	37	0.2	—	—	(1)
Boise, Idaho	75	—	—	75	1.3	2	—	—	24	9	—	—	23	141	—	78	—	—	0.01	0.08	(2)
Well 4, Szentes, Hungary	85	—	—	378	18	7.8	—	1.4	2.8	13	—	0.03	13	1049	0.03	55	—	6.2	—	—	(3)
Spring, Carlsbad, Czechoslovakia	73	7.65	3.3	1718	104	102	47	0.1	2.4	617	1.4	—	1662	2100	1.0	68	—	0.1	—	—	(4)
Laugarás, Iceland	98	8.8	—	79	2.1	4.1	0.01	—	2.1	56	—	—	62	18	0.32	127	0.7	—	—	—	(5)
Well 1, Tigre, Lagoon Louisiana (3930 m)	—	5.4	—	43000	290	1660	161	10	0.95	67100	65	25	200	645	60	63	1	90	—	—	(6)
Well, Kings County, California	100	5.7	15	13600	404	12200	275	0.1	1.6	44000	238	56	16	80	50	63	0.6	127	0.1	0	(7)

References: (1) Glover (1967); (2) Kunze, et al. (1975); (3) Boldizar (1975); (4) Ovchinnikov (1955); (5) Arnorsson (1974 B); (6) Hankins et al. (1978); (7) White (1965).

analyses of waters in some high-temperature areas associated with power development. The most common type of geothermal water is a sodium, potassium chloride solution containing appreciable concentrations of silica, boron, sulphur, ammonia, fluoride, and various trace metals (see also Table 89). Although most geothermal waters have common features, particular areas may be associated with high concentrations of specific elements. As examples, the El Tatio field in Chile has waters containing up to 45 mg/kg arsenic and Ngawha, New Zealand contains up to 1000 mg/kg boron. It should also be noted that medium- and low-temperature geothermal waters used for non-electrical purposes vary widely in composition depending upon the geological setting of the reservoir and its geochemistry (see Table 90). In this case, the possible pollution effects of each "waste" water need to be assessed individually.

Table 91 gives the outflow of liquid effluents from some hot-water geothermal power stations.

Table 91. Outflow of Liquid Effluents per 100 MW(e)y from Three Hot-water Geothermal Sites (tonne/y)

Constituent	Cerro Prieto	Wairakei	Broadlands
Li	320	350	300
Na	150000	35000	30000
K	40000	5500	5000
Ca	10000	500	80
F	40	200	200
Cl	300000	60000	50000
Br	500	150	150
SO_4	170	700	200
NH_4	800	50	500
B	400	750	1100
SiO_2	40000	20000	20000
As	30	100	100
Hg	—	0.025	0.035

Based on the variability in the amount and type of dissolved solids in geothermal fluids, a number of different methods for disposing of wastewater have been tested and used. These include direct release to surface water bodies, evaporation, surface spreading to shallow aquifers, desalination with subsequent water re-use, and reinjection to the production reservoir. A common problem is that of dissolved silica. Silica in the waste water commonly precipitates as amorphous silica in the waste water channels in which case the channels must be cleaned at regular intervals. The selection of a disposal method depends on local hydrologic conditions, water requirements, the quality of the waste water and environmental regulations.

For most power units in water-dominated geothermal fields, direct discharge of waste water is unlikely to be satisfactory. For example, there is concern in New Zealand about the long-term effects on the Waikato River from the waste waters discharged from the Wairakei plants. Coulter (1978) noted that nitrogen is the factor limiting phytoplankton growth in the Waikato River in summer, and that further input from the Broadlands station would be undesirable. Weissberg and Zobel (1973) showed that mercury concentrations varied up to about 1 mg/kg in bottom sediments of the Waikato River following the entry of effluent waters from several geothermal areas. Fish in parts of the river had average mercury concentrations ranging up to about 0.8 mg/kg, i.e. above the accepted 0.5 mg/kg level for human consumption. Arsenic discharge from the wells in the reduced trivalent form is rapidly oxidized to the pentavalent form when the waste waters are diluted by river water. Although some arsenic is taken up by weed growth in waterways (Reay, 1972), its behaviour downstream is predominantly that of a soluble element. Further studies on the transport mechanism, fate and effects of chemical elements in waste waters directly discharged to surface waters in other geothermal fields are needed.

Ponding of water effluents to evaporate and to allow the possible recovery of dissolved salts, presents interesting possibilities in dry climates. Ponding also lowers the possibility of thermal pollution of receiving waters. At Cerro Prieto, an 8 km^2 pond allows the concentration of the 2000 tonne/hour water flow by a factor of 2-3, with some precipitation of silica and heavy metals, before the effluent enters the Rio Hardy (Mercado, 1977). The recovery of salts is also being examined. Fernelius (1975) described the operation of a desalination plant with geothermal brines at East Mesa, California. In arid areas where irrigation and drinking water is valuable an integrated desalination, a salt recovery/water supply industry may be usefully associated with geothermal power production. Such a scheme was proposed for the El Tatio, Chile, geothermal area.

Moderate temperature waters (up to 130^{0}C) used for various domestic, commercial and agricultural heating purposes do not present unique problems and should be considered as any other waste water whose environmental significance is related to its salinity and chemical compositions. However, large volumes of water are involved in heating schemes, e.g. a 1 MWt heating system based on an 85^{0}C water source requires a water flow of about 18 tonnes/hour. As an example, in Iceland where a community of 110,000 about Reykjavik receives heat from wells tapping a dilute water at about 100^0, there have been no disposal problems from 9000 tonnes/hour of water at 40^{0}C being discharged to waste. The waters are little different in composition from the local surface waters (Arnorsson, 1975). In arid areas, dilute thermal waters may be valuable in supplementing local water supplies. Hydrogen sulphide is often at a low concentration level and can be removed by oxidation.

Reinjection is the favoured means of waste water disposal for most geothermal development. However, it cannot be described yet as well-established technology, except for the simple case of steam condensate reinjection where relatively small volumes of water are involved, and the solutions are dilute. In an area where water supplies are drawn from underground aquifers it is important to have a knowledge of the local hydrology and to carefully monitor any effects arising from the injected geothermal water. In high-temperature water areas it is likely that emergency discharge ponds would need to be provided in case reinjection wells ceased to operate temporarily. For example, at Broadlands, New Zealand, a 40 hectare pond 5 m deep is planned to store 30 days discharge from the proposed station (NZED, 1977). Reinjection has been practiced satisfactorily to date at Ahuachapan, The Geysers, Larderello-Travale, but well scaling and thermal degradation problems have arisen at Otake and Hatchobaru, Japan.

Thermal Discharges

Geothermal power plants have a low thermal efficiency; for dry steam operated plant the efficiency is about 15% whereas for liquid-dominated geothermal field the efficiency is about 8-10%. In other words about 85 to 92% of the total heat energy contained in the geothermal fluid is emitted as waste heat. The considerable outflow of water vapour from the silencers, condensate trap valves, hot-water drains, and from the cooling towers of a power station, may in some climates create a tendency for local fog formation and icing of roads in winter. An average 110 MWe Geysers station cooling tower emits 658 tonnes of water vapour per hour (Reed and Campbell, 1975). The problem is greater for stations with mechanical draught cooling towers than for natural draught cooling towers which have a greater plume rise. The 500 MWe installation at The Geysers could increase the moisture content in the air of the local basin by 50% (Swanberg, 1975).

Water-dominated fields have an appreciable potential for thermal pollution. In future developments this need not be the case since reinjection is becoming common, and the currently-operating stations mark only stages in the development of a new technology. For example, the low efficiency of the Wairakei station is partly due to an early concept of combining power generation and heavy water production. Axtmann (1975 A) reviewed the extreme case of the Wairakei station where once-through river water cooling is used and separated hot water also enters the Waikato River. Of the 1720 MW of energy piped to the surface, some 850 MW finally enter the river through condensate (580 MW) and waste hot water (270 MW). At the average river flow this raises the river water temperature by about 1.5^0C after mixing, but at very low flows temperature rises of several degrees can occur and there is evidence of the fish life being affected (Coulter, 1978). In few situations will river water cooling be permissible, as temperature rises of a few degrees are disastrous for many kinds of fish. For example, trout usually cannot survive above about 23.5^0 and the optimum conditions are 13-16^0. Their eggs do not hatch above 14.4 (McKee and Wolf, 1971). In hot water fields other than Wairakei, stations use a cooling tower/condenser cycle. Combined with waste hot water reinjection (or ponding, as practiced at Cerro Prieto) hazards to local waterways from thermal pollution can be minimized. Far better would be the utilization of separated hot water (115-160^0) for community or industrial heating, before reinjection.

CHAPTER 18

Solar Energy

SOLAR radiation is an inexhaustible source of energy. The solar flux at the outer fringes of the earth's atmosphere is 1.35 kW/m², a value known as the solar constant. As the sun's radiation passes through the earth's atmosphere the energy is depleted by absorption, scattering and reflection. This is due to the air molecules themselves, dust, and the naturally occurring gases such as ozone (which removes much of the ultraviolet), water vapour (which absorbs strongly bands in the near-infrared region) and carbon dioxide (which absorbs strongly bands in the middle-infrared). By the time it has reached the earth's surface, the direct-beam solar radiation has been reduced in magnitude. The annual amount of solar radiation received at the surface of the earth is about 1.2×10^{17}W (or about 1×10^{18} kWh), which is equivalent to more than 20,000 times the present annual consumption of energy of the whole world (UNEP, 1979). However, the extraction of all the energy provided by sun is nearly impossible and only a fraction of this energy can be extracted; the efficiency of extraction depends on the location and the prevailing meteorological conditions.

For the design of systems to utilize solar energy, the most useful information available is the energy received on a horizontal surface (insolation) at the particular location. This is measured hourly at a large number of meteorological stations throughout the world and the data stored both as published records and magnetic tapes for use in computer programmes. Using these data, it is possible to calculate the insolation on collectors mounted at any angle and oriented in any direction, and the information obtained used to determine the output of collectors used, e.g., in heat generating systems either month by month or annually (Morse, 1977). The daily average figures for insolation vary from about 9 MJ/m² for Antarctica to about 25 MJ/m² for the Australian North West Coast, Saudi Arabia or Peru. Although it may be said that our knowledge of the character of solar radiation at the earth is satisfactory and commensurate with the state of the art of solar energy technology, it should be noted that an extension of organized solar radiation monitoring together with an improvement in data quality control is desirable. Some international organizations (e.g. the World Meteorological Organization) and regional ones are undertaking such activities.

The energy from the sun can be directly harnessed by developing passive and active solar systems, and it is in this respect that the term "solar energy" is used in the strictest sense. However, solar radiation contributes through the interaction with natural processes to the development of secondary sources of energy that can be harnessed by suitable technologies (see Fig. 10). In the present study, "solar energy" is used in its broadest sense to encompass all the energy sources that are derived from solar radiation.

Solar energy has been used in many areas for centuries, but mainly in a primitive way (for example, for crop drying in rural areas). Several solar devices for heating water, green houses, etc. have been developed and although at present the use of solar energy is still limited, extensive research and development programmes are under way in many countries to harness solar radiation efficiently for a broad number of applications. These range from heating and cooling of buildings, water heating, desalination, refrigeration, solar drying, telecommunications, irrigation, electricity generation to ovens for high-temperature materials processing*.

PASSIVE SOLAR SYSTEMS

A passive solar energy system is one in which the thermal energy flow is by natural means, that is by radiation, conduction, or natural convection. Passive solar offers a simple, economical, comfortable and reliable means of building heating (for example the Trombe-Michel Solar houses, Trombe,1977). By careful consideration of building orientation, treatment of glazing, storage of excess heat in building mass, shading, and ventilation, a building can be designed to be quite climatically adaptive, absorbing solar radiation during a winter day and storing heat through to the night and yet rejecting summer sun. Passive cooling effects can also be achieved by control of radiation losses, evaporation, ventilation, and especially the use of building mass heat capacity to average diurnal temperature swings thus reducing afternoon overheating (Bahadori, 1978; Balcomb, 1979).

ACTIVE SOLAR SYSTEMS

Any surface becomes heated when exposed to sunlight because solar radiation is absorbed and transformed into heat. If a heat-transfer medium – air, water or a chemical fluid – is allowed to flow over this surface, it will extract heat from sunlight. Several solar collectors have been designed; these are generally classified into: non-concentrating solar collectors (flat-plate collectors) and concentrating solar collectors. The former are further classified as low temperature ($40-60^0$C), medium temperature ($60-100^0$C) and high temperature (above 100^0C)**

Solar Water Heating

The most immediate use of solar energy is for water heating. The systems most commonly used today are solar water heaters for residential and commercial purposes. Several countries have used solar water heaters for many years. Japan had about two million units installed; in Australia, the number of solar water heaters installed now is in the order of 70,000 (Barnes, 1979). Solar

---

* The literature on the different uses of solar energy is quite extensive and it is not the aim of this study to review such uses in detail. (For further information, see: Palz, 1978; UNIDO, 1978; International Solar Energy Congress, 1978, 1979; EEC, 1979; Sørensen, 1979).

** For a detailed description of different solar collectors see, for example, Palz (1978); Cooper (1979) and Giutronich (1979).

Fig. 10 Energy from the sun.

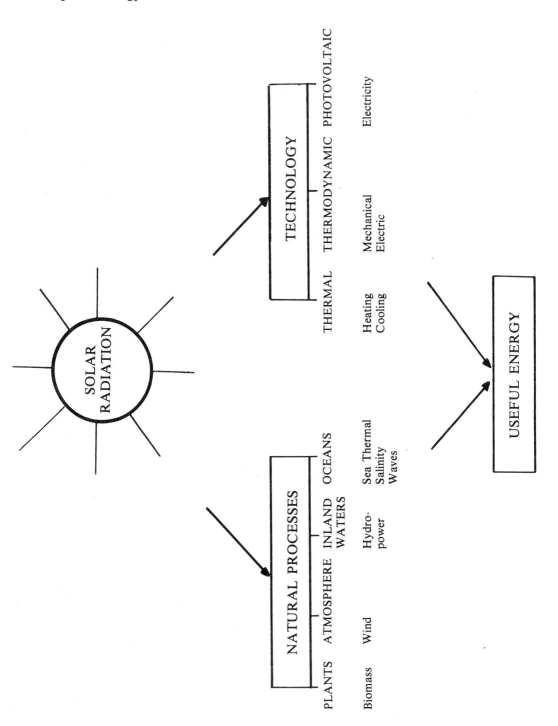

water heaters are being produced in many other countries, for example, in the U.S.A., UK, France, Federal Republic of Germany, Israel, Niger, Mali, Senegal, etc.; in several other countries governments are sponsoring active research and development of solar water heating. Solar swimming pool heating has been developed in some countries, for example, in Australia (Morse, 1977, Barnes, 1979) and in the U.S.A. Considerable potential exists for the utilization of solar energy as an alternative major resource in industrial heating generating systems (Morse, 1977, Morse et al. 1977; Read, 1978, 1979; Barnes, 1979; Sargent et al. 1980). Field tests to demonstrate the technical feasibility of solar industrial process heat systems have been undertaken in some countries, for example in Australia and the U.S.A. These tests aim at determining system performance, reliability, maintainability and economics. In practice, the use of solar energy is considered most cost-effective at low or intermediate temperature (Sargent et al. 1980).

Solar Drying

Of all the direct uses of solar energy, solar crop drying is perhaps the most ancient and widespread. The customary technique involves spreading the material to be dried in a thin layer on the ground to expose it to sun and wind. Copra, grain, hay, and fruits, and vegetables are still being dried in this manner in many parts of the world.

In recent years innovations have been adopted, particularly for fruit drying, in which fruit is placed in carefully designed racks to provide controlled exposure to solar radiation and wind and to improve material handling. Solar drying is of special interest in the case of soft fruits, meat and fish which are particularly vulnerable to attack by insects, as the sugar concentration increases during drying. For dried fruits and vegetables, sun drying is the cheapest and simplest way to dry crops in regions having abundant sunshine and where the post-harvest season is characterized by low relative humidity and little or no rainfall. Although there is no significant commercial manufacture of solar crop dryers, a number of experimental designs are now in use (see, for example, UNIDO, 1978, Keener et al., 1978). These range from the use of solar-heated air in more or less conventional air dryers to a combination of direct drying and air drying by placing the materials to be dried in flat-plate collector-dryers. Among the former are various designs developed in the United States, Turkey, Canada, Brazil, and Australia. Combination collector-dryers have been designed and used successfully in India and Trinidad. Development of solar drying can conceivably benefit from further development of collector-dryer combinations and flat-plate air heaters and energy-storage systems to supply hot air to dryers. Research in the design and control of these processes, for particular crops or other materials to be dried, could lead to other practical applications that could result in improved utilization of food supplies in developing countries (NAS, 1976).

Space Heating and Cooling

The technology of solar heating where, for example, water is the medium for transfer of heat is essentially an extension of the technology employed in solar water heating, except that heat has to be recovered from the tank through a hot water pump. Collectors and storage units much larger than those employed in solar water heaters are necessary to provide a substantial portion of the heat requirements. Systems employing air as the heat-transfer medium between the collector and a storage bin containing small rocks have also been used successfully. Solar heat is stored as heat in the rocks, and is recovered when needed by passage of air over the rocks and thence to the rooms. Several experimental solar houses have been built and operated with the heating system comprising the collector, a heat-storage unit, an auxiliary heater, and appropriate heat-distribution and control systems. There has been much diversity in concept and design and some of the systems are associated with heat pumps. An extensive review of the possible methods of solar cooling for air conditioning is given by Duffie and Beckman (1974). The most intensively studied system to date is the lithium bromide/water absorption process, in which the generator receives hot water from the solar system at temperatures in the order of 80–90^{0}C. The development of collectors of reasonable cost that can operate efficiently at temperatures well above the boiling point of water will allow consideration of the well-known ammonia/water system.

Solar Refrigeration

Closely related to air conditioning, solar refrigeration is generally intended for food preservation or for storage of biological and medical materials. There have been experiments in several countries including the U.S.A., Sri Lanka, France, and the U.S.S.R. on solar-operated coolers, using absorption cooling cycles. Most of these are aimed at household-scale food coolers or small-scale ice manufacture.

Thermal Storage

Storage is needed in all solar heating systems to avoid wasting the heat generated by the collector when this exceeds the instantaneous system load. With current technology, it is not practicable to store more than a few days input to a system, but this is sufficient to enable the solar input to be matched to the load for periods of about a month when the average weekly load is approximately constant (such as domestic hot water services). Small hot water systems use insulated copper or glass-lined steel tanks, and recent technologies using large systems for seasonal storage are now being developed (Torrenti *et al.*, 1976).

Copper, concrete, or stainless steel has also been used. Rock storage has been used experimentally with air heaters, and although satisfactory is somewhat bulky. Where the product itself can be stored, e.g. hot water, it is merely a question of providing an adequately sized, insulated storage tank. Off-peak electrically-heated hot water services have for many years used this principle, and such systems are readily available commercially. It is a simple matter, therefore, to incorporate such a storage into a solar water heating system, it being only necessary to increase the size of the tank to provide one and a half to two days' supply and perhaps increase the amount of insulation provided. For industrial process heating and other applications where heat rather than hot water is required, storage presents a greater problem in that the tanks become rather bulky when only the sensible heat of water changing in temperature by a few degrees centigrade is available. Nevertheless, large insulated tanks are being used satisfactorily for this purpose (Morse *et al.,* 1977).

Salt-gradient ponds (solar ponds) have been proposed for collecting and storing thermal energy from the sun for low-temperature applications, such as space heating, water heating, etc. (Tabor, 1963, Tabor and Matz 1965; Nielsen, 1975). In such ponds, the non-convective character of the dense highly saline water at the bottom leads to storage of high temperatures (as high as $95^{0}C$) and the heat can be exploited by suitable heat exchangers installed at adequate depth. Solar ponds have been constructed in some countries, and extensive experimental work is underway, for example, in Israel, Chile, U.S.S.R., U.S.A., and other countries (Nielsen, 1975), to solve the different technological and economical problems of extracting thermal energy from such ponds (see also, Mehta, 1979).

Solar Distillation and Desalination

The use of solar energy for desalting sea-water and brackish water has been demonstrated in several moderate-sized pilot plants in the U.S.A., Chile, Brazil, the U.S.S.R., and several other countries. This century-old process consists of a shallow pool of brine from which water, slowly evaporated, condenses on the underside of the cooler glass covers and runs to troughs at the lower edges to storage. Excess brine that has not evaporated is run to waste as salt water is supplied to the basin. The idea was first applied in 1892 at Las Salinas, Chile, in a plant supplying drinking water for animals working in nitrate mining and transport. The Las Salinas plant reportedly operated for 30 years (NAS, 1976). Modern developments in solar distillation have been directed to the use of materials and designs for economical and durable constructions, to reduce the distilled water cost. Among recent developments is water desalination by reverse osmosis which makes use of thermodynamic solar energy converters (Maurel, 1979). New solar distillation processes utilizing a filtration glazing, distillation during daylight and recirculation at night, have been developed (Balligand, 1979).

Solar distillation requires relatively large capital investment per unit of capacity but, in properly designed and constructed systems, a minimum of operating and maintenance costs. Product-water costs depend primarily upon still productivity service life, capital cost of the installation, and amortization and interest rates. Productivity of a solar still is dependent on the intensity and duration of the sunlight it receives. Experience shows that a still will yield about 1300 $1/m^2$ annually, with some variations dependent on climate and design. A typical lifetime for a still constructed of concrete, glass, and other durable materials is 20 years or more, other still designs involve less durable materials that must be renewed periodically. The lack of more general use of solar stills is almost entirely the consequence of high capital investment required and the resulting high cost of water produced. There seems to be little prospect that large solar distillers will be competitive with large desalting plants supplied with conventional energy sources unless fuel prices escalate greatly. However, in situations where a community or an industry requires small quantities of water – say less than 200,000 litres per day – the solar still may be more economical than conventional desalting plants. This is particularly true: (a) in small communities where potable water is unobtainable except at very high cost; (b) in certain industrial and commercial applications where materials must be processed in a region where all available water is brackish; or (c) for watering of livestock in areas where grazing is possible if water is supplied. It is for these moderate-volume requirements that solar stills have been built.

Solar Cookers

Different types of solar cookers have been described since the beginning of this century. Simple solar cookers may be classified as (a) concentrating parabolic and spherical dish or trough collectors where the heat at the focus of the collector directly heats either a vessel containing the food or the food itself and (b) ovens or food warmers which are insulated boxes with transparent covers in which solar energy is collected by direct radiation or by radiation from some type of reflective surface. A solar cooker with parabolic mirror was developed and marketed in India as early as 1950. Although it was sold for a low price, it did not attract much interest. Few housewives could become accustomed to the "new" cooking process which was not something done by generations of ancestors. Also the cloudy weather (which reduces the efficiency of cooking) and the need to warm food at times when there is no sun, added to the frustration of the housewives (see Walton et al., 1978 for evaluation of solar cookers). The principal requirements for the successful use of a solar cooker in rural areas in developing countries may be summarized as follows: (a) The unit should cook food effectively; it should therefore provide energy at a sufficient rate and temperature to properly cook desired quantities and type of food. (b) It should be sturdy enough to withstand rough handling, wind and other hazards. (c) It should be sociologically acceptable and fit in with the cooking and eating habits of the people, i.e. provide for cooking to be done in the shade and if possible at times when the sun is not shining. (d) It should be capable of manufacture with local materials and by local labour. (e) It should be possible for the user to obtain a cooking unit at a sufficiently low cost. In order to overcome the problems of cooking in the direct sunshine and of cooking only when the sun is shining, two advanced solar cooking devices have been proposed. The first involves the use of a heat transfer system to permit cooking to be done inside the house. The second involves the use of some type of energy storage system which would permit the cooking to be done in the evening or at other times when the sun was not shining. Such types of cookers are considerably more expensive than the simple open-air cookers.

MECHANICAL ENERGY AND ELECTRICITY FROM SOLAR ENERGY

The term "solar engine" designates an engine operated by solar energy. The thermodynamic cycle of such an engine may be as follows: Vapour is obtained when a liquid working fluid is heated by solar radiation. This vapour expands in a reciprocating or rotating engine, doing work. From the engine it flows to a heat exchanger, in which it condenses. The condensate is reinjected by a pump (usually operated by the solar engine itself) to another heat exchanger, in which it evaporates, closing the cycle. "Low temperature" solar engines make use of flat-plate collectors and work with temperatures below 100°C, while "medium" and "high" temperature machines work with temperatures above 100° C and in this case focussing solar collectors, which track the sun are used. The most extensive application of "solar engines" is for water pumping. Several solar irrigation systems have been developed in many countries; solar pumps have been installed in Senegal, Upper Volta, Mauritania, Niger, Mali, Sudan, Chad, Cameroon, Mexico, U.S.A., Australia and other countries. The first generation of these pumps are rated at 1 kW; later generations are rated at up to 100 kW (e.g. in Mexico, 25 kW, 900 m^3/day in 1976; in Mali, 75 kW, 9000 m^3/day in 1979), and such pumps are commercially available. The present trend in solar irrigation systems is to utilize collectors with selective surfaces or concentrators in order to increase the temperature of the working fluid and thus increase the overall efficiency of the system (Phéline, 1979).

Solar Thermal Electric Conversion

Solar-thermal-electric conversion systems (STECS) collect solar radiation, convert it to thermal energy first and then to mechanical energy via a thermodynamic cycle to drive an electro-mechanical energy conversion device to generate electrical energy. STECS have the potential to convert up to 25% of the incident solar radiation into electricity. There are two broad categories of

STECS: (a) distributed collector systems and (b) central receiver system. In distributed collector central generation systems, solar energy is collected throughout the collector field and the heat is transported to a central energy conversion plant via a pumped fluid or chemical through a piping network. In another approach, electricity is generated at each collector and brought to a central point for transmission. This concept is known as distributed collection with distributed generation. In the central receiver system, a large field of steered mirrors (called heliostats) reflect solar radiation to a single receiver mounted on a central tower. The solar collector field, in effect, is a large parabolic steerable reflector of focal length equal to the tower height (typically 200 to 600 m). With a proper design for the receiver, high-temperature steam is generated which is then used in a steam turbine driving a generator to produce electrical energy.

An active development programme on STECS is being pursued in the U.S.A., with plans for the construction of units varying in size from 5 MWe to a 100-MWe unit to be constructed in the mid-1980s. Similar development plans for STEC systems in France, Italy and other EEC countries are underway (Etievant, 1979; Phéline, 1979). A typical 100-MWe STEC unit might be envisioned as consisting of 12,500 heliostats, each having a reflecting surface of approximately 40 m^2, with a central receiver tower 250 m in height, supporting an absorber to provide steam or hot gas to a turbine for periods ranging from six to eight hours per day. Present designs incorporate either conventional fuel storage for operation of the plant in a hybrid mode or thermal storage sufficient to extend operation to the intermediate power-generating mode. Overall efficiencies are expected to be in the range of 15-20%. Reliability of energy supply by means of STEC systems will be a function, in part, of the siting strategy chosen and the nature of the total utility system with which STEC units would be interconnected. By the end of the present decade, there should be sufficient experience with operating units in the range of 50-100 MWe so that a meaningful estimate of STEC's economic attractiveness can be made (Auer et al., 1978). It should be noted that STEC's are most suited for areas with high solar radiation and low cloudiness such as those located in many developing countries in arid and semi-arid zones.

Photovoltaic Conversion

A second way of using sunlight for electricity production is by photovoltaic conversion (Durand, 1975; Rodot, 1977; Palz, 1978). Photovoltaic cells, containing a thin slice of crystalline silicon or a thin film of cadmium sulphide protected from the weather by a suitable transparent material, generate direct electric current when light shines on them. The stronger the light, the more electricity is generated. With present technology, silicon cells are more efficient than those using cadmium sulphide — up to 15% — but are more expensive. Solar cells are generally rated in terms of peak power — how much electrical wattage they produce in clear day, normal incidence sunlight. The energy can be stored in batteries and used as needed. If the power is required evenly around the clock, about 5 to 10 peak watts of capacity will be required for every average watt desired. Photovoltaic cells are expensive at present and research and development are underway to increase the efficiency of the cells and reduce their costs. In 1976, the cost per peak kilowatt was about 15,000 US$; in 1978 it was about 11,000 US$, this year the cost dropped to about 5,000 US$/kW and is expected to be less than about 500 US$ in 1985 (Auer *et al.,* 1978). This corresponds to about 60$ per m^2 photovoltaic array at 12% efficiency. The potential advantages of photovoltaic conversion are impressive. There need be no moving parts; lifetimes can (in principle) exceed 100 years, although cell performance may be expected to degrade continuously over its operating lifetime; maintenance involves little skill; both direct-beam and diffuse solar radiation can be utilized effectively; the system is inherently modular and readily lends itself to the design of virtually any system size, small or large.

At present, photovoltaic solar converters are restricted to few kW range, in which their use is developing rapidly. In remote or developing rural zones, extensive applications are for refrigeration (preservation of vaccines, food, etc.) water pumping and electric lighting. Some typical examples are: solar powered navigation lights around the airport of Medina, Saudi Arabia, navigational lighthouse in Indonesia, solar panels for battery charging, solar water pumps, educational televisions in Niger (Palz, 1978; NAS, 1976), telecommunications (Holderness, 1978; Phéline, 1979).

Energy Storage

Since energy storage is an essential component of most of the renewable energy systems presently under development (solar, wind, sea thermal, etc.,), it is essential that storage mechanisms be developed which will operate economically at the multi-megawatt level if such energy sources are to have a significant impact on the power systems of the future. Thermal storage systems have been dealt with on previous pages. Pumped storage in case of hydropower is dealt with in chapter 21. Energy can also be stored in chemical form in secondary batteries, the most well known of which is the lead-acid battery. In their present state of technology, lead-acid batteries cannot satisfy the requirements of either electric cars or storage of off-peak electrical energy of utilities. Other commonly available batteries such as nickel-cadmium and silver-zinc batteries are too expensive to be used on a large scale and are severely limited by material availabilities. Advanced battery systems under development (e.g. lithium-sulphur, sodium-sulphur, aqueous zinc-chlorine, etc.) are predicted to have round-trip efficiencies of 70 to 85%. Hydrogen holds promise of being a versatile and efficient energy carrier. Because of its flexibility, hydrogen is considered as the fuel of the future and hydrogen energy storage systems are being increasingly recognized and developed for a variety of applications. Electrical energy from a variety of sources (off peak utility power, electricity generated by solar cells and wind-electric systems are some of the commonly suggested inputs) is used to dissociate water into hydrogen and oxygen gases. The hydrogen can then be stored as compressed gas or as metal hydride for use when required.

ENVIRONMENTAL ASPECTS OF SOLAR ENERGY SYSTEMS

Solar energy devices (whether for electricity generation or for non-electric applications) have a number of environmental impacts. Decentralized small units (e.g. water heaters, solar dryers, cookers, photovoltaic-operated equipment, solar refrigeration, space heating and cooling etc.) do not only reduce the demand for fossil fuels leading to conservation of such non-renewable energy sources, but also would lead to the reduction of the bulk of pollutants emitted by burning such fuels. In industrialized countries when, for example, space heating can take up to 20% of the energy budget of the country, the use of solar heating can drastically reduce the dependence on oil products for such purposes. In the developing countries, the use of decentralized solar devices can lead to substantial improvements in the quality of life in remote areas. It can minimize the consumption of the generally subsidized kerosene, firewood and other non-commercial energy sources, the use of which in many cases has created unacceptable environmental impacts. It should be noted, that although several systems for solar energy utilization are now commercially available, there is still need to demonstrate their feasibility for wide applications. Several governments, regional and international bodies have recently embarked on different demonstration projects related to different applications of solar energy; such projects will contribute to our knowledge of the technology, economic, social and environmental aspects of these energy systems.

The use of solar energy for water heating for domestic or industrial purposes is environmentally benign. The same is true for space heating and cooling. No land is required since roof-top solar collectors are used, there are no emissions or wastes to be disposed of since the solar thermal systems are closed systems. Accidental leakages from the systems are not hazardous if water is used as the heat-transfer medium. If organic liquids are used, certain precautions should be taken to avoid the undesirable effects of these liquids in case of leakages.

Some studies (OTA, 1978, EPA, 1976) point out that orienting new buildings to maximize opportunities for the collection of solar radiation for space heating and cooling might create design problems and possibly increase the cost of housing in the community. However, orientation of buildings is not so important, especially if the system used for heating or cooling has an adequate storage component. The use of solar systems could influence residential and commercial architectural styles. The design of collector configurations as integral parts of roofs and the use of concentrating surfaces could radically change the shape of buildings. These effects would probably be mitigated, and the public become accustomed to the change.

Solar thermal power plants do not emit gaseous, liquid or solid effluents like fossil fuel or nuclear power plants. They are relatively "neutral" as far as excess heat rejection is concerned: that is, heat left at the site is similar to what would have been deposited if no plants were there. A central solar thermal plant is estimated to create an excess of about 0.25 MW (t) per 1MW (e) which is very low in comparison to 2.1 MW (t) per 1 MW (e) for a light water reactor or 1.7 MW (t) for a fossil fuel thermal plant (Caputo, 1977). The heliostat field associated with a solar thermal power plant could produce localized changes in the net albedo, energy balance, moisture-balance, low-level wind-flow patterns, and air/surface temperatures. The impact of such changes on microclimate within and immediately surrounding the solar thermal power plant remains to be adequately assessed (DOE, 1977; Williams, 1979).

It is commonly stated that a major draw-back of solar power plants is the extensive land requirement. However, the amount of land required for a solar plant is comparable to what is required by the system which provides electricity from fossil fuels or nuclear power. The area requirement for 1 MW(e)y of solar thermal power plant over its lifetime is about 2000 m^2; that of coal, for example, is about 3000 m^2 (Caputo, 1977). While the area requirements are comparable in the two cases, operation of the solar unit would actually involve a more benign use of the land. However, it is claimed (DOE, 1977) that heliostat fields might disrupt the ecology of the area, especially if located in desert areas. Although this might be the case in certain locations, the location of heliostat fields in some deserts (e.g. in North Africa) would have beneficial effects by creating centres of activities in such barren areas, attracting, for example, nomads to settle down near such centres.

Some attempts have been made to quantify the environmental impacts (with particular reference to health effects) associated with emissions resulting from the manufacture of solar equipment (Caputo, 1977; Inhaber, 1979). It should be noted, however, that these emissions are similar to those encountered in many industrial activities. The studies of Caputo, Inhaber and others aimed at comparing the health effects and risks from fossil fuels, nuclear energy and non-conventional sources of energy. However it should be pointed out that such comparative assessment is debatable and does not take into consideration all the environmental dimensions physical as well as socio-economical. Other attempts have been made to quantify the global impacts of an intensive use of solar energy in the future, for example on the global albedo of earth or on the climatic changes (Gilbert, 1974; Mitre, 1975; Peyches, 1977). These studies indicate the beneficial effects of a substitution of fossil fuels or nuclear energy by solar energy on the heat balance between the earth and the atmosphere.

The development and use of solar systems, either separately or in combination with other energy sources, provide opportunities not only for energy conservation but also for employment. This is particularly true in the developing countries where the manufacture of solar equipment can be labour-intensive. Yet, the wide scale use of solar energy depends on a large extent not on the technology but on the attitude of policy-makers, which is an important factor in promoting the public acceptance of solar energy as a part of the national energy "mix".

CHAPTER 19

Wind Energy

WIND energy is a renewable source of energy manifested by the power of the sun. The source of wind is in the atmospheric temperature differences generated by the sun, which, in turn give rise to pressure differentials. The wind is a mechanism for dissipating, as kinetic energy, the potential energy accumulated in those pressure differences.

Wind energy has been used for thousands of years to propel boats and ships and to provide mechanical power for grinding grain and processing other agricultural products. The significance of wind power peaked in the sixteenth century, and in 1850, windmills provided about 1×10^9 kW (McGowan and Heronemus, 1975). By the turn of the century there were more than 30,000 windmills in Denmark and more or less similar numbers in the Netherlands and other countries, used mainly for grinding and for pumping water.

Different authors have estimated the global wind energy resource base to be in the range of 1200 TW (Sørensen, 1979), but the harnessing of this power is very site dependent. Average wind velocity, at a height of 20–30 m above ground level should be high enough so that the energy flux (that is, the "power" in the wind) through a properly oriented vertical section would reach an annual average suitable for conversion. A wind machine located at a site where the annual average wind power is in the range of 500 W/m² (7 m/sec wind speed) can convert to electricity about 175 of the 500 W.

The energy flux contained in a moving air stream is proportional to the cube of the wind velocity. Not all the energy in the wind can, however, be extracted, even by an ideal machine. Theoretically, the maximum fraction of the power in the wind that is extractable is 59.3% (Merriam, 1978). In practice, the best power extraction efficiency reported for a real machine is about 0.50: and this is not achieved at all windspeeds, but only at the single windspeed for which the wind machine design has been optimized. Also, the conversion of mechanical to electrical energy takes place at less than 100% efficiency; typically 75–95%. All these factors together, the electrical power delivered by a real wind generator is perhaps 30–40% of the power in the wind when the wind generator is operating smoothly at some windspeed within the design range. However, windspeeds occur that are not within the design range. There is a windspeed below which the machine will not operate at all ("cut-in speed") and there maybe also one above which the machine is designed to feather, turn out of the wind or otherwise protect itself ("furling speed"). If the windspeed exceeds rated windspeed some of the extractable power in the wind is wasted in order not to exceed the electrical rating of the generator. Considering these factors the electrical output over a year's time might be 15–30% of the energy in the wind, or even less, depending on the site and the characteristics of the machine.

The technology of wind machines, and also their history, is exhaustively discussed in the literature (see for example, Burke and Meroney, 1975; Torrey, 1976; Simmons, 1975; Merriam, 1978; Noel, 1979). The physics and engineering of the machines is well understood. Testing and modification are in progress and mass production may be established in the near future. The very large wind generators to date have been prototypes. All have operated for limited periods.

Failure or shutdown has occurred; however, this could have probably been avoided with proper engineering work. There are no substantial technical problems that would limit large scale development of wind machines in the near future. The dominant technology in most national programmes today, is the two blade propeller turning about a horizontal axis. The vast majority of large wind-electric generators constructed in the past have been two blade or three blade propellers. Other technologies, such as the vertical axis Darrieus rotor, single blade horizontal axis rotors, and others, may offer advantages, but the advantages are not likely to be crucial for the development of wind energy.

In spite of the lack of extensive wind data and problems of site selection, it is possible to make gross worldwide estimates of possible wind contribution to the energy budget. The crucial factors that will determine whether or not wind energy contributes in a substantial way to human energy needs over the next half century are not technological in nature, and relate mainly to national energy policies, cost and public acceptability. Table 92 gives an estimation of the possible contribution of wind energy to world energy supplies until the year 2000.

Table 92. Estimated Possible Contribution of Wind to World Energy Supplies

Year	Installed capacity (kWh)	Energy output kWh/year
1985	5×10^5	1.5×10^9
1990	1×10^7	3×10^{10}
2000	2×10^8	9×10^{11}

After Merriam (1978). For comparison, the total electricity production in the world in 1976 was 7.6×10^{12} kWh (UN, 1979).

The potential wind power extraction along continental shore-lines is shown in Table 93.

Table 93. Estimate of Coastline Wind Potentials

Region	Assumed Coastline (10^6 m)	Est. coastline Wind potential (TWh per year)	Present electricity consumption (1978) (TWh per year)
North America	46	754	2400
South America	22	604	300
Oceania	20	780	150
USSR	11	494	1000
Asia (except USSR)	42	701	1000
Europe (except USSR)	24	1051	2000
Africa	27	534	200

Source: Sørensen, 1979; consumption figures have been rounded.

The most probable trend in the development of wind energy in the near future is most likely to focus on a much increased use of wind machines in the 5 to 100 kW range for water pumping and rural electrical systems with two day storage batteries. Machines for generating electric power, into existing utility grids, in the range of 100 kW to 5 MW are also being developed.

In addition to the classical applications of windpower a number of other future applications has been considered. These are briefly described in the following: (a) The production of fertilizer using windpower: Static electricity produced by a windmill system can be sparked in an airstream and the oxides of nitrogen thus produced are absorbed in water forming a nitric acid solution. Prototypes of this system have been investigated. Given the large demand for nitrogen-based fertilizers in the world, the possibility of developing initially small scale systems to produce fertilizers in remote areas, islands and mountainous regions, could reduce the transportation costs. (b) The generation of electricity through wind-power can be applied to existing electrolysis technology for the production of hydrogen and oxygen. (c) The use of windmills to compress air which can then be bubbled at the bottom of river beds and harbours to prevent the buildup of ice, has been discussed as a future possibility in cold regions. (d) The use of windpower to compress air which can then be used for the aeration of various bodies of water, particularly in warm regions and during summer periods, can be envisaged to overcome the problems of stagnation and depleted oxygen in bodies of water. This has an application in aerating fish ponds and other water basins where there are detrimental environmental effects. Several commercial firms are already marketing small wind operated air compressors feeding aeration systems placed in ponds. (e) Several investigations have been proposed and studies for the use of windpower to operate windpowered desalination systems for brackish or seawater conversion into fresh water. Generally the windpower would be used to provide mechanical shaft power to operate modified reverse osmosis or vapour compression desalination plants. (f) Given the fact that there are considerable windpower potentials off the coastline of a number of countries (see Table 93), the problem of generating electricity through the use of windmills located on off-shore platforms has been given some attention and will no doubt be considered more seriously in the future. The electricity that is generated can be transmitted through underwater cables and coupled with the national electric grid on shore. The technologies of platform construction, used in off-shore drilling activities in the petroleum field, as well as the transmission of electric power through underwater cables are already established.

ENVIRONMENTAL EFFECTS OF WIND ENERGY

Environmental concern about wind energy involves such factors as the risk of accidents, noise, interference with telecommunications and the possibility of local climatic alterations.

Since the ground area requirements are small and the land surrounding the windmills can be used with few restrictions, the space problem is mainly one of aesthetics. The use of wind turbines, of all various types, involves the erection of towers and the incorporation of some form of air screw which extracts the energy from the wind. This might constitute an unacceptable physical and visible effect on the environment. In some instances, however, the classical windmills of Holland, the sailmills of Portugal, Crete and Rhodes, the visible effects have been turned to advantage as the windmills constitute a tourist attraction. In certain areas of Latin America for example, there is a widespread, concentrated use of fanmill water pumpers. To date, the spacing between mills has been reasonable and the environment is not affected to any great extent from a visual, noise or physical viewpoint. This is even true of towns in these areas where the extensive use of fanmills is an accepted and necessary component of municipal services. Hence it is possible for the windmill to find acceptance within a specific community without causing undue side effects. The setting up of windmill arrays in mechanically cultivated fields can cause a certain amount of interference due to the presence of towers, guy wires, etc. This effect is reduced if the land is used for grazing purposes only. In the case of wooded areas the tree cover could seriously interfere with the operation of the windmills and may completely preclude their use altogether. All energy transmission systems leading from windmill arrays established in agricultural areas will require overhead installation to avoid interference with traditional agricultural operations. The installation of large windmill arrays (windmill parks) in areas normally inhabited by wildlife or along the traditional migratory paths of wild fowl, birds, could hinder or harm the wildlife.

The large scale generation of electricity through windmill arrays can have some modulating effects on telecommunications etc. Studies of TV interference at a single wind energy converter has indicated the need to place aerials at distances of 5–10 rotor diameters (Sørensen, 1979).

Catastrophic failure of wind energy converters has occured, with fortunately no loss of life or serious injury to anyone. Dangerous scenarios include:

(a) Blade failure — in which a blade might fail, e.g. by fatigue, and then be thrown off its supporting structure.

(b) Tower failure — in which the entire windmill structure would fail and topple to the ground, perhaps because of the action of wind speeds exceeding design values or because of poor engineering design.

(c) Collision by a low-flying aircraft — should a windmill be erected in an area often used by aircraft. This, however, can be mitigated by using adequate warning lights.

(a) and (b) have already occurred.

Considering that very large wind systems will most likely be installed in areas with little population, the chances for loss of human life would indeed be small. Restriction of land surrounding windmill parks could be made to guarantee safe conditions. Also, safety inspections of large systems could be made mandatory. Small wind systems give added problems considering that they are under the control of the people using them. As these windmills are sometimes installed next to dwellings, farm yards, etc., the dangers are obviously greater. Therefore, special considerations would have to be given to: (a) certification of windmill components with regards to mechanical safety; (b) laws concerning erection of windmill towers; (c) special design of systems in areas prone to hurricanes, cyclones and typhoons, etc.

The noise problems associated with harnessing wind energy are of two kinds. One is the noise created by the gear box and generator. This noise can be reduced far below the standards required. The other type of noise is that created by the air-flow off the rotor blades. It increases at least as fast as the power. At distances of 5–10 rotor diameters, the noise of one 23 year old wind energy converter was found to be 5-10 dB (A); much of the noise was, however, of the lower magnitude (Sϕrensen, 1979). Since typical frequencies of windmill rotors are in the range of 1/2 to 3 revolutions per second, one may expect to find infrasound waves created in the wake of a mill. Intense infrasound waves may represent a biological hazard. However, the power carried by such waves, behind a windmill, will normally be very small and will be rapidly dispersed by turbulences (Sϕrensen, 1976). Thus, the possible concern about infrasound waves transmitted from a wind-power plant may be taken care of by placing restrictions on housing behind a windmill. Measurements of infrasound is part of some ongoing national wind energy programmes and preliminary results indicate that the dispersion is rapid.

Another concern is the possible climatic impact of wind energy extraction on a large scale. However, there are not sufficient data and information and the subject requires detailed studies. So far the principal concentrations of windmills have been for the pumping of water for agricultural purposes. The erection of large scale windmill arrays, will generally be undertaken to generate electricity which may be transmitted several hundred kilometers away to the urban centres. This could have some environmental impacts that remain to be assessed in detail.

It would be incomplete if reference were not made to the social acceptability of using windmills. Man has for millenia harnessed the energy of the wind for transportation via sail powered boats. The use of windmills was a logical outcome of this use of sails for motive power — the fixed windmill providing power from the wind in a given location. Classical post and sail windmills have been regarded as artistic monuments. It is not likely that there will be a general public rejection of the use of windmills, given their universal popularity in the past. Decision-makers should nonetheless sound out the public mood before initiating large-scale wind power projects. In this way, the public will participate in the overall development process and appreciate benefits that may accrue through the use of this renewable energy source. Some wind energy converters, however, do not have the same charm as classical windmills and may not be held in the same esteem that was accorded to them. All factors with regard to the acceptability of modern windmill "parks" by the local population should be investigated.

CHAPTER 20

Energy from the Sea

I. WAVE POWER

WAVE power derives from wind energy, which derives in turn from solar energy. Although only recently the subject of serious investigation, wave power is by no means a new concept. Hundreds of wave machine designs have been invented during the last 200 years. It is estimated that since 1856 over 350 patents were granted on wave power utilization by 1973 (Voss, 1979; Clarke, 1979). Today wave energy is only used on a small scale to power buoys; the average power output of these systems ranges from 70 to 120 W (Panicker, 1976; Merriam, 1978; Voss, 1979).

The energy and power in waves can be considerable. A typical mid-ocean has approximately 1.5 m waves at periods of 8 seconds (Isaacs *et al.*, 1976). These waves correspond to a mean flux of wave power across a section of the ocean of the order of 10 kW/m. Summed over the entire ice-free ocean, the total power available would be 2.7×10^{12} W; the amount of power that could be extracted from this would, however, be a fraction of this amount (Merriam, 1978).

Devices for harnessing wave energy may be classified into: propulsion schemes, buoy power supply devices, offshore power plants and shore-based power plants (see Panicker, 1976; Merriam, 1979; Count, 1979; Voss, 1979; Clarke, 1979 for detailed description of different devices). The engineering challenge for harnessing wave energy in power plants of several hundred kilo- and megawatts is to find a cost-effective way to convert a fraction of the large amount of energy contained in wave motion under the inhospitable conditions existing in the open sea. Electric energy extracted from wave power by schemes known today might cost, according to present estimates, about 10 to 20 times the cost of electric energy from conventional generating plants (Count, 1979). World-wide application of wave energy conversion systems will be limited by geographic factors and the economics of energy transport and distribution over large distances. It is difficult to estimate the potential contribution of this source, but because the resource base is about 10 times less than geothermal and 400 times less than wind (Sørensen, 1979), it could never become a major source of power except for a few favourable locations.

Environmental Impacts of Harnessing Wave Power

Because there are no large-scale wave power stations existing today it is difficult to assess the environmental effects of harnessing this energy source. Wave power plants will produce no thermal discharges or emissions or cause changes in water salinity or require fresh water for operation. The most direct environmental impact is to calm the sea, since they will act as efficient wave breakers. This has beneficial effects in several locations near harbours offering safe anchorage in times of storms, and/or protecting shorelines from erosion

(Nath and William, 1976; Zeigler *et al.,* 1976). However, the calming of the sea might have adverse biological effects because of the absence of waves and associated mixing of the upper layers of the sea. On the other hand, wave energy plants are likely to enhance the growth of many marine organisms by providing a more protected habitat. The obvious success of artificial reefs in attracting marine life seems to pose two problems for energy collecting devices: (a) Marine growth may foul machinery or collectors, causing reduced efficiency or actual destruction during storm, and (b) anglers and spearfishers may haunt these fishing "hot spots" in great numbers (Zeigler *et al.,* 1976). Large scale wave energy systems may influence climate, because they will interfere with the wind-driven oceanic circulation and with sea-atmosphere transfer processes.

II. TIDAL POWER

Tidal power is derived from the combined kinetic and potential energy of the earth-moon-sun system. Although discussions of the prospects for the utilization of tidal power have extended over many years, the total resource base is small (Gibrat, 1966). Table 94 lists the main sites which meet the necessary conditions of tidal amplitude and coastal topography. The combined hydraulic power of these sites is about 6×10^{10}W, of which no more than 10-25% could be converted to electricity (Auer *et al.,* 1978; Voss, 1979). Suitable tidal sites have also been reported in India, S. Korea (Clark, 1979) and Mauritius.

To harness the tides for power two physical conditions are necessary. First, the tidal amplitude must be large — several meters at the very least. Second, the coastal topography must be such as to allow the impoundment of a substantial amount of water with a manageable volume of civil works. This means a bay with a narrow inlet, a river estuary, or a set of strategically located islands. When the coastlines of the world are examined for locations meeting both of the above conditions only a few dozen are found and some of these impractical, because of inaccessibility or for some other reasons.

Tidal power can be harnessed using the single or double basin schemes. The simple single basin, single effect scheme has been used for centuries (though on a very small scale). The sea enters the basin through open sluices, then when the level has reached its peak the sluices are closed and the water is impounded in the basin at high tide level. When the level of the sea has fallen the water in the basin is allowed to go out through the powerhouse. Power is generated on the ebb tide only. Alternatively, the generator can be turned around and power generated on the rising tide, but this gives less energy. The civil works required are only a single dam, with movable sluice gates and a powerhouse. Substantially increased utility with only minor increase in cost is possible today with a single basin, double effect scheme. This is the system that has been used in the only large tidal power

Table 94. Tidal Power Sites

Location	Average Tidal Range (m)	Hydraulic Energy (10^9 kWh/year)
North America		
Bay of Fundy		
Passamaquoddy	5.5	15.8
Cobscook	5.5	6.3
Annapolis	6.4	6.7
Ninas-Cobequid	10.7	175.0
Amherst Point	10.7	2.25
Shepody	9.8	22.1
Cumberland	10.1	14.7
Petitcodiac	10.7	7.0
Memramcook	10.7	5.2
(Cook Inlet, Alaska)		
Knik Arm	7.5	6.0
Turnagin Arm	7.5	12.5
South America		
Argentina, San Jose	5.9	51.5*
Europe		
England, Severn	9.8	14.7
France		
Aber-Benoit	5.2	0.16
Aber-Wrach	5.0	0.05
Arguenon/Lancieux	8.4	3.9
Frenaye	7.4	1.3
La Rance	8.4	3.1
Rotheneuf	8.0	0.14
Mont St. Michel	8.4	85.1
Somme	6.5	4.1
U.S.S.R.		
Kislaya Inlet	2.4	0.02
Lumbovskii Bay	4.2	2.4
White Sea	5.7	126.0
Nezen Estuary	6.6	12.0

Source: Auer *et al.* (1978).

* Suarez (personal communication) estimates the actual figure to be $2-3 \times 10^9$ kWh/y.

Notes: 1. The basin area depends on the position of the dams and reservoirs and for some locations the best position is not obvious. Thus the energy figures are open to some variations.

2. The sites have been screened for tidal range and general engineering feasibility but not for economic feasibility in terms of distance from energy load centres, comparison with other local energy sources, and so on.

3. It is likely that there exists tidal power sites other than those listed but not large ones near heavily electrified regions.

4. Of those listed, only La Rance and Kislaya have been developed.

station ever built, at the mouth of the Rance River, in France (Cotillon, 1974; Bonnefille, 1976; Merriam, 1978). The civil works are the same as for the single effect scheme, the difference is the turbines. The turbines used are different in design from the Kaplan turbines usually used in low head hydro schemes in that the turbine axis, and the direction of water flow, is horizontal. This allows utilization of very low head in the presence of a strong tidal current. The vane orientations are controllable hydraulically, and the unit can be operated as either a pump or a turbine in either direction. In fact the vanes can be moved out of the way so that the unit is simply an orifice. Thus there are five modes of operation; two directions of pumping, two directions of flow to generate power, and an open orifice mode. (The orifice can also be blocked). The multiple mode operating flexibility gives the possibility of generating power on both the incoming and the outgoing tide. It is also possible to pump water in or out of the basin in order to increase generating capability a few hours hence. Using this capability, the Rance power station has been operated to maximize the value of power delivered, not to extract maximum energy from the tides.

Single basin schemes can generate power only intermittently, but a double basin scheme can provide power continuously, or on demand, which is an advantage. The drawback is that the civil works become more extensive. In the simplest double basin scheme there must be a dam between each basin and the sea, and also a dam between the basins, containing the power-house. One basin is maintained always at a lower level than the other. The lower reservoir empties at low tide, and the upper reservoir is replenished at high tide. If the generating capacity is to be large the reservoirs must be large, which usually means the dams will be long. The Passamaquoddy project in U.S.A. has been proposed as a double basin scheme.

Environmental Impacts of Harnessing Tidal Power

Works of a type and magnitude associated with exploiting tidal energy will necessarily give rise to social and environmental impacts requiring appraisal on a broad plane. Tidal energy may be pollution free, in that it does not add pollutants either to the atmosphere or to the water, but it will change the ecology of its tidal basin and, to some degree, may also affect the tidal regime on the sea side of the development. The extent of this effect would of course depend on the magnitude of the tidal development. Some of the detrimental effects on eco-systems attributable to river hydro plants would be applicable also to tidal power plants (see Chapter 21). However, one must weigh the benefits of exploiting these two renewable sources of energy for the production of electricity against the regional and global effects of burning fossil fuels and fissionable materials to develop electricity. It should also be noted that the ecological effects associated with the development of tidal power vary from one place to another and should be assessed for each site independently (Shaw, 1975; El-Hinnawi, 1977).

III. SEA THERMAL POWER

The most important location of the sea thermal resource is, roughly, the 2000 km-wide area around the equator between the Tropic of Capricorn and the Tropic of Cancer. In that area deep sea water (750 to 1000 m) may be from 15° to 25°C colder than surface water, (outside such areas similar temperature differences can be encountered in summer only). The concept that the temperature difference between the surface and the deep waters of the ocean could be used as a source of energy dates back for about a century. The first pilot scale plant was built by Claude off the coast of Cuba in the late 1920s (Claude, 1978; Swann, 1976) in which warm sea water was used as the working fluid. Recently several plants have been designed for ocean thermal energy conversion, known as OTEC plants (Avery, 1978; Marchaud, 1979).

It has been estimated that by placing OTEC facilities approximately 15 km apart throughout the oceans between 20°S and 20°N latitude, the theoretical maximum limit for electricity production is of the order of 50 TW. An ultimate installed capacity representing a small fraction of this potential would still represent a significant contribution, especially, to the energy needs of islands and isolated coastal regions (Auer et al., 1978, Lavi and Lavi, 1979).

The OTEC plants that have been investigated make use of a closed Rankine cycle incorporating a working fluid suitable for the temperatures and the temperature differences that would be encountered. Warm sea-water would be pumped through an evaporator, and the working fluid would be vaporized. The vapour would expand through a low-pressure turbine, creating shaft harsepower that could then be used to drive an electric generator. The vapour would then be condensed by thermal exchange with deep ocean water and brought to the condenser by the cold-water inlet pipe. The working fluid would then be pressurized and pumped to the evaporator to continue the cycle (Swann, 1976; Zener, 1976, Berkovsky,1978).

A large number of technical and economic problems are associated with the development of OTEC systems. Some could be resolved through further research and development. These include improving the efficiency and lowering the costs of the heat exchangers, finding a solution to potential environmental hazards created by the large amounts of ammonia used in the system, and examining alternative designs for the outer hull of the plant structure. However, other problems are inherent in the system itself. Essentially a tremendously large system — one consisting of pumps, heat exchangers, and other expensive equipment would be required to generate a relatively small amount of power. Because of the hostile marine environment, significant problems of materials, operation, and lifetimes would be encountered. In addition the power generated would have to be economically transported in one form or another to land-based load centres (see, for exexample, Zener, 1976).

Environmental Impacts of OTEC

Environmental problems posed by the extensive use of ocean thermal power systems range from questions dealing with the biological and ecological aspects of antifouling agents, primarily used for the evaporators, and exchanger erosion, to those dealing with the ecological and environmental impacts due to changes in salinity and thermal redistribution on biota, mixing processes, density structure and climate (Swann, 1976; Lavi and Lavi, 1976). A positive effect which some authors (e.g. Othmer and Roels, 1973) have postulated is that the movement of vast quantities of water from the nutrient rich cold waters could be used for large scale food production.

One problem is that of ocean water mixing. Because of the huge volumes of water involved, there is a potential danger of reduced surface temperature which may have local and global weather effects. Also, such disturbance of the thermal resource may degrade the performance of OTEC plants clustered nearby. Another environmental concern is chemical discharge from the plants. Leakage of working fluid may be objectionable depending on the working fluid selected. Leakage of ammonia in small doses is tolerable but major spills may temporarily harm the ecosystem. Fortunately, should there be a major ammonia spill, the damage to the environment is reversible. On the other hand, leakage of a halocarbon (Freon) or a hydrocarbon (propane, isobutane) working fluid is potentially more serious. Likewise, discharge of biocides which may be required to control macrofouling in water passages may be environmentally objectionable. Some experts (von Hippel and Williams, 1975), have pointed out the possibility of substantial CO_2 release into the atmosphere caused by the operation of many OTECs. Deep cold water is highly saturated with dissolved CO_2. As the water rises to the platform to enter the condenser, the decrease in hydrostatic pressure may upset the equilibrium. It has been conjectured that the pressure drop and the temperature rise will result in CO_2 release. However, Lavi and Lavi: (1979), pointed out that this may not be a serious issue because the condenser discharge will be below the surface. Further, the transit time of the water through the cold water pipe and the condenser is relatively short compared to the diffusion rate of dissolved CO_2. Also, some of the CO_2 will be eliminated from the solution through the formation of carbonates. The presence of numerous moored plants with kilometres of electric cable suspended in the ocean may interfere with shipping and may introduce a risk element to OTEC installations both from submarines and surface vessels.

Among possible large-scale environmental hazards is the lowering of temperatures in the tropics and increasing temperatures in the mid-latitudes, because of new currents in the ocean and altered atmospheric circulation. Large-scale utilization of ocean thermal energy will cause a substantial transfer of heat from the surface to deep layers so that, the cooled surface of the ocean will evaporate less water and radiate less energy back into space. The lowered heat flux from the ocean through evaporation and decreased radiation with the same energy input, through absorption of solar energy, will cause an imbalance. Part of this net input will cover the energy produced by the plant, while part should be dissipated outside the tropics, presumably by new currents warming the higher latitudes. If tropical OTEC power plants produce 65×10^{19} joules of energy (approximately the global energy consumption projected for the year 2000), there could be a drop of less than 1^0C in the water of tropical oceans. (Berkovsky, 1978).

IV. ENERGY FROM SALINITY GRADIENTS

When two aqueous phases having different salt concentrations (i.e. different activities of H_2O) are mixed, a large amount of energy is released at the interface between the two solutions. At constant temperature a greater net transfer of water will occur from the less saline solution (higher activity of water) through a semipermeable membrane into the more saline solution (lower activity of H_2O), and will cause the pressure to increase in the more saline solution and decrease in the less saline if the two solutions are confined. The equilibrium pressure difference across a membrane which results from a greater net transfer of water from the less saline to the more saline side of the membrane until equilibrium is reached is known as the osmotic pressure.

Calculations indicate that the equivalent pressure head between 0.5 molar seawater and fresh water is approximately 240 m or almost 25 atmospheres (Bromley $et~al.,$ 1974). A freshwater flow of 1 m^3/sec in such saline water could provide 2.2 MW of salination power. The Columbia River, for example, has a mean flow of about 6600 m^3/sec, and if the osmotic potential of half that flow were realized at 30% efficiency, about 2100 MW of electricity could be produced (Wick and Isaacs, 1976). Marine termini of rivers are not the only likely locations for salinity power plants but innumerable salt pans along desert coasts may also be considered.

The salinity gradient energy is readily converted to mechanical or electrical energy. In the direct mechanical approach, the river water is allowed to flow through hydroelectric turbines into a reservoir at a level which is some hundreds of metres below sea level. The discharge of the fresh water directly into the salt water through semi-permeable membranes (membranes permeable to water but not to dissolved solids) will produce a pressure head that could operate a waterwheel and generator to produce electricity. Other methods for harnessing salinity gradient energy, for example the dialytic process, have been proposed (Wick and Isaacs, 1976). It should be noted, however, that these methods are still at an early stage of development and the feasibility of one or another depends on solving the different technological and economic problems.

Environmental Impacts:

The environmental impacts of the development of salinity power are not known. However, negative environmental effects could occur in the zone where mixing fresh and saline water occurs. It should be mentioned that estuaries are found at the mouths of many rivers and that in many locations, they are the most productive areas in the marine environment. Thus any development schemes should be designed in such a way as to minimize the stress on these ecosystems. Some important problems that need to be solved are the management of sediments carried by the rivers and the protection of marine animals that might be sucked into inlet pipes from the ocean. Corrosion, biological and sedimentary fouling may constitute serious problems if these membranes are in seawater. There will need to be some sort of filtration on both the seawater and the river water. Perhaps even pretreatment of the water will be necessary in order to prevent fouling and corrosion and also in order to increase the efficiency of the membranes or to increase the voltage in some way (Wick and Isaacs, 1976; Zeigler *et al.*, 1976).

CHAPTER 21

Hydro-Power

Man's earliest extensive use of energy, other than muscle power of man and animals and direct solar energy, was that derived from flowing water. The use of waterwheels of various types extends back to the early civilizations. The size and efficiency of waterwheels increased over the centuries, and in the nineteenth century, water-powered mills of various types ushered in the beginning of the industrial age. The peak of this early water power development phase was reached about the middle of the nineteenth century because favourable mill sites, within the reach of the mechanical transmission of power, were limited. Furthermore, at that time, the more flexible steam engines were improving in economy and dependability. With the advent of electricity in the 1880's, and with alternating current technology making transmission of electric energy more economical, the development of hydroelectric energy was well underway by the beginning of the twentieth century. Developments were rapid and by the 1930's projects such as the 1.3 million kW powerhouse at Hoover Dam in the U.S.A. were completed. Large hydroelectric installation such as this increased the utilization of energy in the industrialized countries and programmes to utilize the large hydroelectric potentials were pushed ahead.

The growth of electricity production from hydro-power has considerably increased since the 1950's. In 1950, the hydroelectricity production was 343×10^9 kWh; in 1978 it reached 1557×10^9 kWh, i.e. an increase of 4.5 times. Table 95 gives the hydroelectricity production as compared to electricity produced from other sources. On the average, hydro-electricity constitutes about 21% of the total world electricity production.

Table 95. World Electricity Production (in 10^9 kWh)*

Year	Thermal	Hydro	Nuclear	Total	% Hydro
1972	4226	1294	144	5664	22.8
1973	4570	1317	197	6084	21.6
1974	4567	1446	255	6268	23.1
1975	4660	1467	351	6478	22.6
1976	5063	1471	410	6944	21.2
1977	5227	1524	509	7260	21.0
1978	5527	1557	530	7614	20.5

* After: World Energy Supplies 1973 1978, (United Nations, 1979).

Table 96 gives an estimate of the hydro potential of the world together with the calculated percentage of developed resources in the different regions. Armstrong (1978) estimated the total potential from hydro resources of the world at 2.2×10^6 MW of installed and installable generating capacity at 50% capacity factor. It should be noted that these estimates are conservative and do not take fully into account the great potential of resources for small hydro-power schemes.

Table 96. World Hydro-power Potential

	Potential available 95% of time (10³ kW) (1)	Potential output 95% of time (10⁶ kWh/y) (2)	Present installed capacity (10³ kW) (3)	Current annual production (10⁶ kWh/y) (4)	Per cent of developed potential (4)/(2)x100 (5)
Africa	145,218	1,161,741	11,437	49,663	4.3
Asia	139,288	1,114,305	59,773	245,096	22.0
Europe					
(incl. USSR)	102,961	827,676	177,797	620,676	75.0
N. America	72,135	577,086	111,402	434,035	75.2
Latin America	81,221	649,763	38,582	176,845	27.2
Oceania	12,987	103,897	9,578	31,669	30.5
World Total	553,810	4,434,468	408,569	1,557,984	35.1

Notes: (1) Columns (3) and (4) are 1978 figures according to World Energy supplies 1973–1978, United Nations (1979).

(2) Columns (1) and (2) are after UN Water Conference Secretariat (see Biswas, 1979).

The operating hydroelectric capacity at present is about 400,000 MW with an annual production of about 1558 x 10⁹ kWh, which is approximately 35% of the total world potential available 95% of the time. Although about 75% of the hydro potential in Europe and in North America has been exploited, only about 22% in Asia, 27% in Latin America and 4.3% in Africa has been harnessed for energy production.

Hydro-power is an important renewable source of energy, and constitutes an integral part of optimum overall water resource utilization. It is a catalyst in socio-economic development, particularly in rural areas of developing countries (El-Hinnawi, 1977, 1979); and its economic justification is improving because of its "inflation proof" characteristics and its long life and low maintenance costs (Armstrong, 1978). It should be noted, however, that in order to maximize the benefits of hydro-power, it should be developed as a part of an overall water development plan. The development of large-scale waterpower *per se* in some developing countries. for example, has created a number of socio-economic problems. One consideration being that the output of electricity requires, for example, expensive transmission lines and a social and economic structure ready to take advantage of this form of energy (Kristoferson, 1977). Low density of population, long distances between energy consumption centres, and low income levels are some factors that often hamper the growth of electricity consumption among the population at large. Because of the large investments necessary for many dams, in many cases only very large-scale industrial operations are regarded as economically profitable. This may result in large, energy-intensive industries being constructed to match the output of large power stations. The influence of this type of industrial development in general on socially developing countries is sometimes questioned. Adequate planning and cost/benefit analysis is, therefore, an important prerequisite in the development of hydro-power installations.

Hydroelectric generating plants vary considerably in size from as small as 3 kW (El-Hinnawi, 1977) to as large as 12×10^6 kW (ICOLD, 1977). Small hydro-power stations (also called mini-hydro, village-size hydro-power installations etc.) are generally those with installed generating capacity less than 1000 kW (El-Hinnawi, 1979)*. Such Stations are normally built as part of water management and rural development strategies. In the People's Republic of China, for example, more than 60,000 small hydro-power stations, with a total generating capacity of about 3000 MW were built in the last 15 years (El-Hinnawi, 1977) in rural areas. This has largely accelerated the development of such areas by providing power for better irrigation schemes, rural industries and domestic use. There is no technical reason why such small hydro-power schemes should not be built in rural areas of other developing countries using any suitable available water course (canals, rivers, reservoirs, falls etc.).

Large hydro-electric power plants can reach more than 12,000 MW (for example, the Itaipu power plant at the border between Brazil-Paraguay which is under construction). According to ICOLD (1977), there are about 100 plants exceeding 1000 MW in the world (29 became operational in the last decade; 36 are under construction or in early phases of development; the rest operational before 1970). The construction of such large hydro-electric power plants occurs on major rivers and the reservoirs resulting from building the dams** vary in area from one place to another. There are about 41 man-made reservoirs that range from 1000 km² to 8730 km² (see Fels and Keller, 1973). The volume of the reservoirs vary from one site to another and can reach about 200×10^9 m³ (Mermel, 1976), Table 97 gives some examples of large hydro-electric power plants and the area of the associated reservoirs.

Conventional hydro-electric developments use dams and waterways to harness the energy of falling water in streams to produce electric power. Pumped storage developments utilize the same principle for the generating phase, but all or part of the water is made available for repeated use by pumping it from a lower to an upper reservoir. There are two major categories of pumped storage projects: (a) developments which produce energy only from water that has

---

*Smil (1976) defines small hydro stations as those with capacities less than 500 kW, whereas in the People's Republic of China, small hydro-power schemes are those having installed generating capacities less than 12,000 kW (El-Hinnawi, 1977). In the U.S.A. small hydro is a generating unit with less than 15,000 kW (Wehlage, 1979).

**Different types of dams ranging from earth dams to concrete ones with different designs have been built. In some cases, particularly in small hydro-power schemes, it is not necessary to build dams. The development of bulb-type self-contained turbine-generators can utilize the normal flow of rivers in some cases. It should be noted that there are about 24,000 dams in the world (Mermel, 1976) and that dams are not necessarily built for harnessing hydro-power. For example, in the U.S.A. dams are being built at the rate of 125 per year, about 5% of which are associated with hydro-power, the rest with water supply irrigation etc. (Mermel, 1976).

Table 97.   Examples of Large Hydro-electric Power Plants and of Large
Reservoirs**

Dam	Present installed capacity (MW)	Ultimate installed capacity (MW)	Reservoir area (km²)
Krasnnoyarsk (U.S.S.R.)	6096	6096	2130
Churchill (Canada)	5225	5225	6200
Bratsk (U.S.S.R.)	4100	4600	5426
Volgograd (U.S.S.R.)	2560	2560	3160
Volga Lenin (U.S.S.R.)	2300	2300	6500
Aswan High Dam (Egypt)	2100	2100	5250
Saratov (U.S.S.R.)	1360	1360	1950
Kariba (Rhodesia/Zambia)	1266	1866	5180
Furnas (Brazil)	1216	1216	1606
Kainji (Nigeria)	320	1000	1243

** Compiled after ICOLD (1977), Fels and Keller (1973).

previously been pumped to an upper reservoir, and (b) developments which use both pumped water and natural runoff for generation. Although pumped storage projects may have conventional hydro-electric generating units and separate pumps, most developments utilize reversible, pump-turbine units. Some plants contain both conventional reversible units. The use of pumped storage enables power systems to obtain large quantities of peaking power on demand. It also permits more efficient use of water power via substantial improvements in the overall system efficiencies in converting potential energy resources into usable and available energy. Several pumped storage schemes have been constructed in many countries; others are being developed in some countries in conjunction with hydro-power schemes (Jeffs, 1979). Another scheme for harnessing hydro-power is to tap the glacial run-off from the Greenland ice cap (Partl, 1977). A large volume of surface ice melts every summer and either runs off into the sea or drains back into the lower ice and freezes at depth. If this water could be collected and directed to a number of high-level natural lakes, a series of large power stations could be built. The main problem would seem to be in the construction and maintenance of the collection channels in the ice. The total installed capacity could range 60 to 120 GW depending on load factor.

Environmental Aspects of Hydro-electricity Generation

Hydro-electricity generation has a number of environmental impacts. No dam can be built and no lake can be created without environmental costs and benefits of some kind. A dam becomes a dominant factor in the hydrological regime, and sets in motion a series of impacts on physical, biological and socio-cultural systems. The many consequences on the environment of the dam and the lake behind it appear to be factors in common regardless of the dam's geographical location. The environmental side-effects of dam construction are generally divided into two categories: (a) the local effects and the reactions within the area of the man-made lake: (b) the downstream effects resulting from a change in the hydraulic regime. Both categories have their physical, biological and socio-economic elements.

The environmental impacts of man-made lakes and the reactions within the area of the lake are summarized in Fig. 11. It should be noted that these effects are interrelated and that they affect each other. The construction of a dam and the filling of its reservoir cause substantial local change by introducing an immense structure into a natural setting. The flooding of the region could have immediate significant impact on the means of communication, historic sites, communities which are inundated, and the local flora and fauna. The dam itself presents an obstacle not only to the free running of water, but also to fish migration and navigation. Continued fish migration may be achieved by the construction of fish ladders, or by collecting fish and transporting them by road. In the case of some very high dams, artificial conditions favourable to spawning have been constructed below the dam, and eggs cultured under ideal conditions in hatcheries (UNEP, 1976). While this is a costly process, it can ensure the survival of larger numbers of eggs and fingerlings than possibly under natural conditions. Care must also be taken to assure migration during dam construction.

Physical Aspects

Although it is designed to store water, a man-made lake immediately begins to store sediment carried by the stream. The amount of sediment deposited in a given reservoir depends on the amount of sediment delivered to it and the reservoir's ability to retain the sediment. Therefore, reservoirs differ greatly in the amount of sediment deposited in them because of the tremendous variability, both in time and space, in the amount and characteristics of the sediment carried by streams and the circumstances causing its deposition. Dendy *et al.* (1973) found that in the U.S.A., the average storage loss was 3.5% annually for small reservoirs (with a capacity less than about 10,000m,^{3}) and that the storage loss decreases with increase in the storage capacity (the loss was about 0.16% for reservoirs with a capacity greater than 10^9 m^3). Cyberski (1973) pointed out that the storage depletion was 0.51% per year for 19 reservoirs in Central Europe (ranging in capacity from 1.5×10^5 to 23×10^6 m^3). In Lake Nasser, Egypt, it has been estimated that the reservoir was losing about 60 million m^3 of storage per year due to siltation in the first few years of its filling. At this rate, the dead storage capacity of 30 km^3 will be filled in about 500 years (Biswas, 1978).

The effects of sediment deposition in reservoirs are evidenced in many ways but perhaps most significantly in terms of the reservoir's ability to perform its intended functions. Water resource functions most commonly served by reservoirs include water supply, irrigation, flood control, hydro-electric power, navigation, recreation etc. To the extent that sediment distracts from the services provided or expected from a reservoir, it is a liability expressible as the lesser of either (a) the cost of services foregone because of the sediment or (b) the cost

Fig. 11 Environmental Aspects of Man-Made Lakes (After El-Hinnawi, 1980)

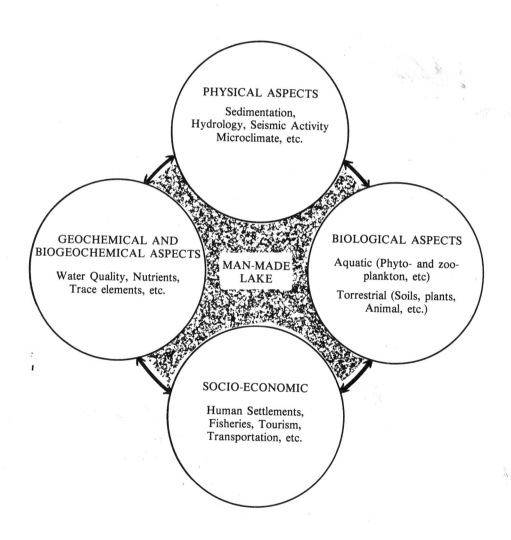

required to remove the sediment from the reservoir or to keep it out in the first place (Glymph, 1973). Depletion of storage capacity is but one of the upstream effects of reservoir sedimentation. The stream channel is likely to aggrade for some distance above the reservoir because of backwater effects on sediment transport. The formation and growth of deltas tend to accelerate and extend the process still further upstream. Thus channel gradients become flatter, channel cross sections are reduced, flooding occurs more frequently, and drainage of floodplain lands is impeded because of reservoir sedimentation. On the other hand, as a result of the siltation in the reservoir, clear water flowing downstream cause channel degradation and stream bank erosion. The sediment-free water passing through the reservoir can entrain another sediment load and proceeds to do so where the material is available. The phenomenon applies downstream from both large and small reservoirs. It is most likely to occur when a dam is built on an alluvial channel that previously had a generally stable relationship between such factors as stream discharge, sediment load, channel gradient, and widths and depths of stream channel. Another effect of the siltation in the reservoir is possible erosion of the river Delta. Prior to the construction of the Aswan High Dam, for example, the Nile Delta used to be built up during the flood season, with the silt carried by the River to the Mediterranean. This situation in the Delta compensated for the erosion that resulted from the winter waves of the preceding year. Without enough siltation, erosion of the Delta has become a major problem (Biswas, 1978).

Changes in quality and quantity of sediments downstream are believed to affect agriculture and fish production in many ways. Whereas, theoretically the principles involved in the foregoing surmises are sound, whether the apprehended effect do, in fact, occur in a given river is debatable and is dependent on various features of the river itself, such as the silt content of the waters, the nutrient status of the silt, the extent of flooding, and so on. It is well-known that due to deprivation of the annual mutrient silt deposits due to the Aswan High Dam, substantial amounts of fertilizers have to be used to recharge the soil downstream with nutrients (El-Hinnawi, 1980). The lack of Nile sediment has also led to the migration of sardines and crustaceans from their habitat in the Mediterraneah north of the Delta. However, there are no precise studies establishing a cause and effect relationship between the Aswan High Dam and the disappearance of the sardines, notwithstanding the fact that the postulates in support of this surmise appear to be theoretically sound. Sardines have disappeared elsewhere in the world in situations completely unrelated to dam construction. They have equally mysteriously reappeared after a lapse of some years (for example, in the Indian Ocean).

Man-made lakes generally alter not only the streamflow regime but also the water balance (and the hydrological cycle). These effects may be of particular significance in arid and semi-arid regions. Comparative studies have shown that the construction of several small and medium size storage reservoirs has reduced the annual flow by 10% in average years and by 25% during dry years in a 2000 km^2 semi-arid river basin in northeast Brazil (Dubreuil and Girard, 1973). In some areas, where the permeability of the substrate is high, vast amounts of water are lost by seepage from the reservoir. Lateral seepage from Lake Nasser, for example, has been estimated to reach 1000 million m³/year (Abu Wafa and Labib, 1973). On the positive side, this seepage has led to changes in volume and direction of ground water flow, facilitating reclamation of lowlying arid lands at considerable distances from the lake (El-Hinnawi, 1980). On the negative side, however, the increase in the water table downstream has led to the production of bog effects and in some cases salinization that adversely affects the agricultural use of the soils (Biswas, 1978).

The filling of a lake imposes new stresses on the earth's crust that, in turn, generate seismic movements and that, in some cases, generate earthquakes that are severe enough (6 on the Richter scale) to cause human losses. These seismic activities may vary in magnitude and time in accordance with a number of factors. Water height of more than 100 m in a reservoir constitutes a factor that may be of major siesmic importance in combination with geologic formation and structure (Rothé, 1973). Moreover, the saturation of sedimentary formations by seepage from the reservoir may cause additional seismic movements. Seismic activities have been recorded for a number of dams and associated reservoirs (Rothé, 1973) and in some cases have led to catastrophes. For example, the earth movements that caused the Vaiont Dam Disaster in Italy in 1963 (which caused the death of more than 2000 persons) were preceded by several years of considerable seismic activity characterized by a clear relation between the frequency of shocks and the progress of filling the reservoir. The 1967 Koyna dam disaster in India, which resulted in some 177 deaths and considerable property damage was due to an earthquake whose epicentre coincided with the dam itself. In general, the seismic disturbances can be traced to the existence of inactive faults and it seems likely that the effect of the added forces contributed by the dam and reservoir liberate orogenic tensions of much greater strength.

Relatively little is known about the precise influence of man-made lakes on weather and climate. At the microclimatic level, most of the evidence comes from comparison with natural lakes and their influence on precipitation, direction and frequency of wind, thunderstorms, hail, snow, and other phenomena. The impact of a man-made lake on temperature and precipitation is related to both local conditions and mesoscale meteorological elements and should be assessed for each site separately (Nemec, 1973).

Geochemical and Biogeochemical Aspects

The physical, chemical and biological properties of the water leaving a lake may differ significantly from the waters entering the lake. A large number of factors affect this quality transformation process, and the water conditions of the lake are of basic importance in understanding, predicting, or influencing these changes. Seasonal temperature fluctuations governed by the energy regime are the most common causes of a density stratification, but other agents such as dissolved or suspended solids can also be influencial. Beyond differences in densities (temperature) at different depths, the rate of flow through a given cross section is the principal factor determining the possibility of the formation of a stagnant layer or distinct water masses in the reservoir. In many cases, hydrogen sulphide is found in these stagnant layers imparting obnoxious odour and creating a fishless void. Water stratification in the reservoir created by the Roseires Dam on the Blue Nile caused heavy fish mortality in 1967, when deoxygenation affected the water temporarily (El-Hinnawi, 1980). In many reservoirs the human discharge of waste has a significant effect on water quality, often as a result of measures not taken into account when the reservoir was planned. Flows of domestic and industrial effluents, of farm waste, and of excess fertilizers and pesticides may drastically modify the quality of stored water and of that discharged downstream.

Biological Aspects

The construction of a dam and the creation of the associated reservoir has a number of impacts on terrestrial and aquatic biota which are of four main kinds: (a) those of short life-span and frequent population turnover (e.g., nutrient-cycling bacteria and many algae), (b) those of intermediate life-span and turnover (e.g. cereal crop plants, some small fish), (c) those of long life-span and slow population turnover (e.g. perennial plants and large aquatic or terrestrial animals or domestic livestock), and (d) people. The impacts of dam construction on these ecosystems should be assessed in detail and the actual and potential benefits of not undertaking varsus undertaking the scheme should be determined. In this regard the elimination of terrestrial production in the inundated land should be carefully assessed. This production may be natural, i.e., culminating in wild plants and animals; however, it may be managed production, i.e., yielding agricultural crops, livestock, or forest crops. Also the elimination of actual and potential riverine production can be compared to anticipated gains from production in the new lake.

When the reservoir begins to fill for the first time, the terrestrial and riverine environments progressively disappear, and the lacustrine environment originates and expands. The inter-relationships between the biological system and other components of the ecosystem are in rapid transition. As the reservoir changes, the plant and animal communities also change. In the lake basin, there is a sequential shift dominance by flowing water species and communities to dominance by those of more quiet water. During the rapid development of the lacustrine system, an increase in bioproductivity is encountered manifested by production of planktonic algae and increase in fish production. During the stabilization process, lake evolution is irregular in speed and direction; after stabilization, future change is slowed and is usually overall in a predictable direction. In the stabilizing period the first benefits in fisheries and agriculture become available. For example, fish catch, as a partial indicator or production, in the section of the Volta River later covered by Volta Lake changed within 5 years from 4,000 to some 60,000 tonnes per year in the new Lake (Ackermann *et al.,* 1973). Raheja (1973) and Latif and Rashid (1973) reported an increase in fish production from lake Nasser, Egypt from 764 tonne in 1966 to 4,560 tonne in 1969; FAO (1974) reported 8,343 tonne in 1972. At this time, explosive development of nuisance weeds has also occurred. In the tropics, nuisance weeds include the water hyacinth, water fern, and water chestnut. One of the more evident characteristics of a lake in the process of stabilization can be a rise or decline in the area covered by emergent aquatic weeds and the occurrence of dense phytoplankton blooms.

The stabilized stage in the file history of a man-made lake is reached when fluctuations in its biological parameters of production exceed only slightly, if at all, those in a natural lake of similar physical characteristics and like latitude and elevation, as an example, stabilization would be characterized by a seasonally cyclic balance in the oxygen budget of the lake. Stabilization is also shown by the emergent aquatic plants when their rapid initial spread has ceased, attained an extent from which there is little annual change, or even retreated from an initial maximum extent as it has at Lake Kariba (Ackermann *et al.,* 1973). The development of the terrestrial ecosystem around the lake depends on the overall development plans of the region. Agriculture and livestock management activities are important elements of such development. In Egypt, for example, FAO (1974) reported that about 100,000 hectares around Lake Nasser are suitable for agriculture.

Once the impoundment has been completed, the succession of plant growth colonizing the shorelines, and at times even the body of the lake, will influence the incidence and development of vector-borne diseases. Excessive weed or algal growth downstream as a result of the biogeochemical changes in the water quality of the river or due to changes in the irrigation system (e.g. from basin to perennial) will also influence the incidence of such diseases. In Egypt, for example, the replacement of simply primitive irrigation with perennial irrigation has caused a high incidence of both *S. mansoni* and *S. haematobium.* Infection rates in four selected areas, within 3 years of introduction of perennial irrigation, rose from 10 to 44%, 7 to 50%, 11 to 64% and 2 to 75% (Biswas, 1979). Not only will such vegetation promote the breeding of schistosome-bearing snails and malaria mosquitoes, but it could also encourage the development of the filariasis vectors (Brown and Doem, 1973). Available evidence from the major man-made impoundments in Africa clearly indicates that transmission of schistosomiasis is taking place in the main body of every lake as well as in the existing irrigation works. It is important to emphasize, in this context, that the definitive host, man, is responsible for the dissemination of schistosomiasis by contaminating the aquatic environment, where he in turn becomes infected. The snail is only a passive intermediate host. Thus the containment and abatement of schistosomiasis calls for management of the impounded water by means of shoreline sanitation, education of the human population to improve its habits in disposing wastes and excreta, and sanitary engineering to minimize the contacts between man and lake water. These measures are particularly important in view of the impossibility of applying molluscicides over an entire lake to control the snail population chemically (Brown and Doem, 1973). Devestating epidemics of malaria have been recorded following construction of some dams. However, many recently created man-made lakes in Africa have not shown any patent malaria resurgence. Onchocerciasis in Africa is mostly transmitted by *simulium damnosum,* whose larvae breed in the rapid sections of streams and rivers. The effect of impoundments is beneficial insofar as they drown out the breeding places for several kilometers above the dam. Although breeding may subsequently take place on the spillways and below the dam, the larvae may be artificially flushed away by opening the sluice gates.

Socio-Economic Aspects

When a man-made lake is created, the members of the lake basin population are displaced, crowded or supplemented by new migrants. Within the lake basin the human population can be divided into four general categories, of which the first two pose the most problems. These categories are (a) those who must relocate because their homes and fields will be partially or totally inundated by the reservoir (the relocatees), (b) those among whom most of the relocatees must be resettled (the hosts), (c) those lake basin inhabitants who are neither relocatees nor hosts and (d) immigrants who move into the lake basin and seek new opportunities that accompany dam construction and reservoir creation. Most of the displacement exercises of the population has created several human problems. Thus the Volta Dam in Ghana has inundated an area of about 85,000 km², and the resulting lake has a shoreline of over 6,400 km. As a result of the development, some 78,000 people and more than 170,000 domestic animals had to be evacuated from over 700 towns and villages of different sizes. Eventually, 52 new settlements were developed to house 69,149 people from 12,789 families (Jones and Rogers, 1976). It was a major problem since a large number of people coming from small villages (600 of the 700 original villages had less than 100 people, and only one had a population of over 4,000), and having different ethnic backgrounds, traditions, religions, social values and cultures, had to be resettled into only 52 locations. The complex emotional relationships between the different tribes and their lands were not properly understood. The development of a socially cohesive and integrated community, having a viable institution infrastructure became hard to achieve. Similarly, the Kariba Dam on the River Zambesi (Zambia and Zimbabwe) displaced approximately 57,000 Tonga tribesmen, who had to pay a major price for this progress. Technology transfer at that level was a major problem, since many of the planners were from outside Africa. The resettlement programme for the Tonga tribesmen left much to be desired; not only did they suffer great cultural shocks when being thrust into communities as different from their own, as theirs from Great Britain, but also it took two years to clear sufficient land to meet their subsistence needs. The government had to step in to avert famine and very serious hardships and, ironically, this good-intentioned step became one of the most destructive parts of the process. The food distribution centres also became transmission sites for the dreaded sleeping sickness disease. Similar results from water development projects have, unfortunately, not been unique. Approximately 100,000 people had to be relocated for the Aswan High Dam without sufficient planning, and the World Food Programme had to rush in famine relief for the Nubians*. Other examples of lakes and populations displaced are the following: Lake Kainji in Nigeria — 42,000; Keban Dam in Turkey — 30,000; and Ubolratana Dam in Thailand — 30,000 (UNEP, 1977). Resettlement of population due to water development projects in many developing countries has not been a satisfactory experience. Inadequate planning, insufficient budget, incomplete execution of plans and little appreciation of the problems of technology transfer have all contributed to the failure of plans. The fact that much of the population to be resettled were rural and illiterate, and thus had very little political power, did not help either. The direct beneficiaries of the projects were often the educated elites, who are in power, whereas the direct social costs were mostly attributable to the rural poor (Biswas, 1978).

On the other hand the building of dams and the creation of reservoirs have a number of socio-economic benefits: better water management and development of irrigation systems to increase agricultural land and production of power needed to accelerate industrialization and socio-economic development. The lake itself provides opportunities for a number of socio-economic activities ranging from agriculture on the sides of the lakes, to fisheries, tourism and the development of small industries.

Transmission Lines:

The environmental impacts of high voltage transmission lines have been consider in Chapter 5 from six aspects: aesthetic considerations, land requirements, communications hazards, ozone and habitat effect; see Chapter 5.

---

* The High dam created a vast reservoir, having a shore-line length of 9,250 km, surface area of 6,220 km^2 and volume of 156.9 km^3 at 180 m elevation. Though much of the land inundated was thinly populated, it contained areas rich in historical monuments; two main temples (Abu Simbel and Philae) had to be dismantled and moved to higher locations.

CHAPTER 22

Energy from Biomass

For the purpose of this study, biomass is defined as renewable organic matter produced by photosynthesis — a mechanism that converts solar energy into stored chemical energy — directly in the case of plants and indirectly in animals whose ultimate food source is plant material. Biomass can be trees, crops, aquatic plants, natural ecosystems (i.e., plant communities not subject to human management) and organic wastes of different kinds. Biomass also includes fuel crops specifically cultivated on energy farms for their fuel content (see section on energy farms). Fuel crops could be conventional annual crops, fast-growing trees, aquatic plants, or even multi-plant communities. Biomass is an important source of energy; perhaps the most important in terms of actual users, for it is the principal fuel for the vast majority of people in most developing countries. Since biomass is a renewable resource, with proper management the energy from this source could be increased and partially fill the gap created by the eventual exhaustion of fossil fuels (EEC, 1979; Chartier, 1979; UK ISES, 1979).

Photosynthesis is theoretically and practically the least efficient of any of the direct solar energy conversion processes currently being considered as alternative energy sources, but it is the only solar technology that is already deployed by mankind on a large scale and capable of directly providing storable fuels. The average conversion of solar energy by the biosphere is 133 TW of which 76 TW is terrestrial, and 57 TW is marine (Sørensen, 1979). Photosynthetic conversion efficiencies vary from less than 0.5 to about 5% and represent the efficiency of the total process, sunlight-fixed chemical energy (Hall, 1979; Benemann, 1978). Under optimum field conditions values between 3 and 5% conversion are achieved by plants; however, often these values are for short-term growth periods, and when averaged over the whole year they fall between 1 and 3%. In practice, photosynthetic conversion efficiencies in temperate areas are typically between 0.5 and 1.3% of the total radiation when averaged over the whole year, while values for sub-tropical crops are between 0.5 and 2.5% (Hall, 1979); the average for the biosphere is about 0.25%.

The dry weight of all living biomass (plant matter) on the earth's land surface has been estimated as nearly 2400×10^9 dry tonnes, with a primary production rate of 172×10^9 dry tonnes/y (Rodin et al., 1975). The latter is equivalent to more than 10 times the present total world energy consumption* from all commercial sources (assuming that the average calorific value of dry biomass is about 1900 kJ/kg). However, only a modest fraction of the biomass enters the human economy, and much of that fraction leaves the economy as waste or residues. Of the forest biomass, for example, which constitutes an estimated 98% of all living terrestrial plant matter, estimates of the fraction of the annual increment that is currently harvested range from 45% (Arnold 1979) to 75% (Openshaw 1978) of which about half is used for fuel. However, this use is unevenly spread over the world and in many areas especially in developing countries more than the increment is being removed and the forest capital is being

---

* The world energy consumption in 1978 was $8,755 \times 10^6$ tonnes coal equivalent, which is equivalent to $2,574 \times 10^{14}$ kJ (United Nations, 1979).

destroyed. The sources of biomass are varied, and a vast difference exists between the amount of biomass that is readily available and that which is hypothetically available.

THE RESOURCE BASE

Trees are an important source of biomass. Forests and woodlands cover approximately 30% of the world land area — 3,800 million ha., of which 1,460 million ha. are classified as tropical forests, 220 million ha. as sub-tropical forests, 1,000 million ha. as open savannah woodlands, 450 million ha. as temperate forests and the remaining 670 million ha. as boreal forest. The total volume contained in this forest area is between 340 and $360x10^9$ m^3, which is equivalent to about $270x10^9$ tonnes air dry wt (12% moisture content) or $175x10^9$ tonnes of coal equivalent. This is about 20 times the world's current energy consumption from all sources, but of course the consequences of using it as an energy source would be catastrophic. Only the annual increment, an estimated 6,610 million m^3, can be used with safety (Openshaw, 1978).

Another source of biomass is grass. Earl (1975) estimated the world's grassland area to be $26x10^8$ ha. In the tropics as in other regions, a sizeable fraction of the natural grass biomass is grazed by wild animals and by domestic livestock. Another and smaller fraction is harvested for use as fodder. Highly productive grass, for example, elephant grass (*penonisetum purpureum*) offers great possibilities as a source of forage for domestic animals and for energy (in case of energy farms; see later). However, the extensive development of this source of biomass is still being studied to determine its viability as a source of energy.

Practically all agricultural crops produced by farmers throughout the world are technically capable of being converted into some form of energy, whether the primary purpose of the crop be for food, animal feed, fibre, or other products. However, the emphasis for energy use has focussed on some crops, for example, sugar cane, cassava etc. and on agricultural and agro-industrial residues. Crop residues vary and range from none in alfalfa to 6.7 tonne per hectare in corn (air dry weight). Only corn and rice exceed the crop average of 4.4 tonne of residue per hectare (Pimentel *et al.*, 1978). Throughout the world, farm crops leave substantial residues, the extent and scale of which is rarely realized. Wheat, with a yearly crop production of 355 million tonnes, rice with 344 million tonnes, corn (maize) with 322 million tonnes, sorghum with 55 million tonnes, millet with 46 million tonnes, and several other less widely-grown grain crops all contribute to a grand annual total of 1,700 million tonnes of cereal straw (UNEP, 1978) much of which is at present unutilized. Agro-industries also produce vast quantities of residues. The sugar cane industry, for example, creates each year 50 million tonnes of sugar cane tops and 67 million tonnes of bagasse, as well as molasses and press mud. Although molasses wastewater is used in Indonesia to fertilize rice paddies, the press mud is widely used as a soil conditioner and as a component of animal feed, and the bagasse is often burned as a low-grade fuel in the sugar mills themselves, the sugar cane residues are generally greatly under-utilized. Pineapple is typical of many fruits for which much of the crop is wasted. Where pineapples are canned, less than 20% of the whole fruit is used; the remainder, often in the form of a highly-polluting liquid, can cause considerable disposal problems. In Malaysia in 1974, for example, 250,000 tonnes of pineapples generated only 40,000 tonnes of canned fruit together with 210,000 tonnes of residues. The core, skins and fresh wastes of pineapples are often crushed for juice, and in Kenya and elsewhere the resulting bran is dried and used as cattle feed. In the Philippines, the residues are converted into wine but in other countries they are not utilized at all (UNEP, 1978).

Agricultural residues are used extensively in many rural areas of the developing countries as a source of energy (through direct combustion). It is indeed difficult to give an estimate of the amount of agricultural residues used as fuel, since they are of non-commercial nature and have not so far been accounted for in any global, regional or national statistics. Earl (1975) estimated that the world annual consumption of agricultural residues, e.g. bagasse, cotton sticks, and hay, as fuel is about 10 million tonnes of coal equivalent. However, Zumer-Linder (1976), referring to India alone, quotes an estimate of the consumption of agricultural residues as totalling 27 million tonnes/y (or about 17 million tonnes coal equivalent). A wide range of proposals exists concerning the utilization of crop remains as a source of energy, but analysis of the agricultural, environmental, and energetic aspects of the use of crop remains suggest that little or none of these remains should be used for biomass energy conversion (Pimentel et al., 1978; Chartier, 1979). Crop remains left on land prevent sediment runoff, conserve soil and water, maintain soil carbon ratios and soil structure, and prevent nutrient (N, P, K, etc.) loss. The beneficial role of crop residues in agriculture, including their potential as feed for livestock, outweigh their use as potential fuel. On the other hand, food processing wastes (pods, hulls, cobs, chaff, and other by-products) have value as an energy source (UNEP, 1978).

Aquatic vegetation (fresh water and marine) has recently attracted attention as a source of biomass for energy conversion. In the warm waters of the tropics aquatic weeds exist and multiply in abundance and have been used to a varying extent for food, animal feed and fertilizer. However, in many countries, the nuisance value of water weeds exceeds their benefits: they impede navigation, irrigation and fish culture projects and may contribute to the spread of tropical diseases (see Chapter 21). Because most aquatic plants contain large quantities of water (about 90%), they are unsuitable for most energy-conversion processes unless they are predried by pressure (which itself is energy-consuming) or by spreading in the sun (which is time and space-consuming). At present, anaerobic fermentation appears to be the only feasible process for converting this biomass into fuel (see later).

The only biomass of animal origin that has any significance for energy production is excrement. Besides its great importance as a fertilizer, animal dung

(dung-cakes) has been used for centuries as fuel in many developing countries. Earl (1975) suggests that the total world consumption of cow dung for fuel is not less than 150 million dry tonnes/y of which 40% is in India; other estimates vary from 80 million tonnes/y to 200 million tonnes/y. As in the case of agricultural residues, it is difficult to estimate the amounts of dung used as fuel, since most of it is a non-commercial, and has not been accounted for in any national statistics. It should be noted that the increased use of dung as fuel (by direct combustion) has deprived the land from a good fertilizer. Each tonne of cow dung burned may mean a loss of about 50 kg of food grain. The expansion in introduction of biogas technology can contribute largely to the alleviation of this problem. Table 98 gives the estimated domestic animal populations of the world and Table 99 gives the estimated annual production of dung.

Table 98. Representative Domestic Animal Populations (in million head)*

Animal	World Total	Developing Countries
Cattle	1201	685
Buffalo	132	99
Horses, Mules	121	73
Camels	14	13
Pigs	674	120
Sheep, Goats	1447	745

* After FAO (1975).

Urban refuse or municipal solid wastes constitute also an important source of biomass. The characteristics and quantity of the urban refuse are factors in not only the amount of energy that is recoverable, but also the type of energy recovery system that can be employed. The quantity of solid waste generated in most industrial countries is substantial. In the U.S.A., for example, about 115×10^6 tonnes of urban refuse are generated annually; in the UK, about 15×10^6 tonnes/y while in Spain the figure is about 9×10^6 tonnes/y (Pfeffer, 1979). Assuming that the energy content of U.S., refuse is about 11 GJ per tonne, the annual amounts of refuse in the U.S.A. could produce 1.3×10^{18} J/y.

Biomass as an energy source has positive and negative effects on the environment. On the positive side, it acts as a renewable energy store, as a sink for atmospheric carbon dioxide and it contributes to soil stabilization, reduction of water run off and desertification. Biomass provides a low-sulphur fuel and an inexpensive source of energy, especially for rural areas of developing countries. The conversion of organic wastes into fuel, in addition to providing a source of energy, reduces the environmental hazards associated with these wastes. On the other hand, if biomass is not properly managed (e.g. through excessive cutting of trees, slash and burn practices, etc.), soil erosion, desertification and related detrimental environmental consequences will occur. Moreover, land used for biomass production may be partially lost for other use—such as for food production.

Table 99.   Estimated Annual Production of Dung and its Energy Content

Source	Fresh Manure* production per head kg/y	Calculated** amount of manure t/y	Estimated*** amount of biogas production m^3/y	Energy content**** MJ (kcal)
Cow	5,400	64.9 x 10^8	19.5 x 10^{10}	42 x 10^{11} (10.2 x 10^{14})
Buffalo	7,000	9.2 x 10^8	2.8 x 10^{10}	6.2 x 10^{11} (1.5 x 10^{14})
Horses, Mules	1,700	3.3 x 10^8	0.66 x 10^{10}	1.5 x 10^{11} (0.35 x 10^{14})
Pigs	1,300	10.1 x 10^8	55.1 x 10^{10}	11.2 x 10^{11} (2.7 x 10^{14})
Sheep, Goats	500	7.2 x 10^8	3.6 x 10^{10}	7.9 x 10^{11} (1.9 x 10^{14})

* Fresh dung production (ESCAP, 1980); the amount of manure produced varies according to the weight of the animal, amount and nature of animal feed. The values given here are average values nearly representative to conditions in developing countries in general.

** Calculated figures using data in Table 98.

*** Estimated figures using data in Table 101.

**** Calculated figures assuming an average calorific value of 22 MJ/m^3 (5,300 kcal/m^3) of biogas.

CONVERSION TECHNOLOGIES AND THEIR
ENVIRONMENTAL IMPACTS

Two main processes are known to convert biomass into energy or synthetic fuel: thermochemical and biochemical. The first includes burning or complete combustion in open air, pyrolysis, gasification or liquefaction, and is generally applicable to terrestrial biomass with moisture content less than about 30%. Biochemical processes include anaerobic digestion and fermentation to produce synthetic fuel.

A. THERMOCHEMICAL CONVERSION

1. DIRECT COMBUSTION

Today direct combustion of biomass provides energy for cooking and heating to the majority of the rural population of the world. Biomass provides process heat and/or electricity throughout the pulp, paper and forest products industry. In addition, some utilities are producing small quantities of electricity from wood fired power plants. In rural areas, the main biomass source may be wood, crop residues and/or manure. The fuels used in the pulp, paper and forest industries often include the bark and other residues from sawmills and pulpmills.

Firewood:

Wood has been a primary fuel since man discovered fire. Wood is composed principally of cellulose and ligno-cellulose together with gums, resins, inorganic matter, and a variable amount of moisture, the amount depending on the kind of wood, the seasons in which it is cut, and the extent to which it has been allowed to dry. The approximate proportions by weight of the main elements present in wood are: carbon 50%, hydrogen 6%, oxygen 44%, and a trace of nitrogen. Fuelwood can be obtained from any tree whether occurring naturally or planted, either directly from the forest or from waste material produced at sawmills and wood-using industries. It is very simple to prepare and use the raw material for secondary energy sources such as charcoal, methyl alcohol etc. The most important controllable factor influencing the efficiency of wood as a fuel is the moisture content; the less the moisture content, the higher the calorific value would be. The gross calorific value of oven-dry wood averages

19.7 kJ/g, but in practice it ranges from 14.7 to 16.8 for wood dried naturally over a period of 4–8 months.

The use of wood as fuel accounted for about 47% of world wood consumption in 1977 — 1,184 million m³ out of some 2,500 million m³*. Wood fuel consumption in developed countries was about one-tenth of their total use of roundwood, while consumption in developing countries was about four-fifths of their total roundwood use (FAO, 1979). The importance of wood as a primary energy source varies widely among different parts of the world. Table 100 gives the consumption of fuelwood in different regions, from which it is clear that most consumption takes place in the developing countries.

Table 100.   Consumption of Fuelwood (in 10⁶ m³)

	1975	1976	1977	1978
Africa	285	292	297	298
N. America	17	18	18	18
Latin America	192	196	196	198
Asia	533	542	549	556
Europe (incl. USSR)	124	121	115	113
Oceania	7	6	6	6
Total	1158	1175	1181	1189

Source: United Nations, Statistical papers J–22 (1979).

In rural areas of the developing countries where wood is readily available, nearly 95% of households use it as a primary source of energy; the per capita consumption varies from 1.3 m³ to 2.3 m³/y, with an average of 1.5 m³/y (Openshaw, 1978). About 50% of the wood consumed as fuel is used for cooking; 30% for domestic heating and the remaining 20% for processing agricultural products, industry etc.

Environmental Impacts:

Firewood comes overwhelmingly from local sources, and this puts growing pressure on the trees, bushes and shrubs near to centres of population. Long before the demand for firewood leads to complete destruction of the tree cover, it

---

* It should be noted that these figures may be considerably underestimated. No accurate records are kept by many countries of self-collected or self-produced products such as fuelwood, charcoal (the so-called non-commercial sources of energy) and these are not accounted for in official statistics. Openshaw (1978) estimated the total world consumption of wood products to be nearly twice the official consumption figures, the largest difference being accounted for by woodfuel which is estimated to be three times greater than reported consumption. Accordingly, Openshaw (1978) estimated the woodfuel consumption in 1976 to be 3,050 million m³ instead of the official figure of 1,184 million m³.

can have a markedly degrading environmental effect. Excessive pruning of the branches may reduce a tree's capacity for growth; removal of the more easily-felled younger trees may reduce the regenerative ability of the forest; excessive opening of the canopy through the removal of too many trees can render the forest susceptible to damage from wind and sun and can affect wildlife generations; the removal of all residues, even to the point in some areas of sweeping up the leaves, removes the nutrients that should return to the soil to maintain its fertility; removal of stumps, bushes and shrubs can destroy much of what remains of the soil's protective cover and binding structure. And eventually, the whole forest may be felled and disappear (UNEP, 1977).

The uncontrolled and indiscriminate collection of firewood for cooking and heating, therefore, can have the most serious implications. Until the 1940s, forests had completely disappeared from most of China, because the trees had been cut down for fuel; in recent years, though, this trend has been reversed and vast areas have been successfully reafforested. In much of India, however, the forests are still rapidly declining (Eckholm 1975). With the growth of human populations, the forests are being cut down faster than they can grow, partly to make room for new farmlands, and partly for use as fuel. As a consequence, the upland area — the watersheds for the great rivers which flow through the plains — are subjected to destructive erosion, while the resulting sediments cause rapid filling of reservoirs and destructive floods downstream severely reduce the cover of cultivable soil and the food which can be produced from it. The forest areas of the developing countries are being destroyed at a rate of 16 million hectares a year (Sauma, 1978). This means that every three years a forest area is being lost that could supply in perpetuity, if properly managed, the firewood requirements of about 480 million people.

In semi-arid regions the ecological consequences of firewood consumption contribute to the process of desertification. Firewood is a scarce and expensive item throughout sub-Saharan Africa, all the way from Senegal to Ethiopia. A manual labourer in Niamey must now spend nearly one quarter of his income on fuel. But the price is higher than even he realizes. The caravans that bring this precious resource into the towns are contributing to the creation of desert-like conditions in a wide band along the desert's edge. Virtually all the trees within 70 km of Ouagadougou have been consumed as fuel by the city's inhabitants, and the circle of land stripped bare for firewood is continually expanding (Eckholm, 1975).

The forest clearing and the burning of firewood add to the atmospheric carbon dioxide budget. Recently, it has become evident that an additional input of non-fossil excess carbon dioxide into atmosphere, due to forest cutting, forest burning, soil management practices, etc., is of the same order of magnitude as the input of fossil fuel CO_2 (estimated at 5×10^{15} g C/y). The burning of fuel wood may inject CO_2 into the atmosphere at a rate of $0.2 - 0.4 \times 10^{15}$ g C/y; forest clearing in the developing countries is estimated to inject 3.6×10^{15} g C/y (SCOPE, 1979). About half of these quantities of carbon dioxide are removed from the atmosphere through different geochemical cycles, and the net contribution of non-fossil fuel sources to the carbon dioxide budget of the atmosphere is not accurately known but considered a substantial one.

The direct combustion of firewood results in emissions consisting mainly of particulates, polycyclic aromatic hydrocarbons and carbon monoxide. While data have long been available on emissions from industrial boilers burning wood residuals, only recently have data become available on emissions from wood stoves (Butcher and Buckley, 1977; Butcher and Sorenson, 1979). However, the available data base and overall understanding is much less than might be desired considering the potential impact of emissions from wood stoves (Morris, 1979).

The question for most developing country villages is not only how to replace firewood with other renewable sources of energy, but also how to supply firewood in an environmentally sound and sustainable way. There are two approaches which seem most promising at present with respect to firewood: using firewood more efficiently, and the more sustainable production of wood via fuelwood plantations. As burnt at present, firewood is not an efficient fuel. Cooking on an open, slow burning fire requires about five times as much energy as cooking on a kerosene stove (Eckholm, 1975). A study in Indonesia found that on the usual type of firewood stove 94% of the heat value of the wood was wasted (Singer, 1971). Simple improvements to stove design cut this loss by about 20%. Cutting the wood some weeks before use and then drying it in the open air reduced the loss by a further 10%, and a new type of cooking pot, partly sunk into the stove, cut heat loss by an additional 30%. In all, the consumption of firewood for cooking was reduced in these ways by about 70%. Several designs of stoves to increase the efficiency of using firewood have been described (see, for example, Evans, 1979). However, a number of political, cultural, historic, economic and technical factors contribute to the rather slow progress in the design and widespread use of efficient stoves (Joseph, 1979).

Trees, unlike oil, are a renewable resource, provided appropriate management and conservation measures are taken. The logical immediate response to the firewood shortage, one that will have many incidental ecological benefits is to plant more trees in plantations, farms, along roads, in shelter belts, and on unused land throughout the rural areas of the poor countries. For many regions, fast-growing tree varieties are available that can be culled for firewood within a few years, especially if the trees are coppiced (cut off at ground level and the stumps allowed to sprout new shoots). Some calculations, for example, suggests that a 40–60 ha plantation could supply the fuelwood needs of a 1,000-family village (Arismunandar, 1980). It has been estimated by the EEC that forest energy plantations on 13% of the area of the 9 countries of the Community could cope with 10% of the 1974 energy requirements. To make such land available in heavily-cultivated areas would face considerable political and cultural resistance, and the experience of the tree-planting programmes of many governments in wood-short countries over the past few decades has not always been encouraging. Nevertheless, if the energy needs of developing country villages are to be satisfied without continuing an increased environmental degradation, some way must be found of growing fuelwood on a sustainable basis in addition to the concerted efforts to harness other renewable sources of energy. One such way would be to plant trees that would provide fodder and food for the farmers as well as fuel. These could be scattered throughout the fields or in blocks. If such species as legumes were planted then they would improve the soil at the same time by fixing atmospheric nitrogen.

Wood and wood wastes generated during the processing of logs into products have been used in different countries in boilers for steam process production in connection with wood industry and/or in small related power plants. It has been estimated that wood wastes in the processing of logs amount to about 1.0 tonne (dry weight) per $2 - 2.5$ m^3 of lumber cut; the estimated annual amount of recoverable forest residue products in the U.S.A. is about 420 million m^3 (Cheremisinoff and Morresi, 1977). The collection of such residues for direct use as fuel for small nearby space-heating applications — especially for peak winter conditions — has been suggested (Burwell, 1978). The large scale use of wood and wood residues for electricity generation can only be realized in areas where the supply of wood is not limited. It is estimated the 1 MW(e)y requires about 6,200 tonnes wood (Earl, 1975). In other words a 100 MW(e)y power station (at 0.70 load) would require 434,000 tonnes of wood (about 600,000 m^3; the forest area to sustain such wood supply would be 30,000 hectares).

2. PYROLYSIS

When wood or other lignocellulosic biomass (e.g. a wide variety of agricultural residues and wastes like cotton gin waste, wood wastes, peanut hulls etc. and municipal solid wastes) is heated in the absence of air it breaks down both physically and chemically into a complex mixture of liquids and gases and a residual char, commonly known as charcoal. This process, known as "pyrolysis" can be carried out under a wide variety of conditions, the two extremes being either to maximize the production of charcoal, or to convert the total organic portion of the biomass into combustible liquids and gases (liquefaction and gasification). Pyrolysis is not a new technology. Charcoal has been produced in primitive earthen kilns from the dawn of civilization. But, using such primitive techniques, valuable by-products (oil/tar and gas) have been wasted. Modern pyrolytic conversion processes provide a means for recovery of these materials. For various reasons, these processes have been primarily applied to municipal refuse and agricultural and forestry process wastes. Thus, pyrolysis in its modern form offers a means for producing clean-burning zero-sulphur, storable and transportable fuels from otherwise wasted materials which are themselves major pollution sources.

Charcoal

The bulk of the world's charcoal is produced from wood by technologically primitive means. Traditional earth kiln methods yield one tonne of charcoal from

12 m³ of wood (8.6 tonne air dry wood). Modern industrial processes in commercial use are capable of producing charcoal from wood at yields close to the theoretical one which is 3.3 tonne of wood per tonne of charcoal product. Thus they are 2–3 times more efficient in their use of the wood than the traditional methods. A more common conversion factor is 6 m³ (or 4.3 tonne air dry wood) of roundwood per tonne of charcoal (Openshaw, 1978).

World charcoal production has been estimated to be about 4.6x10⁶ tonnes in 1978 (United Nations, 1979). However, it should be noted that this figure is based on information supplied by a limited number of countries; many countries do not have adequate information on charcoal production. In many developing countries much charcoal manufacture is carried out by itinerants working for only part of the year and keeping no records of what they produce and sell. Even in developed countries, the statistics of charcoal production are not always readily available and are often published in an ad hoc way. Openshaw (1978) estimated that about 570 million m³ of roundwood were used in 1976 for charcoal production. Assuming that the bulk of charcoal has been produced by traditional earth kilns which use about 12 m³ to produce one tonne of charcoal, the production of charcoal in that year would be about 48 million tonnes, which is nearly 10 times the official estimate.

Charcoal offers a number of advantages over wood as a fuel. Charcoal is easier to transport, store and distribute, more efficient in burning, less polluting, and it has special advantages in some industrial uses. Charcoal has a higher calorfic value than fuelwood. Comparing them both on a dry basis ready to use, charcoal has a calorific value 2–3 times that of wood (the average calorific value of charcoal is 30,400 kJ/kg while that of air dry wood is about 14,700 kJ/kg). Traditional charcoal making produces 1 tonne of charcoal from about 8 tonne of wood; the net efficiency of using that tonne of charcoal as fuel is nearly equal to that of using 8 tonnes of wood (wood stoves are about 6-8% efficient while charcoal appliances are 27–35% efficient). If modern conversion techniques are applied, less wood would be required to produce 1 tonne of charcoal and the net efficiency of the charcoal system will be much higher. In addition, appropriate techniques make it possible to transform into charcoal large volumes of wood of all dimensions which would otherwise have been destroyed in land clearing or left on the ground in exploited areas. Charcoal is traditionally preferred for a number of domestic uses as well as industrial, mainly metallurgical and chemical.

Environmental Impacts

The environmental impacts of production of charcoal are essentially the same as those described above under firewood, from which the bulk of charcoal is obtained. However, the use of charcoal is less polluting than firewood. It is smokeless, an essential requirement for cooking and heating in closed places, with much less emissions of particulates and hydrocarbons. Without adequate ventilation carbon monoxide poisoning may, however, occur.

Other Products of Pyrolysis

A wide variety of modern pyrolytic processes have been recently developed and applied to a variety of biomass resources: wood, agricultural and agro-industrial residues, urban refuse etc. These processes aim at maximizing the conversion of this material into, charcoal, oil/tar and/or gases (for detailed description of pyrolytic processes, see Weinstein and Toro, 1976; Tatom *et al.,* 1975; Golneke and McGaughey, 1976; Anderson and Tillman, 1977; Tillman *et al.,* 1977; Jones, 1978; Jones *et al.,* 1978).

The primary variables regulating the amount and quality of product yields are reaction time, reaction temperature, pressure, the presence of catalysts, the reactor type, etc. Considering the wide variety of processes developed or in development, the choice of a system for a particular application presents a difficult problem and involves many factors, including the type of feed to be utilized, its proximity to the conversion equipment, the desired product yields and other site-specific requirements. For example, a unit to convert agricultural wastes might emphasize char and oil production since a gas would likely be difficult to utilize. On the other hand, a system applicable to urban wastes might favour gas and/or oil production, since equipment to use these fuels may already be in use in many homes and factories. For rural applications where charcoal, oil/tar are suitable products, a low to medium temperature, vertical bed reactor using dried feed is most appropriate, while for urban uses where gas is more suitable, a high temperature convertor or a "flash pyrolysis" system would perhaps be most desirable.

Besides their use as industrial fuels, the products of pyrolysis have numerous applications. Recent US EPA supported work (Polk, 1977) indicates that the pyrolytic oils contain as many as 14 organic compounds, several of which are valuable compounds. Computation indicates that by separating these compounds from the remaining oil which would be used as a fuel, the value of the combined product would be increased by almost 200%. The charcoal produced by pyrolysis, depending on the feed characteristics, the reaction temperature etc., may be activated, thus allowing uses in water and perhaps air filtration systems. Moreover the charcoal, through densification could provide a useful fuel for mobile gasification plants to power heavy equipment such as tractors, trucks, etc. Pyrolytic gases have been used as fuel for a slightly modified spark ignition engine which operated at about 65% of its rated power (Tatom, 1978).

Environmental Impacts

Pyrolysis provides a means for efficient, economical conversion of residues (both urban and rural) which themselves are current major sources of pollution, into transportable, clean burning renewable fuels and chemicals. Additionally, pyrolysis provides an appropriate conversion technology for biomass, which today is an important source of energy for the developing countries. In addition, the world wide shortage of capital is forcing the re-examination of technologies such as pyrolysis which can be made to be relatively inexpensive and labour intensive. In the developing countries, where firewood and agricultural residues are used in a most inefficient way, pyrolysis offers several advantages by converting this biomass into high energy, dense fuels that could replace current charcoal and kerosene used for cooking and lighting (Chiang and Tatom, 1976). Moreover, the improved efficiency of utilization of the charcoal and oil — even after the conversion losses — reduces the amount of biomass required to the extent that perhaps 2.5 times as much food can be cooked with a given quantity of biomass after pyrolytic conversion than before (Tatom, 1977).

Because biomass and most residues are sulphur free, pyrolytic conversion of these materials results in fuels also free of this pollutant. These fuels can be burned by themselves to produce essentially pollution free emissions or in combination with high sulphur coal and/or oil to reduce SO_x and NO_x off-gas concentrations due to dilution of this element in the mixed fuel (Tatom et al., 1978). Tests at the Pittsburgh Energy Research Centre indicate that the charcoal and oil can be easily burned with pulverized coal and petrol respectively in unmodified fired boilers, and that the pulverized charcoal can be readily burned in an oil dung in unmodified oilfired boilers. In the latter case, the low ash and zero sulphur content of the char makes this an especially attractive application and in effect allows the substitution of a cheap renewable solid fuel for a significant fraction of petroleum; thus reducing pollution levels. The process of pyrolysis, in its various forms, can also be made to be environmentally clean. Since there is much great variation in the processes currently in development, it is difficult to make specific comments regarding the emissions from these separate methods. However, it appears that there are essentially no emissions that cannot be practically controlled to meet local air, water and ground pollution standards.

3. DISTILLATION

Methanol (CH_3OH) was originally produced on a limited scale in wood distillation plants as a byproduct of the production of charcoal. It has always been regarded as an excellent fuel because it burns clearly and can be readily transported and stored, although it has only about 50% of the calorific value of gasoline on a volume and weight basis. Methanol was used as a fuel during most of the eighteenth and part of the nineteenth century (for cooking, heating and lighting), until it was replaced by kerosene and other fuels. Interest in the use of methanol as alternative motor fuel dates back to the early 1910's. Methanol has a high heat of vaporization compared to gasoline which results in cooler fuel mixtures and a lower temperature of combustion. These lower temperatures may cause starting problems but they also result in lower nitrogen oxide emissions. This is important because these compounds are the major contributors to photochemical smog. Methanol also has a much lower calorific value than gasoline so fuel consumption will be higher on a weight or volume basis and larger gas tanks will be required. However, specific energy consumption (per km) will be lower because of higher compression ratios and the use of simpler emission control systems. Because of its slower burning characteristics methanol is a high octane fuel and therefore should not require aromatic supplements which are normally added to low-lead and lead-free gasoline for improving octane quality. Thus even the partial use of methanol should reduce the carcinogenic risk caused by the presence of these aromatic compounds in the fuel and in the exhaust gases.

One potential problem with gasoline-methanol blends is the possibility of phase separation in the presence of small amounts of water. This could cause problems in the car engines and therefore some method of keeping water out, or some additive to dissolve it in the fuel mixture, would have to be developed. Test data by a number of organizations have indicated that up to 20% methanol can be added to the gasoline without requiring any engine modifications but above this level some minor adjustments to the carburettor and fuel system are necessary (Bene et al., 1978). Another potential problem is toxicity. However, methanol has been used extensively during the past 150 years and there is no medical evidence to indicate that the addition of methanol to gasoline would materially change the hazards associated with the handling of this fuel.

During the 1920's a process was developed for producing "synthetic" methanol by passing a mixture of hydrogen and carbon monoxide over a catalyst at high temperatures and pressures. The hydrogen and carbon monoxide were produced by the reaction of steam with coke; charcoal produced from biomass would have served equally well except for its much higher cost. This process was used until the 1940's when natural gas became available in quantity and the process was then changed from using "water gas" produced from coke to "synthetic gas" produced by the reforming of methane. This is easily the most economical and most direct process known at the present time and all of the current production of methanol is based on the use of natural gas as a raw material. However, several recent studies have focussed on the use of forest and municipal wastes for the manufacture of methanol (Bene et al., 1978). The question at present is that of cost, and it is expected that the economics of producing methanol from biomass will become more favourable in the future as the costs of coal and oil increase.

B. BIOCHEMICAL CONVERSION

1. ANAEROBIC DIGESTION

The formation of methane and carbon dioxide by the anaerobic digestion of animal and vegetable matter was first recognized and reported during the latter part of the 19th century (Tietjen, 1975). This mixture of gases (essentially about 2/3 methane and 1/3 carbon dioxide with small amounts of hydrogen, nitrogen, organic sulphides and hydrocarbons) is commonly known as biogas (gobar gas or Marsh gas). During the past 50 years, some countries in Europe and North America have built anaerobic digesters to treat their sewage and in many cases have used the resulting biogas "sewage gas" as a source of energy. In 1951, for example, 48 sewage treatment plants, in the Federal Republic of Germany provided more than 16 million m^3 of biogas, 3.4% of which was utilized for power production, 16.7% for digester heating, 28.5% was delivered into the municipal gas supply system, and 51.4% was converted to vehicle motor fuel (Tietjen, 1975). Several small installations for production of biogas from manure were built in several European countries (e.g. in Federal Republic of Germany, France etc.) to solve the problem of fuel shortage during the Second World War. Apart from a few scattered small biogas plants in some developing countries (e.g. in India, People's Republic of China) in the 1950's, the main thrust to develop biogas plants came in the late 1960's and early 1970's. In India more than 36,000 biogas plants ("gobar" gas) were built up till 1975–76; now the figure reaches about 80,000, mostly small family scale plants. The largest number of biogas installations is in the People's Republic of China, where more than 7 million plants have been constructed; the largest number is in the province of Szechuan (UNEP, 1979). Nearly 27,000 biogas plants have been installed in Korea, 7,500 in Taiwan and a few hundreds in Pakistan, Nepal, Bangladesh, the Philippines, Thailand, Japan, Indonesia, Sri Lanka, etc. (Barnett et al., 1978). A few biogas plants have also been installed in a number of African countries (e.g. in Tanzania, Botswana, Ethiopia, Kenya etc.).

Anaerobic digestion of biomass* is a two-stage process. In the first stage, acid-forming bacteria act upon complex organics and change the form of complex fats, proteins, and carbohydrates to simple soluble organic material, commonly known as organic or volatile acids. The second stage involves the fermentation or gas-generation phase which produces the desired methane gas. In this step, the methane-forming bacteria use the organic acids produced in the first stage as substrate and produce the end products: carbon dioxide, methane and traces of other gases (see, for example, NAS, 1977; UNITAR, 1976).

The quantity and composition of the gases produced during anaerobic digestion are a function of the fraction of the total waste that is available to the anaerobic bacteria, i.e., the bio-degradable fraction, and the operating environmental conditions of the process. The more bio-degradable the waste material, the greater the quantity of methane generated per quantity of waste added to a digester. Not all wastes are equally bio-degradable and effective in producing methane. The bio-degradable fraction of waste will vary, being a factor of the characteristics of the material, the food ingested to generate the human or animal wastes, and how the wastes were handled prior to digestion. Untreated

---

* A wide variety of material has potential for biogas production, the most important are animal wastes, crop wastes, agro-industrial residues, aquatic weeds (e.g. water hyacinths), etc.

human wastes or fresh manure are highly bio-degradable material. When exposed to the natural environment for some time, natural degradation of the organic fraction takes place and the bio-degradable fraction will decrease (Loehr, 1978).

To attain continuous digestion, in a one-chamber set-up, a proper balance between the acid-forming bacteria and the methane-forming bacteria is required. Optimum levels of some environmental parameters are essential to the establishment and maintenance of this balance; these parameters are: temperature, pH, nutrients, and toxicity of input. Temperature is an important operational parameter in an anaerobic digestion process. As temperature increases, biological reactions proceed much faster, and this results in more efficient operation and lower retention-time requirements which may vary from 7 to 60 days. Two temperature levels have been established: in the mesophilic level, the temperatures range from 20–35°C; in the thermophilic level they range from 35 to 60°C. Although rates of reaction in the thermophilic level are much faster than those in the mesophilic level due to increased bacteria formation, the economics of most digestion systems have indicated operation in the mesophilic level. Another environmental requirement for anaerobic digestion is the maintenance of anaerobic conditions in the digester. The methane formers are strict anaerobes and even small amounts of oxygen can be quite detrimental to them. This necessitates a closed digestion tank which excludes oxygen while also facilitating collection of the methane produced. Another environmental requirement for optimum operation is proper pH control. Anaerobic digestion can proceed quite well under slightly acidic conditions, with a pH of 7 (± 0.5). Beyond these limits, anaerobic digestion proceeds with decreasing efficiency. Under more acidic conditions, a pH of 6.2 or lower, waste stabilization ceases. The bacteria responsible for waste fermentation in the anaerobic process require nitrogen, phosphorus, and other materials for optimum growth. Therefore, another important environmental condition is the presence of the required nutrients in adequate quantities. These nutrients are measured against a carbon-nitrogen (C–N) ratio. If there is insufficient nitrogen to allow the anaerobic bacteria to reproduce, carbon dioxide will be the principal gas produced. On the other hand, if the waste material contains excessive nitrogen, ammonia concentrate will increase and will inhibit the action of bacteria and gas production will slow or even cease. For optimum methane production, the C:N ratio should be in the range of 25–30 : 1 (ESCAP, 1980). If the C:N ratio in the waste material contemplated for methane generation is high (as in the case of crop residues etc.), nitrogen rich manure is added to adjust the ratio. On the other hand, if the C:N ratio is low (as in the case of poultry manure), straw or crop residues are added.

The composition of biogas produced depends on the type of material fermented, the design of the biogas plant* and the above-mentioned factors but it is generally in the range of (ESCAP, 1975):

* The basic design of all the biogas plants provides for an anaerobic digester (tank) and a gasholder to collect the biogas. Several designs have been constructed, each has its own performance characteristics, (for details see, for example ESCAP, 1975; NAS, 1977; Barnett et al., 1978; El-Hinnawi, 1977; van Buren, 1979).

Methane:	55–65%
Carbon dioxide:	35–45%
Nitrogen:	0– 3%
Hydrogen:	0– 1%
Oxygen:	0– 1%
Hydrogen Sulphide:	0– 1%

The calorific value of biogas varies with the methane content, and is generally in the range of 20–26 MJ/m^3.

The amounts of biogas produced depend on the above-mentioned factors and on the nature of the material digested (Zelter, 1979). Several estimates have been given in the literature for the biogas output. However, it is difficult to compare these estimates since in many cases the conditions under which the yields are given are not specified. Table 101 gives the biogas yields possible per unit mass of *fresh* matter (wet) fed to the digester and completely destroyed.

Table 101. Biogas Yield per unit Mass of Fresh Matter Completely Digested

Material	Biogas yield (m^3/tonne)
Cattle dung	22 – 40
Pig manure	40 – 60
Poultry	60 –115
Human faeces	20 – 28
Crop residues	30 – 40
Water hyacinth	40 – 50

Source: ESCAP (1980)

On the other hand, Table 102 gives the actual yield of biogas per unit mass of dry matter.

Many options exist for utilizing biogas. It can be used directly in gas-burning appliances for cooking, lighting and refrigeration or it can be used as fuel for internal combustion engines (after reduction of hydrogen sulphide content and converting the internal combustion engine to use biogas). Another possible use is the use of biogas for the production of electricity. It is estimated that 0.28 to 0.42 m^3 of biogas are required per person per day for cooking, 0.11 to 0.15 m^3 per hour for lighting a gas lamp equivalent to 60 W bulb and 0.6 to 0.70 m^3 of biogas for production of 1 kWh of electricity (NAS, 1977; ESCAP, 1978). Biogas is extensively used in India, the People's Republic of China, Korea, Nepal, and several other countries for cooking and lighting. It is also used for running irrigation pumps and machinery and for electricity production. For example, biogas from sewage in Fushan (Guangzhou, China) is used to operate a power plant with a capacity of 630 kW (FAO, 1979).

Table 102.   The Biogas Yield of Some Input Materials

Input material	Biogas yield (m³/tonne solid contents)	Methane contents %
Livestock dung:		
cattle dung	260–280 ⎫	
pig dung	561 ⎬	50–60
horse dung	200–300 ⎭	
Plant Residues:		
fresh weeds	630	70
wheat straw	432	59
green leaves	210–294	58
rice husks	615	60
Wastes:		
sewage wastes	640	50
liquid wastes from wine or		
spirit making factory	300–600	58

Source: after ESCAP (1980).

The effluent and sludge remaining, after digestion has taken place, is a rich and effective fertilizer. Anaerobic digestion of plant residues and animal wastes conserves nutrients needed for the continued production of crops; the only materials removed from the system for purposes other than putting back on the land are the gases generated — methane, carbon dioxide and hydrogen sulphide. The fate of nitrogen is particularly important; a major advantage of anaerobic digestion of biomass is the conservation in organic or ammonium nitrogen forms, of practically all the nitrogen present in the material used. The sludge produced by anaerobic digestion is believed to have a fertilizer value greater than that of the original raw material (perhaps due to more concentration in relation to volume of material). It has no offensive odour when spread on land and rodents and flies are not attracted to the remaining solid or liquid residues. It is claimed that the use of biogas slurry as a fertilizer had led to an increase in the yield of some crops by 10 to 28% in the People's Republic of China and India as compared to the use of excreta (van Buren, 1979). The use of biogas slurry can also give protection to crops during prolonged drought and also reduce the weeding cost as fertility of weeds is destroyed in the fermentation process. Experiments in the Philippines have shown that rice fertilized with commercial urea had an average yield of 6.5 tonne/ha whereas rice fertilized with manure sludge produced 8.3 tonne/ha (Eusebio and Rabino, 1978). The effluents can be used also for growing algae that can be used as animal feed and input material to the digester; the water from the algae ponds can be used for fish farming.

Environmental Aspects of Biogas

The environmental impacts of production and use of biogas should be considered in the context of the "organic waste — biogas — bioproductivity package (El-Hinnawi, 1977). Such package is illustrated in Fig. 12. Biogas technology provides a means of organic waste management that is beneficial. Not only agricultural and agro-industrial residues can be managed beneficially but also animal manure. This is particularly important in rural areas of the developing countries which lack adequate disposal systems. Manure is a major source of different parasites (e.g. schistosomes, hookworms, etc.), and anaerobic fermentation eliminates to a large extent these organisms (UNEP, 1979). This could lead to effective reduction in the incidence of parasitic diseases in rural areas. However, detailed research work is needed to determine the behaviour of different pathogenic organisms in the fermentation process and the amounts remaining in the slurry.

Like other renewable sources of energy, biogas can provide an appreciable fraction of the energy needs, especially in rural areas of developing countries. This will not only lead to accelerated development of such areas, but also to environmental production through the appropriate management and recycling of organic wastes. The use of biogas can also reduce the demand for firewood and charcoal in different regions (as it is the case in some areas in China), leading to the conservation of wood resources and reduction of the process of desertification. To illustrate the potential of large scale use of biogas, Eusebio and Rabino (1978) pointed out that if 60% of the organic wastes of the farm animals in Southeast Asia had been collected and utilized for methane production as energy source in 1975, the energy produced would have been equivalent to about 8.9×10^9 litres of petroleum. This amount would have cut down 15 per cent of the Southeast Asia total imports of mineral fuels, lubricants and related products. In addition, the amount of fertilizer that would have been produced could have saved about 207 million US$ worth of chemical fertilizer.

Biogas is a cleaner fuel than firewood, coal, etc. and its use for cooking in rural areas of the People's Republic of China and other countries has eliminated the pollution and health hazards associated with the use of wood or coal. The use of biogas has also discouraged wood cutting decreasing, thereby, deforestation and consequent soil degradation.

Several schemes of the "organic waste — biogas — bioproductivity" package have been experimented in some countries on a small scale or on a village level (for example, in China, Philippines, India, etc). The problems to be encountered in large-scale schemes are different from those encountered in small scale (or family-size) plants. These include land and water requirements; collection, storage and handling of manure and other wastes; handling of slurry; distribution systems of biogas, safety questions in relation to handling large quantities of biogas, etc.

Kuo and Jones (1978) studies the environmental impacts of large scale production of biogas. Assuming that biogas will be used for electricity generation,

Fig. 12   Biomass— Biogas– Bioproductivity

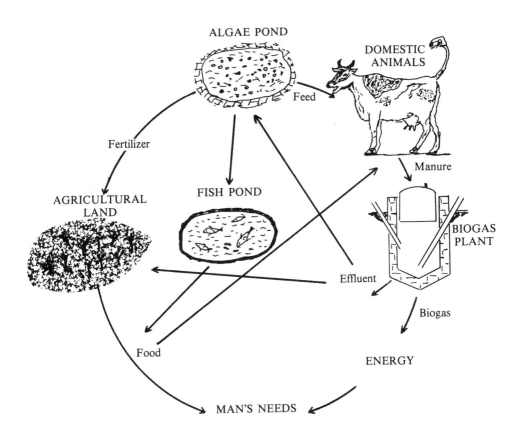

Table 103 gives the material balances for production of 100 MW (e)y, as calculated from the date given by Kuo and Jones (1978).

The water quantities required for digestion are zero when fresh manure is used and waste water is recycled. In case of no recycle, evaporation ponds may be used for liquid disposal to avoid discharge to surface water. Evaporation ponds could produce odours at certain times and become a source of ammonia and H_2S; however, photosynthetic organisms could oxidize the H_2S and minimize its release. This water effluent might be also used for irrigation.

Solid residues from anaerobic digestion of biomass are used as fertilizer and have less environmental impact than undigested manures that are spread directly on agricultural land.

Table 103. Material Balances in Production of Biogas to Generate 100 MW(e)y*

Organic waste Input (cattle manure, fresh)	$\left[\begin{array}{l}\text{1.48 x } 10^6 \text{ tonne solid} \\ \text{5.92 x } 10^6 \text{ tonne water}\end{array}\right]$	
Biogas produced	$\left[\begin{array}{l}\text{Gross amount:} \\ \text{Gas used for heating digester:} \\ \text{Net biogas:}\end{array}\right.$	$\left.\begin{array}{l}\text{4.1 x } 10^8 \text{ m}^3 \text{ (8.97 x } 10^{12} \text{ kJ)} \\ \text{5.3 x } 10^7 \text{ m}^3 \text{ (1.17 x } 10^{12} \text{ kJ)} \\ \text{3.5 x } 10^8 \text{ m}^3 \text{ (7.8 x } 10^{12} \text{ kJ)}\end{array}\right]$
Sludge	$\left[\begin{array}{l}\text{0.91 x } 10^6 \text{ tonne solids} \\ \text{2.77 x } 10^6 \text{ tonne water}\end{array}\right]$	
Wastewater	$\left[\begin{array}{l}\text{3.17 x } 10^6 \text{ tonne}\end{array}\right]$	

* Calculated from data given by Kuo and Jones (1978) assuming that 9,000 kJ are the minimum required per kWh and that the calorific value of biogas is 22 x 10^3 kJ/m³. A part of the biogas produced is used to heat the digester system. Emissions from boilers consist mainly of sulphur and nitrogen oxides; these are calculated as 180 tonne (S) and 163 tonne (N) to produce the amount of biogas necessary for production of 100 MW(e)y.

2. FERMENTATION

Production of ethanol through fermentation can be accomplished in at least three ways: (a) directly using naturally available sugars such as sugarcane; (b) indirectly using carbohydrate or starch sources such as cassava; (c) in combination with acid hydrolysis or enzyamatic hydrolysis of cellulose sources such as wood which produces sugar that then can be fermented. While alcohol production from sugar cane or cassava are familiar techniques, this is not generally so in the case of cellulose hydrolysis even though several plants using this process have operated in the U.S.S.R. for about 40 years. However, the economies of these processes, up till recently, have not been attractive; and the primary source of ethanol has been through chemical modification of petroleum derived ethylene which was more economical prior to the recent rapid increase in oil prices. In 1974, 75% of all industrial alcohol produced in the U.S.A. came from this source (Bene *et al.,* 1978). Apart from the more favourable economics it has been more legical to use petroleum as a source of alcohol rather than cereal grains and molasses, i.e. starch and sugars, because these materials have a more important use as a source of food for both humans and animals. However, the cost gap between these two methods will diminish as the supply of petroleum dwindles and the price increases, and, for these and other reasons, there will be a tendency to revert back to the fermentation technology.

The fermentation process is accomplished with yeast or bacteria, and the weight of the material is reduced by a factor of two in the process; the form of the material. has been changed from an unwieldly solid to an easily used liquid alcohol. Sugarcane is an excellent raw material for the manufacture of ethanol; sweet sorghum, corn and cassava have also been used. Sugarcane is the most efficient biomass that collects and stores solar energy in a useful form. In areas where the rate of photo-synthesis and carbon fixation are high (e.g. along the Amazon Basin, Equatorial Africa and in Equatorial Southeast Asia), the yield of sugarcane is as high as 60 tonnes per hectare per year. Table 104 gives the alcohol yield of some selected crops, from which it appears that on a weight basis (litre per tonne of crop), sugarcane yields the lowest amount of alcohol. However, sugarcane has a higher yield per hectare than other crops and hence the alcohol yield per hectare of sugarcane is the highest (Jackson, 1976).

Table 104.   Alcohol Yield of Selected Crops

Crop	Crop yield per ha/year (tonne)	Alcohol yield per ha (litre)	Alcohol yield l/t
Sugarcane (Brazil)	54.2	3630	67.0
Sweet Sorghum (US)	46.5	3554	76.0
Cassava (Brazil)	11.9	2137	180.0
Corn (U.S.A.)	5.7	2200	386.0

Source: After Brown (1980)

Several pilot-scale studies are underway in several countries for the production of alcohol from biomass; production programmes have already started in some others, the largest so far is the Brazilian National Alcohol Programme (Yang and Trindade, 1978; CTP 1979). In 1975, the alcohol consumption in Brazil was 414 x 10^3 m^3 (of which 35.6% was used as fuel alcohol, the rest in chemical and pharmaceutical industry) and the initial target of the Brazilian programme was to increase this amount to 5 x 10^6 m^3 (of which 73% will be used as fuel alcohol) by 1985 (Yang and Trindade, 1978). This target has been recently reformulated and the official supply of alcohol in 1985 was revised to 10.7x10^6 m^3; more conservative estimates indicate a figure of 7.5x10^6 m^3 (CTP, 1979).

Environmental Impacts

The production of alcohol is not without negative environmental impacts. Large areas of land should be allocated for growing sugarcane or cassava and in several countries such land may not be available or is available but the question whether land should be used for energy farms (see later) or for food production will arise. In the fermentation process, large amounts of effluents are produced. These include the raw material washwaters, stillage (bottom slop or waste residue drawn off the bottom of the first distillation column) and flegmass (bottom effluent from the second distillation column). Stillage is the most important effluent. It is generated at a rate which is 12 to 13 times in volume that of the alcohol production. The estimated production of 7.5×10^6 m^3 of ethyl alcohol in Brazil in 1985 will generate about 94×10^6 m^3 of stillage which if not suitably managed could create serious negative environmental impacts.

Because of its high content of soluble organics and inorganics, stillage has a high pollution potential if discarded into rivers. For example, the total Brazilian stillage output in 1977 (20 million m^3) was roughly equivalent to the sewage corresponding to 50 million inhabitants (Yang and Trindade, 1978). Since stillage normally does not contain pathogenic organisms or toxic compounds, recovery of its minerals and organics is a potentially attractive undertaking. It is technically feasible to convert stillage into marketable products such as fertilizers and feed additive or into methane as a supplementary energy source. Economics will ultimately determine the choice of stillage treatment process and the product form. Presently there are two stillage treatment processes being used in Brazil. The most widespread practice is lagooning. The other process takes advantage of the fertilizer value of stillage. Raw stillage is sprayed over the sugarcane fields, thereby reducing or eliminating the mineral fertilizer requirements (Yang and Trindade, 1978; CTP, 1979).

The energy requirements to produce ethyl alcohol from sugarcane, cassava and sweet sorghum were studied in detail by Da Silva *et al.,* (1978). It has been found that sugarcane is the more efficient crop for ethyl alcohol production, followed by sweet sorghum and cassava from a net energy viewpoint. Sugarcane shows a large net gain per year of 89.4×10^6 kJ/ha, which is 1.43 times the total energy consumed in the alcohol system production (agriculture, industry, etc.).

Up to 20% ethanol can be blended with gasoline without any changes in present-day internal combustion engines. Minor modifications, however, are required for engines running on 100% alcohol. Although ethanol has a lower calorific value than gasoline, ethanol has a higher density and a motor running on ethanol is 18% more powerful than a motor running on gasoline. When added to gasoline, alcohol increases the value of the octane rating. Instead of simply selling gasohol (90% gasoline and 10% alcohol) as a higher grade fuel, refineries could take advantage of this effect by producing a less refined gasoline to combine with alcohol. This would save oil at the refinery (Brown, 1980). Alcohol has the marked advantage of eliminating the use of lead anti-knock additives and of reducing the amount of hydrocarbon and nitrogen oxides emissions. From a toxicological point of view ethanol is much les toxic that methanol (Bene *et al.,* 1978).

ENERGY FARMS

The term "energy farming" means the growing of biomass for its fuel value. Traditionally one thinks of energy farms as forests; recently, however, several alternatives have been suggested, for example, weeds, agricultural crops, grasses and algae (fresh-water and marine). The feasibility of such energy farms is being studied in several countries.

Fast-growing trees have been introduced in some countries as a source of fuelwood. Eucalyptus has been selected in Australia, some parts of the U.S.A. and in some suitable areas of other countries as a source of wood biomass (Mariani, 1978; Hall, 1979). In the Philippines, the best species of fast growing trees seems to be the "giant ipil-ipil" (*Leucaena acidophila*) which fixes nitrogen to ammonia. A feasibility study has shown that a 9,100 ha fuel wood plantation would supply the needs of a 75 MW steam power station if it were not more than 50 km distant (Hall, 1979). Trial plantings of alder, willows, poplars, etc. are being also undertaken in Europe in a number of countries. Improved tree growth via short-rotation forestry is under intensive study in some countries. Some projections indicate that yields of dry organic matter can be substantially increased by coppiceing techniques and genetic improvements (Klass, 1978). Improved methods of harvesting, logging and utilization of residues (in other words improved forest management) would prevent deforestation and associated negative environmental impacts while providing for an adequate supply of fuelwood (to be used directly or converted to other fuels).

Biomass production techniques are being optimized for warm season grasses, sugarcane, sweet sorghum and other species. Improvements in production, by way of higher yields and/or lower costs are being developed through closer plant spacing, improved fertilization and irrigation methods, and higher harvesting frequencies. Species improvements via genetic manipulation and nitrogen fixation are future options under study to increase the yield of biomass for energy. Several problems are, however, inherent to this type of energy farm, the most important are: land areas required, fertilizer and water requirements, pesticide requirements and pressures on soil productivity. The numerous programmes to divert agricultural resources to the production of fuel crops should be carefully evaluated. Even without this diversion, efforts to expand world food production have been losing momentum for nearly a decade (Brown, 1980). Care should, therefore, be exercised not to encourage the production of energy crops at the expense of food production.

Thoughts of using algae and bacteria in biological solar energy systems are not new but have received more attention over the last few years. Algae ponds are open shallow ponds in which algae and bacterial populations work together to utilize sunlight and nutrients to produce cell mass. Under good conditions rapid growth with about 3–5% solar conversion efficiency can be obtained (Hall, 1979). The harvested algae may be fed directly to animals, fermented to produce methane, or burnt to produce electricity. Many liquid and semi-solid domestic and industrial wastes are ideal for the growth of photosynthetic algae and bacteria. The harvested biomass can be fermented to methane (equivalent to 11,600 kJ/kg algae) while the residues would contain virtually all the N and P of the algal biomass, so providing a good agricultural fertilizer — one hectare of algal ponds could supply the fertilizer required by 10–50 hectares of agriculture (Hall, 1979). In California, average yields of algae in excess of 100 kg dry weight/ha/day are obtained, with peak production in summer reaching three times this figure. Yields of 50–60 tonnes dry weight/ha/day would produce 74,000 kWh of electricity.

Through the mechanism of the photosynthetic carbon cycle, the green plant captures the carbon dioxide from the atmosphere and, with the aid of sunshine, separates hydrogen from the water to reduce the carbon dioxide first to carbohydrate. Some plant species can take the carbohydrate and reduce it all the way to hydrocarbon. The rubber tree (*Hevea braziliensis*) and other species like *Euphorbia* and *Avalois sp.* are examples of these plants. The idea being studied is that energy farms from such plants could be a useful source of hydro-carbons (Calvin, 1979; Hall, 1979).

References

Abu Wafa, T. and A. H. Labib (1973): See page losses from Lake Nasser In *Ackermann et al. : Man-Made Lakes*.
Amer. Geophysical Union. Monograph 17.

Ackermann, W. C. *et al.* (1973) : *Man-Made Lakes : Their problems and Environmental Effects*.
Amer. Geophysical Union. Monograph 17.

Allen, G. W. and McCluer, H. K. (1975) : Abatement of hydrogen sulphide emissions from The Geysers geothermal power plant. *Proc. 2nd U.N. Symp. Develop. Use Geothermal Resources,* San Francisco, 2, 1313–1321.

Andersen, S. O. (1975) : Environmental impacts of geothermal resource development on commercial agriculture. *Proc. 2nd U.N. Symp. Develop. Use Geothermal Resources,* San Francisco, 2, 1317–1321.

Anderson, L. and D. Tillman (1977) : *Fuel from waste.* Acad. Press. New York.

Arismunandar, A. (1980) : Personal Communication.

Armstead, H. C. H. (1975) : Environmental factors and waste disposal. *Proc. 2nd U.N. Symp. Develop. Use Geothermal Resources,* San Francisco, 1, 1xxxvii.

Armstrong, E. L. (1978) : Hydraulic Resources. In *World Energy Resources 1985–2020.* World Energy Conference IPC Science and Technology Press.

Arnold, J. E. M. (1979) : Wood Energy and Rural Communities. *Natural Resources Forum,* 3, 229.

Arnorsson, S. (1974 A) : The utility of waters from high-temperature areas for space heating. *Geothermics* 3, 127–141.

Arnorsson, S. (1974 B) : Composition of thermal fluids in Iceland. In *"Geodynamics of Iceland and the North Atlantic Area"*, D. Reidel Publ. Co., Dordrecht, Holland.

Arnorsson, S. (1975) : Geothermal energy in Iceland: utilization and environmental problems. *Naturopa,* No. 23, 23–26.

Auer, P. L. *et al.* (1978) : Unconventional Energy Resources. In *World Energy Resources 1985–2020.* World Energy Conference IPC Science Tech. Press.

Avery, W. H. (1978) : OTEC Status and prospects. *MTS Journal,* April-May.

Axtmann, R. C. (1974) : An environmental study of the Wairakei power plant. *N.Z. Dept. Sci. and Ind. Res. Phys. Eng. Lab. Rept.* No. 445.

Axtmann, R. C. (1975 A) : Environmental impact of a geothermal power plant. *Science* 187, 795–803.

Axtmann, R. C. (1975 B) : Emission control of gas effluents from geothermal power plants. *Environ. Letters* 8, 135–146.

Bahadori, M. N. (1978) : Passive cooling systems in Iranian Architecture. *Sci Amer.* 238, 144 (1978).

Balcomb, J. D. (1979) : Passive Solar Systems. *Proc. Solar Energy Today Conf.,* Melbourne.

Balligand, P. *et al.* (1979) : *Cong. International sur le dessalement de l'eau de mer et le recyclage des eaux usées,* Nice, Octobre.

Banwell, C. J. (1975) : Geothermal energy and its uses: Technical, economic, environmental and legal aspects. *Proc. 2nd U.N. Symp. Develop. Use Geothermal Resources,* San Francisco, 3, 2257–2267.

Barnes, F. J. (1979): Solar Water Heating. Proc. Solar Energy Today Conf. Univ. Melbourne. paper No. 3.

Barnett, A. *et al.* (1978) : *Biogas Technology in the Third World.* IDRC Report – 103e, Ottawa.

Bene, J. G. *et al.* (1978) : *Energy from Biomass for Developing Countries.* IDRC Report.

Benemann, J. R. (1978) : Biofuels, a survey. Electric Power Research Institute. ER–746–SR Special Report. Palo Alto, California.

Berkovsky, B. (1978) : Ocean Thermal Energy – Prospective Renewable Source of power. *Natural Resources Forum* 2, 337 (1978).

Biswas, A. K. (1978) : Environmental Implications of Water Development for Developing Countries. *Water Supply and Management,* 2, 4, 283–297.

Biswas, A. K. (1979) : Water; a perspective on global issues and politics. *J. Water Res. Planning and Mang.,* ASCE, 105, WR2, *Proc. Pap.* 14815, p. 205.

Biswas, A. K. (1980) : Non-radiological Impacts of Nuclear Energy; in El-Hinnawi, E. E. (Editor) *"Nuclear Energy and the Environment".* Pergamon Press, Oxford, 310 pp.

Boldizar, T. (1975) : Research and Development of geothermal energy production in Hungary. *Geothermics* 4, 44–56.

Bonnefille, R. (1976) : Les réalisations d'E.D.F. concernant l'énergie des marées. *La Houille Blanche,* No. 2.

Bowen, R. G. (1973) : Environmental impact of geothermal development. In *"Geothermal Energy",* ed. P. Kruger and C. Otte, Stanford University Press.

Bridges, J. E. (1975) : Biologic Effects of High Voltage Electric Fields. Electric Power Research Institute. Palo Alto, California. Report 381–1.

Bridge, J. E. (1977) : Environmental considerations concerning the Biological effects of Power Frequency (50 or 60 Hz) Electric Fields. *Proc. IEEE Paper.* F 77–256–1.

Bromley, L. *et al.* (1974) : Thermodynamic properties of sea salt solution. *Amer. Inst. Chem. Eng. Journal.* 20, 326.

Brown, A. W. A. and J. O. Doem (1973) : Health Aspects of Man-Made Lakes. In : *Ackermann et al* : Man-Made Lakes. *Amr. Geophysical Union.* Monograph 17.

Brown, L. R. (1980) : Food or Fuel; *Worldwatch paper* 35. Worldwatch Institute, Washington, D. C.

Burke, B. L. and R. M. Meroney (1975) : Energy from the Wind: Annotated Bibliography. Fat Collins, Colo: Colorado State Univ.

Burwell, C. C. (1978) : Solar Biomass Energy. *Science,* 199, 1041.

Butcher, S. S. and D. J. Buckley (1977) : *J. Air Pollution Control. Assoc.,* 27, 346.

Butcher, S. S. and E. M. Sorenson (1979) : *J. Air Pollution Control. Assoc.,* 29, 724.

Calvin, M. (1979) : Petroleum Plantations and synthetic chloroplasts. *Energy,* 4, 851.

Caputo, R. S. (1977) : Solar power plants : dark horse in the energy stable. *Bull. Atomic Scientists.* May 1977 p. 47.

Chartier, P. (1979) : European community biomass programme *Conf. Biomass for Energy,* London, June.

Chartier, P. and Meriaux, S. (1979) : La Biomasse, Passé, Présent, Futur. *Revue Génie Rural,* Octobre.

Cheremisinoff, P. and A. Morresi (1977) : Energy from Wood Wastes. *Environment,* 19, 25.

Chiang, T. T. and J. W. Tatom (1976) : Pyrolytic Conversion of Agricultural and Forestry Wastes in Ghana, a feasibility study. *Georgia Institute Techn.* Atlanta, Georgia. July 1976.

Clarke, F. J. P. (1979) : *Wave Energy Technology.* UNITAR Conf. Long-Term Energy Resources. UNITAR, UN, New York.

Clark, R. H. (1979) : *Prospects for Tidal Power.* UNITAR Conf. Long-Term Energy Resources, UNITAR, UN, New York.

Claude, G. (1928) : *L'énergie thermique des mers.* Conférence faite à la Sorbonne, February.

Cooper, P. I. (1979) : Non-concentrating collectors. *Proc. Solar Energy Today Conf.* Melbourne.

Cotillon, J. (1974) : La Rance, six years of operating a tidal power plant in France. *Water Power,* p. 314.

Coulter, G. W. (1978) : A preliminary appreciation of effects on aquatic environments of geothermal power development in New Zealand. *Geothermal Res. Council Trans.* 2, 119–120.

Count, B. M. (1979) : Exploiting Wave Power. *IEEE Spect.,* Sept. 79, p. 42.

CTP (1979) : *Alcohol: the Brazilian way out of the Energy Crisis* : Centro de Technologia Promon. Contrib. to UNITAR Conf. Long-Term Energy Resources. Paper CF7/VII/5, United Nations, New York.

Cyberski, J. (1973) : Accumulation of debris in Water storage reservoirs of Central Europe. In *Ackermann et al. : Man-Made Lakes.* Amer. Geophysical Union. Monograph 17.

D'Amore, F. (1975) : Radon-222 survey in Larderello geothermal field, Italy. *Geothermics* 4, 96-108.

Da Silva, J. G. *et al.* (1978) : Energy Balance for Ethyl Alcohol Production from crops. *Science,* 201, 903.

Dendy, F. E. *et al.* (1973) : Reservoir Sedimentation Surveys in the United States. In *Ackermann et al., Man-Made Lakes.* Amer. Geophysical Union. Monograph 17.

DOE (1977) : *Solar Thermal Power Systems.* DOE–EDP–0004; U.S. Department of Energy, Washington, D.C.

Dubreuil, P. and G. Girard (1973) : Influence of a very large number of small reservoirs on the annual flow regime of a tropical stream. In *Ackermann et al. :* Man-Made Lakes. Amer. Geophysical Union, Monograph 17.

Duffie, J. A. and W. A. Beckman (1974) : *Solar Energy Thermal Processes.* Wiley, New York.

Durand, H. (1975) : L'avenir des cellules solaires au silicium à usage terrestre. *L'onde Électrique,* 55, 3.

Dutcher, L. C. *et al.* (1972): U. S. Geol. Survey circ. 649, 57 p.

Earl, D. E. (1975) : *Forest Energy and Economic Development,* Clarendon Press, Oxford.

Eckholm, E. (1975) : The Other Energy Crisis: Firewood. *Worldwatch Institute Paper* 1.

Eckholm, E. (1979) : Planting for the future; Forestry for human needs. *Worldwatch Institute Paper* 26.

EEC (1979) : *International Conference on Solar Energy and Development,* Varese.

Einarsson, S, *et al,* (1975) : Disposal of Geothermal Water by Reinjection. *Proc. 2nd U.N. Symp. Develop. Use Geothermal Resources,* San Francisco. 2, 1349.

El-Hinnawi, E. E. (1977) : Energy, Environment and Development. 10th *World Energy Conference.* Istanbul 1977, paper 4. 8–1.

El-Hinnawi, E. E. (1977) : China Study Tour on Energy and Environment. Technical Report. *United Nations Environment Programme,* Nairobi.

El-Hinnawi, E. E. (1979) : Small Hydro-Electric Schemes and Rural Development. *Proc. Inter. Workshop on Energy and Environment in East Africa.* Nairobi, May 1979.

El-Hinnawi, E. E. (1980) : The State of the Nile Environment. *Water Supply and Management,* 4, 1-11.

Ellis, A. J. (1975) : Some geochemical problems in the utilization of geothermal waters. *Proc. Grenoble Symp., Int. Assn. Hydrol. Sci.,* Publ. 119, p. 100–109.

Ellis, A. J. (1976) : Geothermal fluid chemistry and human health. *Simp. Int. Sobre Energia Geothermica en America Latina,* Guatemala City.

Ellis, A. J. and Mahon, W. A. J. (1977) : *"Chemistry and Geothermal Systems",* Academic Press. New York, 392 pp.

Ellis, A. J. (1978) : *Environmental Impact of Geothermal Development.* Report prepared for UNEP. October 1978.

ENEL (1970) : Larderello and Monte Amiata : *Electric Power by Endogenous Steam.* Ente Nazionale per l'energia Elettrica, Rome.

Energies Nouvelles (1978) : *La Géothermie en France.* Collection Energies Nouvelles; B.R.G.M., Orléans, Février.

EPA (1976) : *Potential Environmental impacts of solar heating and cooling systems.* EPA—600/7–76–014. Agency, Washington, D.C.

EPA (1977) : *Western Energy Resources and The Environment* : Geothermal Energy. EPA–600/9–77–010 Washington, D.C.

ESCAP (1975) : Report of the Workshop on Biogas Technology and Unitilization. E/CN.11/1 HT/L.18 Bangkok.

ESCAP (1978) : Report of the Expert Group on Biogas Development., E/ESCAP/NR.5/1. Bangkok.

ESCAP (1980) : *Guidebook on Biogas Development,* Bangkok.

Etievant, C. and Pharabod, F. (1979) : Les Centrales Solaires de puissance. *Revue de l'Energie,* Mars 1979, p. 313.

Eusebio, J. and B. Rabino (1978) : Recycling System in Integrated Plant and Animal Farming. *Compost Sci.* March/April. 24.

Evans, E. (1979): Report prepared for FAO; December 1979.

FAO (1974) : Lake Nasser Area, Agricultural Potential. FI: DP/EGY/66/558. *Techn. Report* 4. Rome.

FAO (1974) : Fish Yield Projections on the Nasser Reservoir. FI: DP/EGY/66/558. *Techn. Report* 5. Rome.

FAO (1975) : *Production Yearbook.* Rome.

FAO (1979) : China, azolla propagation and small-scale biogas technology. *Soils Bull.* 41, Rome.

FAO (1979) : Issue Paper on Fuelwood and charcoal. Techn. Panel on Fuelwood. *UN Conf. New and Renewable Sources Energy.* Rome, January 1980.

REFERENCES

Fels, E. and R. Keller (1973) : World Register on Man-Made Lakes. In *Ackermann et al. : Man-Made Lakes*. Amer. Geophysical Union. Monograph 17.

Fernelius, W. A. (1975) : Production of salt water by desalting geothermal brines. *Proc. 2nd U.N. Symp. Develop. Use Geothermal Resources* 3, 2201–2208.

Gibrat, G. (1974) : L'énergie en l'an 2000. *Communication à l'Académie des sciences morales et politiques*, Paris, 12 Mars.

Gilbrat, R. (1966) : *L'énergie des marées*. Presses universitaires de France, Paris.

Giutronich, J. E. (1979) : Concentrating Collectors. *Proc. Solar Energy Today Conf.* Melbourne.

Glover, R. B. (1967) : The chemistry of thermal waters at Rotorua. *N.Z.J. Sci.* 10, 60–96.

Glymph, L. M. (1973) : Sedimentation of Reservoirs. In *Ackermann et al : Man-Made Lakes*. Amer. Geophysical Union. Monograph 17.

Golneke, G. and P. H. McGauhey (1976) : Waste Materials. *Annual Review of Energy Vol. 1*. Annual Reviews Inc. Palo Alto, California.

Goodman, N. and D. Pimentel (1979) : Biomass Energy Conversion as an alternate energy source. *Compost Science*, Jan./Feb. 28.

Gudiksen, P. H., Axelrod, M. C., Ermak, D. L., Lamson, K. C., and Nyholm, R. A. (1978) : Air quality assessment studies of geothermal development in the Imperial Valley. *Geothermal Res. Council Trans.* 2, 235–236.

Hall, D. O. (1979) : Solar energy use through biology. *Solar Energy*, 22, 307.

Hankins, B. E., Chevanne, R. E., Ham, R. A., Karkalits, O. C., and Palermo, J. I. (1978) : Chemical analysis of water from the world's first geopressurized geothermal well. *Geothermal Res. Council Trans.* 2, 253–256.

Haseler, A. E. (1975) : *"District Heating – an Annotated Bibliography"*. Property Service Agency, Dept. of Environment, London.

Hickel, W. J. (1972) : *Geothermal Energy*. Univ. of Alaska.

Von Hippel and R. H. Williams (1975) : *Bull. Atom. Sci.* 31, 25.

Hirschmann, J. R. (1970) : Salt flats as solar heat collectors for industrial purposes. *Solar Energy* 13, 83–97.

Holderness, A. L. (1978) : Solar Power for Telecommunications Search *Aust. Assoc. Adv. Sci.* 9, 143.

Howard, J. H. (1975) : Principal conclusions of CCMS non-electrical applications project. *Proc. 2nd U.N. Symp. Develop. Use Geothermal Resources*, San Francisco, 3, 2127–2141.

ICOLD (1977) : Contributions of Damns to the solution of energy problems. International Commission on Large Damns. *10th World Energy Conference*. Istanbul, 1977.

Inhaber, H. (1979) : Risk with energy from conventional and non-conventional sources. *Science*, 203, 718.

Inman, R. E. (1977) : Silvicultural biomass farms. Vol. I–VI *Mitre Technical Report* No. 7347. US. ERDA, Washington.

Isaacs, J. D. *et al.*, (1976) : Utilization of the Energy from ocean waves. *Proc. Workshop Wave and Salinity Gradient Energy Conversion. U.S. ERDA Report* No. COO–2946–1, Paper F–1.

ISEG (1978) : *International Solar Energy Congress*, New Delhi, 1978, 3 volumes.

ISEG (1979) : *International Solar Energy Congress*, Atlanta, Georgia, 1979. Abstracts.

Jackson, E. A. (1976) : Brazil's National Alcohol Programme Process Biochemistry, June 1976.

Jain, G. C. (1973) : Heating of solar pond. *Proc. Int. Congr. Semi. in Service of Mankind*, UNESCO, Paris. Paper EH 61.

Janes, D. E. (1977) : Background Information on High Voltage Fields. *Envir. Health. Persp.* 20, 141.

Jeffs, E. J. (1979) : The Application Potential of Hydro-power. *Energy*, 4, 841.

Jhaveri, A. G. (1975) : Environmental noise and vibration control at geothermal sites. *Proc. 2nd U.N. Symp. Develop. Use Geothermal Resources*, San Francisco, 2, 1375–1378.

Jones, J. (1978) : Converting solid wastes and residues to fuel. *Chem. Engin.* 2, 87.

Jones, J. L. *et al.*, (1978) : Pyrolysis, thermal gasification, and liquefaction of solid wastes and residues – Worldwide status of Processes. *ASME 8th Biennial National Waste Processing Conference*, Chicago, Illinois, May 1978.

Jones, J. O. and P. Rogers (1976) : *Human Ecology and Development of Settlements*. Plenum Press, New York.

Joseph, S. (1979) : Problems and priorities in developing wood stoves. *FAO Report*, December 1979.

Kaufman, G. E. and S. M. Michaelson (1974) : Critical Review of the Biological Effects of Electric and Magnetic Fields: In *"Biological and Clinical Effects of Low Frequency Magnetic and Electric Fields"*. Thomas Publ. Co. Springfield. Ill.

Keener, H. M. *et al.*, (1978): Simulation of Solar grain drying. *Agric. Engin. Series* 102, OHIO Agricultural Research Centre.

Klass, D. L. (1978): Energy from Biomass and Wastes — 1978 update. *Symposium on Energy from Biomass and Waste*, paper No. 1, p.1, Instit. Gas. Techn., Chicago.

Koga, A. (1970): Geochemistry of the waters discharged from drillholes in the Otake and Hatchobaru areas. *Geothermics* (Sp. Issue 2), 2(Pt 2), 1422–1425.

Kristoferson, L. (1977): Water power — A short Overview. *Ambio* VI, p.44.

Kruger, P. (1976: Geothermal Energy. In *Annual Review of Energy*, Vol. 1, p. 159. Annual Reviews Inc. Palo Alto, California.

Kunze, J. F.; Richardson, A. S.; Hollenbaugh, K. M.; Nicholas, C. R. and Mink, L. L. (1975): Non-electric utilization projects, Boise, Idaho. *Proc. 2nd U.N. Symp. Develop. Use Geothermal Resources*, San Francisco 3, 2141–2145.

Kuo, M. C. T. and Jones, J. L. (1978): Environmental and Energy output analysis for the Bioconversion of agricultural residues to methane. *Symposium on Energy from Biomass*, paper No. 10, p. 43. Instit. Gas. Techn., Chicago.

Laszlo, J. (1976): Application of the stretford Process for H_2S abatement of the Geysers geothermal power plant. 1976 *Intersociety Energy Corrosion Conf.* (A.I.Ch.E., HT and EC Div.), 7 pp.

Latif, A. and Rashid, M. (1973): Catch of Fish from Lake Nasser. In: *Ackermann et al., : Man-Made Lakes*. Amer. Geophysical Union. Monograph 17.

Lavi, A. and Lavi, G. H. (1979): Ocean Thermal Energy Conversion. *Energy*, 4, 833.

Leibowitz, L. P. (1977): Projections of future H_2S emissions and geothermal power generation: The Geyser region, California. *Geothermal Res. Council Trans.* 1, 183–185.

Loehr, R. C. (1978): Methane from Human, Animal and Agricultural Wastes. In *N. L. Browne: Renewable Resources and Rural Applications in the Developing Countries*. AAAS selected Symposium 6, Washington, D.C.

Makhijani, A. (1875: *Energy and Agriculture in the Third World.* Ballinger Publ. Co. Cambridge, Mass.

Makhijani, A. (1976): *Energy Policy for the Rural Third World.* IIED, London.

Marchaud, P. (1979): L'énergie thermique des mers. *La Recherche,* No. 105.

Mariani, E. O. (1978): The Eucalyptus Energy Farms as a Renewable Source of Fuel. *Symposium on Energy from Biomass and Waste.* paper No. 2, p. 29. Institute Gas. Techn., Chicago.

Maurel, G. (1979): Energie solaire et dessalement, *Conférence Inter. Dessalement,* Nice, Nov. 1979.

McGowan, J. G. and Heronemus, W. E. (1975): Ocean Thermal and Wind Power. *Environmental Affairs,* vol. 4, No. 4, 629.

McKee, J. E. and Wolfe, H. W. (1971): *Water Quality Criteria.* California State Water Resources Control Board, Publ. 3A.

Mehta, G. (1979): Salt stratifical solar ponds, paper CF7/XIII/5. *UNITAR Converences on Long-Term Energy Resources,* United Nations, New York.

Mercado, S. (1967): Geoquimica Hidrotermal en Cerro Prieto, B.C. Mexico. Commission Federal de Electricidad, Mexicali.

Mercado, S. (1975): Cerro Prieto geothermoelectric project: pollution and basic protection. *Proc. 2nd U.N. Symp. Develop. Use Geothermal Resources,* San Francisco 2, 1394–1398.

Mercado, S. (1976 A): Ceppo Prieto geothermal field, Mexico: wells and plant operation. *Proc. Int. Congress on Thermal Waters, Geothermal Energy and Volcanism of Mediterranean Area.* 1, 394–408, National Technical University, Athens.

Mercado, S. (1976 B): Disposicion de desechos geothermicos. *Simp. Int. Sobre Energia Geotermica en America Latina,* Guatemala City.

Mercado, S. (1977): Salmuera de desecho de la planta geotermica de Cerro Prieto. Problematica y possibles beneficios. Boletin Inst. Investig. *Electricas* 1, 12–14.

Mermel. T. W. (1976): International Activity in Dam Construction. *Water Power and Dam Construction,* April 1976, p. 66.

Merriam, M. F. (1978): Wind, Waves and Tides. *Annual Review of Energy,* Vol. 3, p. 29. Annual Reviews Inc. Palo Alto, California.

Mitre Corporation (1975): Energy use and climate. Possible effects of using solar energy instead of stored energy. M 74–66 Rev. 1, NSF–RA.N74.185.

Morris, S. C. (1979): *Health Aspects of Wood Fuel Use.* Brookhaven Natl. Lab. (Draft).

Morse, R. N. (1977): Solar heating as a major sources of energy for Australia. *10th World Energy Conference.* Istanbul Paper 4. 2–3.

Morse, R. N. *et al.,* (1977): Solar Energy as Heat and for Fuel. *Tech. Conf. Inst. Engin.* Canberra, July 1977).

Muffler, L. J. P. (1975): Present status of resources development. *Proc. 2nd U.N. Symp. Develop. Use Geothermal Resources,* San Francisco, 1, xxxiii.

Muffler, L. J. P. and White, D. E. (1969): Active metamorphism of Upper Cenozoic sediments in the Salton Sea geothermal field. *Geol. Soc. Am. Bull.* 80, 157–182.

Muffler, L. J. P. and White, D. E. (1972): Geothermal Energy. *Sci. Teacher* 39, 40–43.

NAS (1976): Energy for rural development. *U.S. National Acad. Sci.,* Washington, D.C.

NAS (1977): Methane Generation from Human, Animal and Agricultural Wastes. *U.S. National Acad. Sci.,* Washington, D.C.

Nath, J. H. and Williams, R. M. (1976): Preliminary Feasibility study for utilization of water wave energy. *Proc. Workship Wave and Salinity Gradient Energy Conversion.* U.S. ERDA Report No. COO–2946–1, Paper I. 1.

Nemec, J. (1973): Interaction between Reservoirs and the atmosphere. In *Ackermann et al. : Man-Made Lakes.* Amer. Geophysical Union Monograph 17.

Nielsen, C. (1975): Salt-gradient solar ponds for solar energy utilization. Environmental Conservation 2, 289.

Noel, J. M. (1979): Les centrales éoliennes. *Annales des Mines.* April.

Noguchi, T. (1966): Exploration of Otake steam field. *Bull. Volcanol.,* 29, 529–544.

NZED (1977): Broadlands Geothermal Power Development Environmental Impact Report. New Zeland Electricity Dept. Wellington.

Olafsson, J. (1977): Mercury and aresenic in Mamafjall and Krafla drillholes. Jokull (in press).

Olafsson, J. and Riley, J. P. (1978): Geochemical studies on the thermal brine from Reykjanes, Iceland. *Chemical Geology 21, 219–237.*

Openshaw, K. (1978): Woodfuel — a time for Re-assessment. Nat. Resources Forum, 3, 35.

OTA (1978): Application of Solar Technology to today's energy needs. Vol. 1. Office of Technology Assessment, Congress of the U.S.A., Washington, D.C.

Otmer, D. F. and Roels, O. A. (1973): Power, Fresh water and food from cold deep sea water. *Science,* 182, 121.

Ovchinnikov, A. M. (1955): Mineral waters of the health resort Karlovy Vary in Czechoslovakia. Vopr. Kurortol. Fiz. Lecheb. 3, 66. (Chem Abstr. 52, 2311).

Palz, W. (1978): Solar Electricity. Butterworths, London.

Panicker, N. N. (1976): Power Resources Potential of Ocean Surface Waves. *Proc. Workshop Wave and Salinity Gradient Energy Conversion.* U.S. ERDA Report No. COO–2946–1 Paper J–1.

Partl, F. R. (1977): Power from Glaciers. IIASA, R.R.–77–20, Laxenburg, Australia.

Peters, W. *et al.,* (1978): An appraisal of World Coal Resources. In *World Energy Resources 1985–2020.* World Energy Conference. IPC Sci. Tech. Press.

Peyches, Y. (1977): Répercussions climatiques des utilisations des énergies primaires, cas de l'énergie solaire. *Revue générale de thermique,* No. 184.

Pfeffer, P. (1979): UNITAR Conf. Long-Term Energy Sources, UNITAR, UN, New York.

Phéline, J. (1977): Développment et Utilisation des Sources Alternatives d'Energie. *Revue Génerale Thermique,* No. 185.

Phéline, J. (1979): Energie Solaire et production décentralisée d'électricité. *Annales Des Mines,* April 1979.

Pimentel, D. *et al.,* (1978): Biological Solar Energy Conversion and U.S. Energy Policy. *Bioscience 28,* 376.

Polk, M. B. (1977): Development of methods for the stabilization of Pyrolytic oil. Annual Report USEPA Grant R–80440010, Atlanta Univ. Atlanta, Georgia.

Raheja, P. C. (1973): Lake Nasser. in *Ackermann et al: Man-Made Lakes.* Amer. Geophysical Union. Monograph 17.

Read, W. R. (1978): Solar water heating for domestic and industrial use. *Search,* 9, 130.

Read, W. R. (1979): Proc. Solar Energy Today Conf. Melbourne.

Reed, M. J. and Campbell, G. E. (1975): Environmental impact of development in The Geysers geothermal field, U.S.A. *Proc. 2nd U.N. Symp. Develop. Use Geothermal Resources,* San Francisco 2, 1399–1410.

Reay, P. F. (1972): The accumulation of arsenic from arsenic-rich natural waters by aquatic plants. J. Appl. *Ecology* 9, 557–565.

Robertson, D. E., Fruchter, J. S., Ludwig, J. D., Wilkerson, C. L., Crecelius, E. A., and Evans, J. C. (1978): Chemical characterisation of gases and volatile heavy metals in geothermal effluents. Geothermal Res. Council Trans. 2, 579–582.

Rodin, L. E. *et al.* (1975): Productivity of the World's Main Ecosystems Proc. 5th General Assembly, Special Committee IBP. National Acad. Sciences, Washington.

Rodot, M. and Vasseur, P. (1977): Cahier No. 2 de l'Afedes, Paris.

Rothé, P. (1973): Geophysics Report. In *Ackermann et al.: Man-Made Lakes.* Amer. Geophysical Union. Monograph 17.

Sargent, S. L. et al. (1980): Solar Industrial Process Heat. Env. Sci. Techn. 14, 518 (1980).

Sauma, E. (1978): Speech at 8th World Forestry Congress, Jakarta.

Sax, N.I. (1975): "Dangerous Properties of Industrial Materials" (4th Ed.), Van Nostrand Reinhold Co., New York.

Schroeder, H. (1971): "Air Quality Monograph". Amer. Petroleum Institute, Washington.

Scope (1979): The Global Carbon Cycle. *SCOPE* 13, J. Wiley, New York.

Shaw, T. (1975): Tidal power and the environment. New Scientist, October 1975 p. 202.

Siegel, S. M. and Siegel, B. Z. (1975): Geothermal hazards: mercury emission. Environ. Sci. Technol. 9, 473–474.

Simmons, D. M. (1975): Wind Power. New Jersey, Noyes Data Corp. 270 pp.

Singer, H. (1971): Improvement of Fuelwood Cooking Stoves and Economy in Fuelwood Consumption. FAO, Rome.

Smil, V. (1976): Exploiting China's Hydro potential. Water Power and Dam Construction. March 1976 p. 19.

Sørensen, B. (1976): Wind Energy. *Bull. Atomic Scientists.* September 1976 p. 38.

Sørensen, B. (1979): Wind Energy. UNITAR Conf. on Long-Term Energy Sources, UNITAR, UN, New York.

Sørensen, B. (1980): *Renewable Energy.* Acad. Press, London.

Stoker, A. K. and Kruger, P. (1975): Radon in geothermal reservoirs. *Proc. 2nd U.N. Symp. Develop. Use Geothermal Resources,* San Francisco 3, 1797–1803.

Suter, G. W. (1978): Geothermal development and wildlife behaviour. Geothermal Res. Council Trans. 2, 627–629.

Swanberg, C. A. (1975): Physical aspects of pollution related to geothermal energy development. *Proc. 2nd U.N. Symp. Develop. Use Geothermal Resources.* San Francisco 2, 1435–1443.

Swann, M. (1976): Power from the Sea. *Environment* 18, p. 25.

Tabor, H. (1963): Large Area Solar collectors for power production. *Solar Energy,* 7, 189.

Tabor, H. (1966): Solar Ponds. *Science Journal* 66, 66–71.

Tabor, H. and R. Matz (1965): A Status report on Solar pond projects. *Solar Energy, 9,* 177.

Tatom, J. W. et al. (1975): A Mobile Pyrolytic System. In *W. Jewell: Energy, Agriculture, and Waste Management.* Ann Arbor Science, Ann Arbor, Michigan.

Tatom, J. W. (1977): Demonstration of alternate fuel production through pyrolysis of agricultural wastes at the UNEP Rural Energy Centre in Senegal. Atlanta, Georgia 1977.

Tatom, J. W. *et al.* (1978): A prototype Mobile System for pyrolysis of agricultural and/or silvicultural wastes. Georgia Techn. Final Report. under EPA Grant R–803430. Atlanta, Georgia.

Thompson, C. R. and Kats, G. (1978): Effects of continuous H_2S fumigation on crop and forest plants. *Environ. Sci. and Technol.* 12, 550–553.

Tietjen, C. (1975): From Biodung to Biogas. In *W. Jewell: Energy, Agriculture and Waste Management.* Ann Arbor Science, Ann Arbor, Michigan.

Tillman, D. A. *et al.* (1977): Fuels and Energy from Renewable Sources. Acad. Press, New York.

Torrey, V. (1976): Wind-Catchers. Brattleboro, Vt: Stephen Greene.

Torrenti, R. *et al.* (1976): Dimensionnement d' un bassin de stockage intersaisonnier COMLES, Marseille. *Revue d'heliotechnique,* 2e Semestre, 1976.

Trombe, F. and Michel, J. (1977): Revue chauffage, Ventilation, conditionnement, Mai. Publications C.N.R.S.

UK ISES (1979): Biomass for energy. Conf. UK Sect. Inter. Solar Energy Soc., Royal Society, London, July.

Umarov, Y. U. *et al.* (1971): Investigating the heat regime in a saltwater solar pond. *Heliotechnology* 7, p. 24.

United Nations (1979): World Energy Supplies 1973–1978. Statistical Paper J 22. UN New York.

UN Committee Natural Resources (1977): E/C.7/64. United Nations, New York.

UNEP (1977): Environmental Issues in River Basin Development UN Water Conference, E/CON. 70/A.26. United Nations, New York.

UNEP (1977): The State of the Environment. Nairobi.

UNEP (1978): The State of the Environment. Nairobi.

UNEP (1979): Environmental Impacts of Nuclear Energy. Energy Report series ERS–2–79. Nairobi.

UNEP (1979): Environmental Impacts of Fossil Fuels. Energy Report Series ERS–1–79. Nairobi.

UNEP (1979): Environmental Impacts of Fossil Fuels. Energy Report Series UNEP–ERS–1, Nairobi.

UNEP (1979): Solar-2000. UNEP Report 8. Nairobi.

UNEP (1979): Biogas utilization and comprehensive Development in China. ESCAP/UNEP Regional Seminar on Alternative Patterns of Development, Bangkok.

UNIDO (1978): Technology for solar energy utilization. Development and Transfer of Technology Series No. 5 UNIDO, Vienna.

UNITAR (1976): Microbial Energy Conversion. New York.

Van Buren, A. (1979): A Chinese Biogas Manual. Intern. Techn. Publ. London.

Velker, J. A. and Axtmann, R. C. (1978): Sulfur emission control for geothermal power plants. M/S in preparation. Princeton University, Chemical Engineering Dept.

Voss, A. (1979): Waves, Currents, Tides — Problems and prospects. *Energy,* 4, 823.

WAES (1977): Workshop Alternative Energy Strategies. Edit. C. Wilson: McGraw Hill, New York.

Walton, J. D. *et al.* (1978): A state of the art. Survey of solar powered irrigation pumps, solar cookers, and wood burning stoves for use in sub-sahara Africa. Georgia Institute Techn. Atlanta, Georgia.

Wehlage, E. F. (1979): Hydro and Hydro electric Power, UNITAR Conf. Long-Term Energy Resources, Paper CF7/XVI/2, UNITAR, New York.

Weinstein, N. J. and R. F. Toro (1976): Thermal Processing of Municipal Solid Waste for Resource and Energy recovery. Ann Arbor Science, Ann Arbor, Michigan.

Weissberg, B. G., and Zobel, M. G. R. (1973): Geothermal mercury pollution in New Zealand. *Bull. Environ. Contamination and Toxicol.* 9, 148–155.

White, D. E. (1965): Geothermal Energy. *US Geol. Survey Circular* 519.

White, D. E. (1965): Saline waters of sedimentary rocks. In *Fluids in subsurface environments — a Symposium.* Am. Soc. Petrol. Geologists Memoir No. 4, 342–366.

White, D. E. (1968): Environments of generation of some base-metal ore deposits. *Econ. Geol.* 63, 301–355.

Wick, G. L. and J. D. Isaacs (1976): Utilization of the energy from salinity gradients. Proc. Workshop Wave and Salinity Gradient Energy Conversion. U.S. ERDA Report No. COO–2946–1, paper A–1.

Williams, J. (1979): Climatic impact of alternative energy sources. *Energy,* 4, 933.

World Bank (1979): World Development Report 1979, Washington, D.C.

Yang, V. and S. C. Trindade (1978): The Brazilian Gasohol Programme. Symposium on Energy from Biomass and Waste, paper No. 42, p. 813. Instit. Gas. Techn., Chicago.

Zeigler, J. M. *et al.* (1976): Environmental effects arising from salinity gradient and ocean wave power generating plants. Proc. Workshop Wave and Salinity Gradient Energy Conversion. U.S. ERDA Report No. COO–2946–1 paper D–1.

Zelter, S. Z. (1979): Methanogenese et déchets agricoles. *Revue Genie Rural,* Octobre.

Zener, C. (1976): Solar Sea Power. *Bull. Atomic Scientists* Jan. 1976, p. 17.

Zumer-Linder, M. (1976): Unprocessed Wood for Housing, Building and Fencing. OECD/CSTP 1976.

SUMMARY

THE rising global demand for energy has been met to an increasing extent by the use of fossil fuels, especially oil, which were cheap and plentiful. Recently, many countries have realized that non-renewable sources of energy are finite in extent and that diversification of energy sources is a must for future development. Concern for future energy supplies is reflected in the programmes of many national governments and in the efforts being made by a number of international organizations to assess global energy resources and possible rates of supply and substitution. Energy policies are nowadays influenced by several factors: population growth, level and nature of socio-economic activity, the relative costs of energy, the adequacy and reliability of supply, the availability of technology and supporting infrastructure, the success of energy conservation programmes and concern about the environmental and safety aspects of production and use of energy.

At local, national and in some cases regional levels, the environmental aspects of energy production and use have become the subject of wide-ranging debate. Environmental awareness and anti-pollution campaigns have affected the formulation of energy policies in many countries, and it has recently been realized that nations are not isolated in this respect: the actions of one country may affect the environment in a neighbouring one. Environmental objectives should not, however, be seen as being inconsistent with, or as imposing constraints upon, energy policy. A balance should be maintained between the need to preserve and improve the quality of the environment and the socio-economic goals and needs which depend on the availability of energy. Nowadays, energy policy decisions are dictated less by technological than by social, environmental and political factors. Although some countries are sensitive to the environmental aspects of energy production and use, there is still need for a comprehensive and more coherent consideration of the subject.

1. Fossil Fuels

Fossil fuels (coal, oil and natural gas) are used for many purposes: domestic, industrial, transportation and for electricity production. The production, transportation, processing and use of fossil fuels have several environmental impacts; some are substantial and others small, some important and others of little consequence, some of short duration and others with long term effects and they might occur in different geographic areas and might affect different communities in different ways. For many environmental changes which are identified as impacts, the state of knowledge and technology will often only permit a qualitative assessment. Only in a few cases it is possible to evaluate an impact quantitatively. Since fossil fuels account for about 75% of the world electricity generation, the environmental impacts of the coal, oil and natural gas fuel cycles have been quantified in this report — when feasible — on the basis of 1 GW(e)y for easy comparison between the different cycles.

Coal mining has a number of environmental impacts. The occupational hazards, especially the incidence of black-lung disease (coal worker's pneumoconiosis) are more marked in underground than in surface mining. The estimated fatal and non-fatal accidents in coal mining are about 0.6 and 45 per 1 GW(e)y in underground mining; the figures for surface mining are 0.3 and 16, respectively. In underground coal mines the number of occupational deaths per 1 GW(e)y is of the order of 0.21. It should be noted that these estimations would vary from one country to another depending on the safety measures adopted to reduce such occupational hazards.

Coal mining, especially surface strip mining, has a potential impact on land. The land area disturbed vary according to the geological occurrence of the coal-bearing formation and its characteristic. Where surface mining is carried out in densely populated areas, it has a direct effect on human settlements and the total infrastructure in the area. Reclamation of strip mined areas has been successfully achieved in several countries. Both underground and surface coal mining have an impact on the water resources in the area of mining. Acid mine drainage has altered the water quality of more than 10,000 km of streams in the U.S.A. and has reduced or eliminated aquatic life in many of them. Different methods for treating acid mine drainage are known and stringent control measures are implemented in some countries to avoid the hazard of acid mine drainage.

Coal conversion processes are dually oriented: (a) to convert plentiful coal of different grades into scarce liquid and gaseous fuel and (b) to remove or treat, during processing, environmentally unacceptable or health — endangering compounds. However, coal conversion involves the handling of large amounts of coal often at high temperatures and pressures; it requires containment and control of both highly corrosive process materials and those that pose a possible health threat; and it calls for treatment and disposal of a voluminous solid waste and a possibly hazardous liquid and gaseous waste. Knowledge of these wastes and emissions and their environmental impacts is still incomplete.

Exploration and production of oil and natural gas, whether carried out on land or offshore, have a number of environmental impacts. Besides occupational hazards, offshore operations are of concern because of the possibility of accidental oil spills. Although some field studies have shown that there is no ecological damage due to offshore operations, further studies are necessary to determine the long-term effects of hydrocarbons, accidentally spilled or resulting from normal operations, on the marine ecosystem. This is particularly important for sensitive eco-zones such as tidal marshes, coastal wetlands, river swamps and sheltered bays.

Marine oil transportation is the most important means of transporting oil from oil-exporting to oil-importing countries. Although the world tanker fleet generally has a good safety record, the fact that ships today are so much larger than they were 20 years ago means that the consequences of an accident are potentially much greater. Experience has shown that such incidents can have serious effect upon the environment and marine life, damaging such important resources as fisheries and tourism. Although oil spillage from transportation (including normal and terminal operations) account for about 35% of all oil discharges in the sea, the consequences of a major spill near coastal zones are far reaching. The environmental impacts of such spill depend on the composition of the cargo (whether crude oil or refined products) and on the climatic conditions. The cleaning-up operations depend on a number of factors: ecological and commercial interests at the site of the spill, legal interests, technical factors and the cost of the cleaning-up operations.

Oil refineries are large industrial installations with air and water emissions, large water requirements for processing and cooling — unless air cooling is used extensively — and safety problems due to the risk of explosions and fires. The most important emissions from refineries are airborne effluents (SO_x, organic compounds, NO_x, CO and particulate matter) and liquid effluents. The latter contain a variety of compounds, the most important of which are oil and grease, phenols, ammonia and suspended and dissolved solids. These wastes are normally treated and the effluents discharged from most refineries contain low concentrations of such pollutants.

The exploitation of oil shale raises a number of economic and environmental problems. The environmental disruption associated with oil shale mining is typical of that of any large surface or underground mining operation, except that size of the operation will mean that the scale of the disruption will be much greater. In the case of surface retorting (recovery of oil from shale in a retort vessel) the disposal of spent shale will create land disturbances of large magnitude, accumulation of toxic substances in vegetation, and contamination of ground and surface water from runoffs. The oil shale industry will exert a potential impact on local water resources due to the demand for clean process water, the need for removal of process effluent discharges and mine dewatering.

The environmental effects of large scale tar sands development can be widespread and severe, if insufficient planning and financing accompanies such development. Land disturbance during the mining of tar sands can be appreciable. The mine area can be restored as in the case of a surface coal mine, with one exception, the reclamation and revegetation. The most abundant solid waste material in the process of extraction of oil from tar sands is the mineral matter sent as tailings to the settling pond.

The combustion of fossil fuels (coal, oil, and natural gas) give rise to a number of airborne effluents, the most important of which are sulphur oxides, nitrogen oxides, carbon monoxide, particulates, organic compounds, trace metals and radionuclides and carbon dioxide. The quantities of these emissions vary according to the fuel used, its composition and the measures adopted at the power plant to reduce the emissions.

Man-made sulphur emissions account for about 40% of the global sulphur budget of the atmosphere; 75–85% of these man-made emissions are from fossil fuel combustion. Sulphur is normally emitted from power plants as SO_2 which is rapidly oxidized to SO_3 and converted to sulphuric acid aerosol or to sulphates. These oxidation products affect a wide variety of materials: plants, soil and other components of the biosphere. Sulphur emissions contribute largely to acid precipitation which has led to the acidification of several fresh water bodies in some countries, with consequent detrimental effects on aquatic life.

One of the primary problems in determining the health effects of sulphur dioxide and other emissions continues to be the development of an understanding of the manner in which these pollutants interact with one another in the atmosphere. Knowledge is insufficient to relate each airborne pollutant independently to health. The effects of airborne pollutants have been quantified in studies on animals, plants, in studies on man and in populations occupationally exposed to high levels of mixtures of different pollutants. Exact estimates of the hazards cannot be precisely defined without knowing the interactions creating the damage. Most epidemiological studies show good correlation between particulates and health effects; the correlation between SO_2 and health effects is less sharp, and it has been difficult to show correlation between impaired health and nitrogen oxides or oxidants. Particulates, especially fine ones that cannot be removed by present removal devices, are of particular concern because they can lodge easily deep in the lungs. Trace elements, polycyclic aromatic hydrocarbons, and other compounds absorbed on these particulates increase the damage through synergistic effects.

Concern has been recently voiced at the implications of extensive coal utilization, especially its possible impact on climate. The atmosphere is believed to show a warming primarily due to the greenhouse effect of increasing carbon dioxide emissions. Studies with climate models and of climate observations indicate that regional anomalies will probably occur, but that the magnitude and impacts of these anomalies cannot be reliably predicted at present. The medium and long-term effects of such possible climatic changes are of such fundamental importance that they command serious attention.

The thermal efficiency of modern fossil fuel power plants varies from 38–40%, i.e. about 60% of the heat energy generated has to be rejected into the environment, in the vicinity of the power station. These thermal discharges can have detrimental effects on the water-based ecosystems and/or cause micro-climatic changes. However, thermal discharges have been put into beneficial use in a number of countries and further development prospects are under investigation.

2. Nuclear Energy

Public concern about nuclear power development has focussed on a number of issues, the most important of which are: (a) the effects of radiation on man, both somatic and genetic, which may be associated with some stages of the nuclear fuel cycle; (b) the safety of nuclear installations; (c) the environmental impacts associated with radioactive waste management; (d) the availability of plutonium in case of reprocessing of spent fuel and fear of its theft and misuse and (e) the socio-economic aspects of nuclear power development.

The total installed nuclear generating capacity in the world in 1978 was 110.8 GW(e) from 227 power stations operating in 20 countries, constituting about 6% of the world electrical power capacity. Estimates of the world nuclear power growth by the turn of the century vary between 900 to 2000 GW(e). This wide variation is due to the complexity of the factors involved in energy forecasts, which includes: socio-economic, environmental, geo-political and technical ones. A recent forecast by IAEA estimates the total nuclear power growth in the world to be 1100-1700 GW(e) by the year 2000, contributing about 20-26% of the total electrical generating capacity. Since light water reactors will continue to be the main type of reactors in use, at least until the year 2000, this study deals mainly with the environmental impacts of the fuel cycle of LWR's.

Man has always been exposed to ionizing radiation from various natural sources: external, such as cosmic rays and radioactive substances in the ground and in building materials, and internal sources in the form of naturally ocurring radioactive substances in the human body. In addition, man has been also exposed to radiation from man-made ionizing radiation. On the average, a member of the world population receives a whole body dose of about 100 mrem/y from natural radiation and about 50-80 mrem/y from man-made sources, of which less than 1.0 mrem/y is due to radiation from coal and nuclear power industry. Although our knowledge about the health effects of ionizing radiation has greatly advanced over the past years, the effects of low levels of radiation have not yet been well-defined and have recently been the subject of public concern and controversy.

Although uranium is a fairly common constituent of the earth's crust, and the cumulative uranium ore requirements by the year 2000 are within known world uranium resources, considerable exploration effort is required before such quantities of uranium can be classified as proven reserves. By the year 2025, the cumulative uranium requirements for LWR's without recycle will greatly exceed the presently estimated world resources of 4 million tonnes. Accelerated efforts for the exploration and exploitation of new uranium resources and/or development of alternative fuel cycles to meet the projected increasing demands of the nuclear industry seem inevitable.

Uranium ores are usually mined by underground or surface mining. Like any other mining operations, this leads to land disruption (which is greater in surface mining than in underground mining), possible changes in the hydrological regime and pollution of nearby ground waters. Exposure to radon daughters is considered to be the most important radiological occupational hazard in uranium mining. The total occupational deaths in underground uranium mining are estimated to be 0.23-0.34 per GW(e)y, of which 0.03-0.1 are from radiation exposure (lung cancer).

Milling of uranium ores requires in some regions considerable land areas, most of which are devoted to ponds for the permanent disposal of mill tailings. The water pollution hazards associated with uranium milling operations depend on the milling process; the management of liquid wastes is very dependent on climate and ranges from treatment prior to discharge to almost total retention. Dissolved toxic substances may then have the potential for percolation into the ground water, or for direct seepage to near-by surface waters. The levels of radon are much lower in the uranium mills than in the mines; the occupational dose from milling is insignificant compared to that from mining. The local collective dose commitments from milling operations depend mainly on airborne releases to radon daughters, which contribute to external exposure from deposited material and to internal exposure via inhalation and ingestion pathways. The inferred cancer mortality from such exposure is 0.005 per GW(e)y. Inactive piles mill tailings present a potential for exposure to radiation; the doses to the general population near inactive mill sites can be greatly reduced by adequate design of the tailings pond and by the use of a clay cover. However, the radioactivity in the piles will continue for very long periods of time and sound techniques for accomplishing the long-term stability and isolation of tailings have not yet been demonstrated.

The uranium hexafluoride conversion, the uranium enrichment and fuel fabrication processes do not have major environmental impacts. All of the land involved in these operations can be reclaimed for unrestricted use except for that used for waste burial. Most of the liquid and airborne chemical effluents can be treated by conventional methods. The uranium enrichment process is the largest consumer of electricity in the whole nuclear fuel cycle; about 36 MW(e)y is needed by a gaseous diffusion plant to enrich uranium required for the production of 1 GW(e)y in a LWR. The enrichment process is also a dominant user of cooling water next to the nuclear reactor in the fuel cycle. The radiological impact of the hexafluoride conversion, uranium enrichment and fuel fabrication processes constitute a minor component of the impact of the entire fuel cycle.

The selection of sites for nuclear facilities depends on technical, socio-economic and environmental factors. Land requirements for power plants vary from one plant to another and is generally about 40 ha per 1 GW(e). Land requirements for cooling systems depend on the system used, cooling towers require about 6.5 ha while cooling ponds require about 860 ha per 1 GW(e).

The thermal efficiency of current light water reactors is about 33% which means that almost two-thirds of the heat energy generated in a reactor core has to be rejected to the environment in the vicinity of the power station. As in the case of other thermal power stations, these thermal discharges can have detrimental effects on the water-based ecosystems and/or cause micro-climatic changes. However, thermal discharges have been put into beneficial uses in a number of countries and further development prospects are under investigation.

Normal operation of a nuclear power station gives rise to a number of fission and activation products. A small part of the radioactive material produced is released in airborne and liquid effluents and part as solid wastes. The occupational radiation exposure at LWRs is mainly due to gamma radiation from fission and activation products. Exposures to the individual are usually kept below 5 rem/y which are within the limits of internationally recommended standards. The population around a nuclear power plant receives very low doses of radiation and existing studies have been unable to establish a correlation between nuclear facilities and increased mortality in the general public. More analyses and research are needed to establish any relationship between low levels of radiation and ill health or mortality.

The probability of "hypothetical" reactor accidents of various degrees of severity has been estimated in several studies. The most comprehensive of these studies "The Rasmussen Report" estimates the probability of such accidents to be very low and compares the calculated risks of reactor accidents with other risks. However, a recent review of the Rasmussen study concluded that the ranges of uncertainty in the probability estimates are understated.

The uranium and plutonium in the spent fuel from LWR's are valuable energy sources. When both are re-cycled, reductions in uranium ore requirements and in the processes at the front-end of the fuel cycle (milling, UF_6 conversion and enrichment) can be achieved. Although the techno-economic, health and safety and environmental aspects encountered in re-cycling are more or less similar to those encountered in no-recycle option, they require different solutions. The re-cycling of plutonium would introduce a traffic in purified plutonium, which might require safeguards in addition to those in effect.

The most important airborne effluents from reprocessing plants are carbon-14, krypton-85, iodine-129 and tritium. Caesium-134 and 137, strontium-90 and tritium are important radionuclides released in liquid effluents. The occupational radiation exposure in recent reprocessing plants is estimated to be well below the dose limit of 5 rem/y applicable to occupational exposure of the whole body. On the basis of current technology the individual risk to members of the workforce of a reprocessing plant is predicted as less than 10^{-4} per year. For the critical groups of the general public the individual risk is predicted as less than 3×10^{-6} per year. Both the occupational risk and the risk to the general public are low compared with levels currently accepted.

Radioactive wastes are generated in practically all areas of the nuclear industry and accumulate as either liquids, solids or gases with ranging radiation levels. The bulk of the wastes occur at the front-end of the nuclear fuel cycle which includes mining and milling, while the more radioactive wastes occur at the back-end of the cycle which includes reactor operation and fuel reprocessing (in case of re-cycling). The latter wastes are generally considered as low, intermediate, high-level and wastes contaminated with transuranic elements. Low-level and intermediate wastes are normally disposed of by shallow land burial or by dumping in the deep ocean in specially designed containers. High-level and transuranic wastes must be conditioned and subsequently disposed of, normally in a solidified form, at a suitable repository. Such material has sufficiently persistent biological hazards and require special long-term isolation. Several options have been proposed, the most feasible of which at present is believed to be the disposal in deep geological formations or in the deep ocean bed.

Decommissioning of nuclear installations is technically feasible. However, the issues involved in decommissioning are complex; the various aspects of the problem — technical, economical, radiological, environmental and organizational are conflicting in many ways, and will not be resolved until waste disposal routes are defined. Reactors are likely to present the greatest problems because of their large numbers, the high levels of induced radioactivity and the very large volume of radioactive waste which will be produced.

Under normal conditions, shipments of material required for the annual fuel requirements of LWR (1GW(e)y) expose the general population to very small radiation doses. The estimated maximum annual dose received by a truck driver transporting nuclear fuel cycle material is estimated to be 2.2 rem. The probability of an accident occurring in the transportation of radioactive material is small because of the different precautions taken. In case of an accident, the detriment has been mainly due to reasons other than radiation. No serious injuries from radiation have hitherto been reported from transportation accidents.

3. Renewable Sources of Energy.

The general realization of the finite nature of fossil fuel resources has caused a re-examination of the possibility of using those energy resources which are of a non-depletable nature and, therefore, considered renewable. Such sources of energy are of primary importance both to the developed and developing countries. In the former countries strategies for the exploitation of such sources constitute a part of recent policies which aim at reducing the dependence on fossil fuels to satisfy the growing needs for energy. In the developing countries, particularly those short of fossil fuel resources, the development of technologies to harness renewable sources of energy in a more efficient way constitutes a promise for meeting the future energy needs to accelerate the process of development, particularly in rural areas. In other words, renewable sources of energy have become an important component of national energy supply systems.

The resource base of renewable sources of energy is extremely large; but due to the diffuse nature of most of these sources, only a fraction of the resource base can be exploited at suitable geographical sites. With the present state of technology, it is difficult to estimate how much of the resource base can be technical and economically exploited. Recently, many countries have started major research and development programmes for widescale harnessing of indigenous renewable sources of energy, the results of which would lead to a reliable estimate of the contribution of renewable sources of energy to the overall world energy supply systems.

Renewable sources of energy vary widely in their impact on the environment. Although our knowledge of the environmental impacts of geothermal energy and hydropower is more advanced than that of other renewable sources of energy, little is known of the health impacts and/or quantitative environmental impacts of these sources. In the present study, the assessment of the environmental impacts of renewable sources of energy is carried out on the basis of available information; speculative assessment has been avoided, specially of the impacts of energy systems that have not reached the pilot scale experimentation.

Since geothermal energy must be utilized or converted in the immediate vicinity of the resource to prevent excessive heat loss, the entire fuel cycle, from extraction to transmission, is located at one site. This offers environmental advantages in terms of land area requirements and use and in terms of effluents management. In addition, geothermal power stations do not generally need an external source of water for cooling, since the condensed steam is recycled for that purpose. On the other hand, geothermal energy has a number of negative environmental effects that are site specific, varying according to the geochemical characteristics of the geothermal reservoir. Land subsidence resulting from the withdrawal of geothermal fluids from the reservoir is a potential negative impact of geothermal development is some areas. The problem can, however, be alleviated in a number of ways, especially by reinjection of geothermal fluids to deep wells, following power production. Hydrogen sulphide is the main airborne effluent of real concern in geothermal fields. The main problem encountered is its objectionable smell. It also tends to concentrate around geothermal plants, but so far no health problems have been encountered in communities living near these plants. Liquid effluents from geothermal power stations contain a variety of chemical elements in different concentrations, depending on the geochemical characteristics of the geothermal field. Methods for disposing of wastewater include direct release to surface water bodies, evaporation, desalination with subsequent water reuse, and reinjection to the production reservoir. The selection of one method or another depends on local hydrologic conditions, water requirements, quality of the waste water, economics of recovery of salts and environmental regulations.

At present, the most immediate use of solar energy is for water heating; the hot water can be used for domestic or industrial purposes. Solar drying of products, space heating and cooling, refrigeration, water distillation and desalination, cooking, water pumping, thermal electricity conversion and photovoltaic conversion are other applications of solar energy that are receiving increased attention in many countries. Solar energy devices have a number of environmental impacts. Decentralized small units do not only reduce the demand for fossil fuels leading to conservation of such non-renewable sources of energy, but also to the reduction of the amount of pollutants emitted by burning such fuels. The use of solar energy in developing countries, especially in rural areas, can lead to substantial improvements in the quality of life. Although land requirements for solar power plants are comparable to those of conventional thermal power stations, most of the land can be used for different purposes.

The most probable trend in the development of wind energy in the near future is most likely to focus on a much increased use of wind machines in the 5 to 100 kW range for water pumping and rural electrical systems. Machines for generating electric power in the range of 100 kW to 5 MW are also being developed. Environmental concern about wind energy involves such factors as the risk of accidents, noise, interference with telecommunications and the possibility of micro-climatic alterations.

Several devices have been proposed to harness the energy of the sea. Today wave energy is used on a small scale to power buoys, the average power output of these systems ranges from 70 to 120 W. Because there are no large-scale wave power stations existing today it is difficult to assess the environmental effects of harnessing this source of energy. However, wave power plants will produce no thermal discharges or emissions or cause changes in water salinity or require fresh water for operation. The most direct environmental impact is to calm the sea, since they will act as efficient wave breakers. This has beneficial effects in several locations near harbours offering safe anchorage in times of storms, and/or protecting shorelines from erosion. However, the calming of the sea might have adverse biological effects because of the absence of waves and associated mixing of the upper layers of the sea. Tidal power can be harnessed at some suitable sites. Different schemes have been proposed; some are under development, others are in operation (the largest tidal power station in operation is La Rance in France). The ecological effects associated with the development of tidal power vary from one place to another and depend to a large extent on the magnitude of the project. Several prototype systems of ocean thermal energy conversion (OTEC) have been suggested, and further research and development is needed to solve the technical, economical and environmental problems associated with OTEC. The environmental problems range from questions dealing with the biological and ecological aspects of antifouling agents used in OTEC systems, to those dealing with the ecological and environmental impacts due to changes in salinity and thermal redistribution.

No dam can be built and no lake can be created for hydro-power development without environmental costs and benefits of some kind. The many consequences on the environment of the dam and the lake behind it appear to be factors in common regardless of the dam's geographical location. These environmental impacts are generally divided into: local effects within the area of the man-made lake, and the downstream effects resulting from a change in the hydraulic regime. Both categories have their physical, biological and socio-economic elements. Siltation in the man-made lake and associated effects both in the lake area and downstream; changes in the physical, chemical and biological properties of the water in the lake and downstream and associated health, agricultural and biological impacts and the socio-economic changes in the lake area are among the most important aspects of large hydro-power development projects. On the other hand, the building of dams and the creation of reservoirs provide for better water management and development of irrigation systems to increase agricultural land and for production of power needed to accelerate industrialization and socio-economic development. Environmental impact assessment and cost/benefit analyses taken up at an early stage in the development of hydro-power projects would lead to adequate planning and maximization of the benefits of such projects.

Biomass is a potential renewable source of energy, produced by photosynthesis, directly in the case of plants and indirectly in animals whose ultimate food source is plant material. Biomass can be land plants and crops, aquatic plants and/or organic wastes of different kinds. Biomass can be converted into energy or synthetic fuel by two main processes: thermochemical and biochemical conversion. The first includes direct burning or complete combustion, pyrolysis, gasification or liquefaction, while the second includes anaerobic digestion and fermentation.

Today, direct combustion of biomass provides energy for cooking and heating to the majority of the rural population of the world. The biomass used is mainly wood, crop residues and/or manure (dry cow-dung cakes). The use of wood as fuel accounts for about 47% of the total world wood consumption; in the developing countries it accounts for about 80%. In rural areas of these countries, wherever wood is readily available, nearly 95% of households use it as a primary source of energy (mainly for cooking, heating and in small rural industries). The uncontrolled and indiscriminate collection of firewood lead to soil erosion and degradation and enhance the desertification process. It is estimated that the forest areas of developing countries are being destroyed at a rate of 16 million hectares a year. Afforestation programmes and proper forest management are the main pre-requisites for ensuring an adequate fuelwood supply without causing ecological degradation.

The pyrolytic conversion of wood into charcoal has been known from the dawn of civilization. Charcoal has been produced in primitive earthen kilns, and modern conversion techniques have been developed to increase the charcoal yield from the same amount of wood. Charcoal offers a number of advantages over wood as a fuel. It is easier to transport, store and distribute; it is more efficient in burning, less polluting and it has special advantages in some industries. Charcoal can be also produced, together with oil/tar and gases from the pyrolytic conversion of agricultural and agro-industrial residues, urban refuse, etc. Pyrolysis, therefore, provides a means for efficient conversion of these residues — which themselves are current major sources of pollution — into transportable, clean-burning renewable fuels and chemicals.

Biogas production is one component of a closely interrelated "Biomass — Biogas — Bioproductivity" package. Such package provides a means for the beneficial management of organic waste, especially in rural areas of the ·developing countries. Besides disposing of such waste, biogas is produced and constitutes an important source of energy for cooking, lighting, as fuel for internal combustion engines and for production of electricity. In addition, the effluent and sludge remaining after digestion is a nutrient-rich fertilizer; it can be used directly in the fields and/or for growing algae that can be used as animal feed. It has been found that anaerobic digestion eliminates most of the pathogenic organisms found in the manure, and this could lead to effective reduction in some parasitic diseases in rural areas. It should be noted, however, that the environmental problems encountered in the case of large-scale biogas schemes are different from those encountered in the case of family-size plants. The former include land requirements; collection, storage and handling of organic waste; production, storage and distribution of biogas and related safety questions; handling the large quantities of slurry and waste water produced, etc. Such impacts remain to be accurately assessed.

Ethanol can be produced through fermentation of sugarcane, cassava, corn, sweet sorghum, etc.; sugarcane is the most efficient in terms of net energy yield.

Several programmes for the production of alcohol are underway in some countries; the largest so far is the Brazilian National Alcohol Programme. Up to 20% alcohol is blended with gasoline without any changes in present-day internal combustion engines. This will lead to savings in crude oil requirements in addition to the fact that alcohol increases the value of the octane rating eliminating, thereby, the use of lead anti-knock additives and the associated undesirable environmental impacts. The production of alcohol from fuel crops raises, however, a number of important issues: land area requirement; fertilizer, water and pesticide requirements; the pressures on soil productivity and above all the possible competition with food production. In addition, the fermentation process produces large amounts of effluents (mainly stillage) which if not suitably managed could create serious negative environmental impacts. However, it is technically feasible to convert stillage into marketable products such as fertilizers and feed additives or into methane as a supplementary energy source.

Appendix I

UNITS AND CONVERSION FACTORS

Multiples of Units

tera T	=	10^{12}
giga G	=	10^{9}
mega M	=	10^{6}
kilo k	=	10^{3}
milli m	=	10^{-3}
micro μ	=	10^{-6}
nano n	=	10^{-9}
pico p	=	10^{-12}

Length (metre : m, kilometre : km)

1 ft	=	0.3048 m
1 mile	=	1.609 km

Area (hectare = ha)

1 ha	=	$10^{4}\,m^{2}$
1 acre	=	0.4047 ha

Volume (m^{3})

1 litre	=	$10^{-3}\,m^{3}$
1 barrel (for oil, 42 gal)	=	$0.1589\,m^{3}$

Mass (kg)

1 tonne (t)	=	1000 kg
1 ton	=	907.18 kg
1 lb	=	0.4536 kg

Energy (joule = J)

1 kWH	=	3.6 MJ
1 Btu	=	1.0551 kJ
1 kcal	=	4.1868 kJ
1 Btu/lb	=	2.326 kJ/kg

Power (Watt : W)

1 Btu/h	=	0.29307 W
1 kcal/h	=	1.163 W
1 hp	=	0.7457 kW

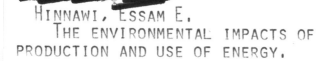

HINNAWI, ESSAM E.
THE ENVIRONMENTAL IMPACTS OF
PRODUCTION AND USE OF ENERGY.